WITHDRAWN

PROGRESS IN

Nucleic Acid Research and Molecular Biology

Volume 18

PROGRESS IN
Nucleic Acid Research and Molecular Biology

edited by

WALDO E. COHN

Biology Division
Oak Ridge National Laboratory
Oak Ridge, Tennessee

Volume 18

1976

ACADEMIC PRESS
New York San Francisco London
A Subsidiary of Harcourt Brace Jovanovich, Publishers

Copyright © 1976, by Academic Press, Inc.
ALL RIGHTS RESERVED.
NO PART OF THIS PUBLICATION MAY BE REPRODUCED OR
TRANSMITTED IN ANY FORM OR BY ANY MEANS, ELECTRONIC
OR MECHANICAL, INCLUDING PHOTOCOPY, RECORDING, OR ANY
INFORMATION STORAGE AND RETRIEVAL SYSTEM, WITHOUT
PERMISSION IN WRITING FROM THE PUBLISHER.

ACADEMIC PRESS, INC.
111 Fifth Avenue, New York, New York 10003

United Kingdom Edition published by
ACADEMIC PRESS, INC. (LONDON) LTD.
24/28 Oval Road, London NW1

LIBRARY OF CONGRESS CATALOG CARD NUMBER: 63-15847

ISBN 0-12-540018-7

PRINTED IN THE UNITED STATES OF AMERICA

Contents

LIST OF CONTRIBUTORS	vii
ABBREVIATIONS AND SYMBOLS	viii
SOME ARTICLES PLANNED FOR FUTURE VOLUMES	xii

The Ribosome of *Escherichia coli**

R. BRIMACOMBE, K. H. NIERHAUS, R. A. GARRETT AND H. G. WITTMANN

I. Introduction	1
II. Ribosomal Components	2
III. Protein–RNA Interactions	8
IV. Topography of Ribosomal Subparticles	13
V. Ribosomal Function	23
VI. Conclusion	36
References	36

Structure and Function of 5 S and 5.8 S RNA

VOLKER A. ERDMANN

I. Introduction	45
II. Primary Sequences of 5 S and 5.8 S Ribosomal RNA	46
III. Secondary and Tertiary Structures of 5 S and 5.8 S RNA	58
IV. Complexes between 5 S RNA and Protein	71
V. Function of 5 S and 5.8 S RNA	79
VI. Outlook	85
References	85
Note Added in Proof	90

High-Resolution Nuclear Magnetic Resonance Investigations of the Structure of tRNA in Solution

DAVID R. KEARNS

I. Introduction	91
II. NMR Studies of Mononucleotides	96
III. NMR Studies of Oligonucleotide Fragments	105
IV. High-Resolution NMR of tRNA	113
V. Application of NMR to the Investigation of Problems of tRNA Structure	130
Summary	145
References	146

* For Addendum to article, see p. 323.

Premelting Changes in DNA Conformation

E. PALEČEK

I. Introduction	151
II. Evidence for Premelting Changes in DNA Conformation	154
III. DNA Conformation in Solution	177
IV. Relation between DNA Conformation and Function	203
V. Concluding Remarks	204
References	206
Addendum	212

Quantum-Mechanical Studies on the Conformation of Nucleic Acids and Their Constituents

BERNARD PULLMAN AND ANIL SARAN

I. Introduction	216
II. The Quantum-Mechanical Methods	217
III. Types of Torsion Angles and Definitions	230
IV. The Glycosyl Torsion Angle χ_{CN}	236
V. Conformation about the Exocyclic C4'—C5' Bond	255
VI. The Backbone Structure of Di- and Polynucleotides	275
VII. Conformation of the Sugar Ring: The Pseudorotational Representation	301
VIII. Related Subjects	308
IX. Concluding Remarks	210
References	313

SUBJECT INDEX 327

CONTENTS OF PREVIOUS VOLUMES 329

List of Contributors

Numbers in parentheses indicate the pages on which the authors' contributions begin.

R. BRIMACOMBE (1), *Max-Planck-Institut für Molekulare Genetik, Berlin-Dahlem, Germany*

VOLKER A. ERDMANN (45), *Max-Planck-Institut für Molekulare Genetik, Abt. Wittmann, Berlin-Dahlem, Germany*

R. A. GARRETT (1), *Max-Planck-Institut für Molekulare Genetik, Berlin-Dahlem, Germany*

DAVID R. KEARNS (91), *Department of Chemistry, Revelle College, University of California, San Diego, La Jolla, California*

K. H. NIERHAUS (1), *Max-Planck-Institut für Molekulare Genetik, Berlin-Dahlem, Germany*

E. PALEČEK (151), *Institute of Biophysics, Czechoslovak Academy of Sciences, 612 65 Brno, Czechoslovakia*

BERNARD PULLMAN (215), *Institut de Biologie Physico-Chimique, Laboratoire de Biochimie, Théorique associé au C.N.R.S., Paris, France*

ANIL SARAN (215), *Institut de Biologie, Physico-Chimique, Laboratoire de Biochimie, Théorique associé au C.N.R.S., Paris, France*

H. G. WITTMANN (1), *Max-Planck-Institut für Molekulare Genetik, Berlin-Dahlem, Germany*

Abbreviations and Symbols

All contributors to this Series are asked to use the terminology (abbreviations and symbols) recommended by the IUPAC–IUB Commission on Biochemical Nomenclature (CBN) and approved by IUPAC and IUB, and the Editor endeavors to assure conformity. These Recommendations have been published in many journals (*1, 2*) and compendia (*3*) in four languages and are available in reprint form from the NAS–NRC Office of Biochemical Nomenclature (OBN), as stated in each publication, and are therefore considered to be generally known. Those used in nucleic acid work, originally set out in section 5 of the first Recommendations (*1*) and subsequently revised and expanded (*2, 3*), are given in condensed form (I–V) below for the convenience of the reader. Authors may use them without definition, when necessary.

I. Bases, Nucleosides, Mononucleotides

1. *Bases* (in tables, figures, equations, or chromatograms) are symbolized by Ade, Gua, Hyp, Xan, Cyt, Thy, Oro, Ura; Pur = any purine, Pyr = any pyrimidine, Base = any base. The prefixes S–, H_2, F–, Br, Me, etc., may be used for modifications of these.

2. *Ribonucleosides* (in tables, figures, equations, or chromatograms) are symbolized, in the same order, by Ado, Guo, Ino, Xao, Cyd, Thd, Ord, Urd (Ψrd), Puo, Pyd, Nuc. Modifications may be expressed as indicated in (*1*) above. Sugar residues may be specified by the prefixes r (optional), d (=deoxyribo), a, x, l, etc., to these, or by two three-letter symbols, as in Ara-Cyt (for aCyd) or dRib-Ade (for dAdo).

3. *Mono-, di-, and triphosphates of nucleosides* (5′) are designated by NMP, NDP, NTP. The N (for "nucleoside") may be replaced by any one of the nucleoside symbols given in II-1 below. 2′-, 3′-, and 5′- are used as prefixes when necessary. The prefix d signifies "deoxy." [Alternatively, nucleotides may be expressed by attaching P to the symbols in (*2*) above. Thus: P-Ado = AMP; Ado-P = 3′-AMP.] cNMP = cyclic 3′:5′-NMP; Bt_2cAMP = dibutyryl cAMP; etc.

II. Oligonucleotides and Polynucleotides

1. Ribonucleoside Residues

   (a) Common: A, G, I, X, C, T, O, U, Ψ, R, Y, N (in the order of I-2 above).

   (b) Base-modified: sI or M for thioinosine = 6-mercaptopurine ribonucleoside; sU or S for thiouridine; brU or B for 5-bromouridine; hU or D for 5,6-dihydrouridine; i for isopentenyl; f for formyl. Other modifications are similarly indicated by appropriate *lower-case* prefixes (in contrast to I-1 above) (*2, 3*).

   (c) Sugar-modified: prefixes are d, a, x, or l as in I-2 above; alternatively, by *italics* or **boldface** type (with definition) unless the entire chain is specified by an appropriate prefix. The 2′-O-methyl group is indicated by *suffix* m (e.g., -Am- for 2′-O-methyladenosine, but -mA- for N-methyladenosine).

   (d) Locants and multipliers, when necessary, are indicated by superscripts and subscripts, respectively, e.g., $-m_2^6A-$ = 6-dimethyladenosine; $-s^4U-$ or $-^4S-$ = 4-thiouridine; $-ac^4Cm-$ = 2′-O-methyl-4-acetylcytidine.

   (e) When space is limited, as in two-dimensional arrays or in aligning homologous sequences, the prefixes may be placed *over the capital letter*, the suffixes *over the phosphodiester symbol*.

2. Phosphoric Acid Residues [left side = 5′, right side = 3′ (or 2′)]

(a) Terminal: p; e.g., pppN . . . is a polynucleotide with a 5′-triphosphate at one end; Ap is adenosine 3′-phosphate; C>p is cytidine 2′:3′-cyclic phosphate (*1, 2, 3*).

(b) Internal: hyphen (for known sequence), comma (for unknown sequence); unknown sequences are enclosed in parentheses. E.g., pA-G-A-C(C$_2$,A,U)A-U-G-C>p is a sequence with a (5′) phosphate at one end, a 2′:3′-cyclic phosphate at the other, and a tetranucleotide of unknown sequence in the middle. (**Only codon triplets are written without some punctuation separating the residues.**)

3. Polarity, or Direction of Chain

The symbol for the phosphodiester group (whether hyphen or comma or parentheses, as in 2b) represents a 3′-5′ link (i.e., a 5′ . . . 3′ chain) unless otherwise indicated by appropriate numbers. "Reverse polarity" (a chain proceeding from a 3′ terminus at left to a 5′ terminus at right) may be shown by numerals or by right-to-left arrows. Polarity in any direction, as in a two-dimensional array, may be shown by appropriate rotation of the (capital) letters so that 5′ is at left, 3′ at right when the letter is viewed right-side-up.

4. Synthetic Polymers

The complete name or the appropriate group of symbols (see II-1 above) of the repeating unit, **enclosed in parentheses if complex or a symbol**, is either (a) preceded by "poly," or (b) followed by a subscript "n" or appropriate number. **No space follows "poly"** (*2, 5*).

The conventions of II-2b are used to specify known or unknown (random) sequence, e.g.,

polyadenylate = poly(A)   or   (A)$_n$, a simple homopolymer;

poly(3 adenylate, 2 cytidylate) = poly(A$_3$C$_2$) or (A$_3$,C$_2$)$_n$, an *irregular* copolymer of A and C in 3:2 proportions;

poly(deoxyadenylate-deoxythymidylate) = poly[d(A-T)]   or   poly(dA-dT)   or (dA-dT)$_n$ or   d(A-T)$_n$, an *alternating* copolymer of dA and dT;

poly(adenylate,guanylate,cytidylate,uridylate) = poly(A,G,C,U)   or   (A,G,C,U)$_n$, a random assortment of A, G, C, and U residues, proportions unspecified.

The prefix copoly or oligo may replace poly, if desired. The subscript "n" may be replaced by numerals indicating actual size, e.g., (A)$_n$·(dT)$_{12-18}$

III. Association of Polynucleotide Chains

1. *Associated* (e.g., H-bonded) chains, or bases within chains, are indicated by a *center dot* (not a hyphen or a plus sign) separating the *complete* names or symbols, e.g.:

poly(A)·poly(U)   or   (A)$_n$·(U)$_m$
poly(A)·2 poly(U)   or   (A)$_n$·2(U)$_m$
poly(dA-dC)·poly(dG-dT)   or   (dA-dC)$_n$·(dG-dT)$_m$.

2. *Nonassociated* chains are separated by the plus sign, e.g.:

$$2[\text{poly}(A) \cdot \text{poly}(U)] \xrightarrow{\Delta} \text{poly}(A) \cdot 2\,\text{poly}(U) + \text{poly}(A)$$
or $\quad 2[A_n \cdot U_m] \rightarrow A_n \cdot 2U_m + A_n$.

3. Unspecified or unknown association is expressed by a comma (again meaning "unknown") between the completely specified chains.

Note: In all cases, each chain is completely specified in one or the other of the two systems described in II-4 above.

IV. Natural Nucleic Acids

RNA	ribonucleic acid or ribonucleate
DNA	deoxyribonucleic acid or deoxyribonucleate
mRNA; rRNA; nRNA	messenger RNA; ribosomal RNA; nuclear RNA
hnRNA	heterogeneous nuclear RNA
D-RNA; cRNA	"DNA-like" RNA; complementary RNA
mtDNA	mitochondrial DNA
tRNA	transfer (or acceptor or amino-acid-accepting) RNA; replaces sRNA, which is not to be used for any purpose
aminoacyl-tRNA	"charged" tRNA (i.e., tRNA's carrying aminoacyl residues); may be abbreviated to AA-tRNA
alanine tRNA or tRNA^{Ala}, etc.	tRNA normally capable of accepting alanine, to form alanyl-tRNA
alanyl-tRNA or alanyl-tRNA^{Ala}	The same, with alanyl residue covalently attached. [*Note:* fMet = formylmethionyl; hence tRNA^{fMet}, identical with tRNA$_f^{\text{Met}}$]

Isoacceptors are indicated by appropriate subscripts, i.e., tRNA$_1^{\text{Ala}}$, tRNA$_2^{\text{Ala}}$, etc.

V. Miscellaneous Abbreviations

P_i, PP_i	inorganic orthophosphate, pyrophosphate
RNase, DNase	ribonuclease, deoxyribonuclease
t_m (not T_m)	melting temperature (°C)

Others listed in Table II of Reference 1 may also be used without definition. No others, with or without definition, are used unless, in the opinion of the editor, they increase the ease of reading.

Enzymes

In naming enzymes, the 1972 recommendations of the IUPAC-IUB Commission on Biochemical Nomenclature (CBN) (*4*), are followed as far as possible. At first mention, each enzyme is described *either* by its systematic name *or* by the equation for the reaction catalyzed *or* by the recommended trivial name, followed by its EC number in parentheses. Thereafter, a trivial name may be used. Enzyme names are not to be abbreviated except when the substrate has an approved abbreviation (e.g., ATPase, but not LDH, is acceptable).

REFERENCES[*]

1. *JBC* **241**, 527 (1966); *Bchem* **5**, 1445 (1966); *BJ* **101**, 1 (1966); *ABB* **115**, 1 (1966), **129**, 1 (1969); and elsewhere.[†]

[*] Contractions for names of journals follow.

[†] Reprints of all CBN Recommendations are available from the Office of Biochemical Nomenclature (W. E. Cohn, Director), Biology Division, Oak Ridge National Laboratory, Box Y, Oak Ridge, Tennessee 37830, USA.

2. *EJB* **15**, 203 (1970); *JBC* **245**, 5171 (1970); *JMB* **55**, 299 (1971); and elsewhere.*
3. "Handbook of Biochemistry" (H. A. Sober, ed.), 2nd ed. Chemical Rubber Co., Cleveland, Ohio, 1970, Section A and pp. H130–133.
4. "Enzyme Nomenclature," Elsevier Scientific Publ. Co., Amsterdam, 1973, and Supplement No. 1, *BBA* **429**, 1 (1976).
5. "Nomenclature of Synthetic Polypeptides," *JBC* **247**, 323 (1972); *Biopolymers* **11**, 321 (1972); and elsewhere.*

Abbreviations of Journal Titles

Journals	Abbreviations used
Annu. Rev. Biochem.	ARB
Arch. Biochem. Biophys.	ABB
Biochem. Biophys. Res. Commun.	BBRC
Biochemistry	Bchem
Biochem. J.	BJ
Biochim. Biophys. Acta	BBA
Cold Spring Harbor Symp. Quant. Biol.	CSHSQB
Eur. J. Biochem.	EJB
Fed. Proc.	FP
J. Amer. Chem. Soc.	JACS
J. Bacteriol.	J. Bact.
J. Biol. Chem.	JBC
J. Chem. Soc.	JCS
J. Mol. Biol.	JMB
Nature, New Biology	Nature NB
Proc. Nat. Acad. Sci. U.S.	PNAS
Proc. Soc. Exp. Biol. Med.	PSEBM
Progr. Nucl. Acid Res. Mol. Biol.	This Series

* Reprints of all CBN Recommendations are available from the NRC Office of Biochemical Nomenclature (W. E. Cohn, Director), Biology Division, Oak Ridge National Laboratory, Box Y, Oak Ridge, Tennessee 37830, USA.

Some Articles Planned for Future Volumes

The Transfer RNAs of Cellular Organelles
   W. E. Barnett, L. I. Hecker and S. D. Schwartzbach

Mechanisms in Polypeptide Chain Elongation on Ribosomes
   E. Bermek

Mechanism of Action of DNA Polymerases
   L. M. S. Chang

Initiation of Protein Synthesis
   M. Grunberg-Manago

Integration vs. Degradation of Exocellular DNA: An open Question
   P. F. Lurquin

The Messenger RNA of Immunoglobulin Chains
   B. Mach

Bleomycin, an Antibiotic Removing Thymine from DNA
   W. Müller and R. Zahn

Protein Biosynthesis
   S. Ochoa

Vertebrate Nucleolytic Enzymes and Their Localization
   D. Shugar and H. Sierakowska

Regulation of the Synthesis of Aminoacyl-tRNAs and tRNAs
   D. Söll

Physical Structure, Chemical Modification and Functional Role of the Acceptor Terminus of tRNA
   M. Sprinzl and F. Cramer

The Biochemical and Microbiological Action of Platinum Compounds
   A. J. Thomson and J. J. Roberts

Transfer RNA in RNA Tumor Viruses
   L. C. Waters and B. C. Mullin

Structure and Functions of Ribosomal RNA
   R. Zimmermann

The Ribosome of *Escherichia coli*

R. BRIMACOMBE,
K. H. NIERHAUS,
R. A. GARRETT AND
H. G. WITTMANN

*Max-Planck-Institut für
Molekulare Genetik
Berlin-Dahlem, Germany*

I. Introduction	1
II. Ribosomal Components	2
A. Ribosomal Proteins	2
B. Ribosomal RNA	4
III. Protein–RNA Interactions	8
A. Protein Binding Sites on 16 S RNA	9
B. Protein Binding Sites on 23 S RNA	12
IV. Topography of Ribosomal Subparticles	13
A. The 30 S Subparticle	14
B. The 50 S Subparticle	20
C. The 30–50 S Subparticle Interface	22
V. Ribosomal Function	23
A. Initiation	23
B. Elongation	28
C. Termination	35
VI. Conclusion	36
References	36
Addendum	323

I. Introduction

The properties of ribosomes from *Escherichia coli* have been far more widely studied than those from any other organism. This review is therefore restricted to a discussion of the *E. coli* ribosome, except where interesting conclusions can be drawn by a comparison with corresponding studies on other organisms. The review is focused on two aspects, namely, structure and function. The structural aspects are considered in three sections concerning, respectively, the primary structure of the ribosomal proteins and ribosomal RNA, the interaction between proteins and RNA, and the topographical arrangement of the proteins. The section on function is subdivided into considerations of initiation of protein synthesis, elongation and termination. In general, the literature cited refers to the most recent work on a particular topic, rather than to the older findings, which are covered in a number of recent reviews dealing in more detail with various aspects of the ribosome problem. These reviews are quoted in their appropriate place in the text.

II. Ribosomal Components

A. Ribosomal Proteins

An essential prerequisite for studies on ribosomes at the molecular level is the isolation and characterization of all ribosomal proteins. Isolation of the proteins has been achieved mainly by combination of several methods, *viz.* separation of the ribosomal subparticles by zonal centrifugation, prefractionation by salt treatment if necessary, column chromatography on carboxymethyl- or phosphocellulose, and gel filtration on Sephadex. A description of these various isolation procedures has been reviewed recently (*1*).

Two-dimensional polyacrylamide gel electrophoresis shows (*2*) that the small subparticle of the *E. coli* ribosome contains 21 proteins, designated S1 to S21, and the large subparticle 34 proteins, L1 to L34.[1] The molecular weights of all the proteins fall within the range from 6000 to 32,000, with the exception of the 30 S protein S1, which has a much higher molecular weight of 65,000. The individual molecular weights are listed in a review by Wittmann (*1*). Most of the ribosomal proteins are basic in nature, and this is reflected in their amino-acid compositions and isoelectric points. Many proteins are very rich (up to 34%) in basic amino acids (*3–5*); protein S21, for instance, has 20% arginine and protein L33 has 21% lysine; in consequence, the isoelectric points of the proteins are very high: pH 10 or more for about 70% of the proteins. Only three proteins (S6, L7 and L12) are mildly acidic, and these have isoelectric points of approximately pH 5 (*6*).

The separation of tryptic peptides on a preparative scale and amino-acid analyses have so far been made for almost 50% of the ribosomal proteins (*7, 8,* and Wittmann-Liebold, unpublished results). Information about the primary structure of ribosomal proteins can also be very efficiently and relatively quickly obtained by means of a protein sequenator. In this way, up to 40–60 amino-acid residues from the N-terminal regions have so far been sequenced for each of 18 proteins from the small subparticle, and 14 proteins from the large (*8, 9*). A few

---

[1] The proteins are numbered according to their positions on two-dimensional electropherograms, number first from left to right (i.e., plus to minus, first dimension) and then from top to bottom (i.e., plus to minus, second dimension).

proteins (e.g., S5, S18, L7 and L11) have blocked N termini, and therefore cannot be analyzed by the sequenator.

The complete primary structure has been determined for ten proteins: S4 (*10*), S5 (*11*), S6 (*12*), S8 (*13*), S9 (*14*), S18 (*15*), L7 and L12 (*16*), L25 (*17, 17a*) and L29 (*18*). Sequences of 15 others are well on the way to completion. The total number of amino-acid residues sequenced to date is about 2500, i.e., 30% of the approximately 8000 amino acids contained in the *E. coli* ribosomal proteins. This large effort is justified for the following reasons.

1. The ribosome, and in particular the *E. coli* ribosome, is the only cell organelle for which a thorough understanding at the molecular level of both structure and function can be expected in the near future. The detailed determination of the spatial distribution of ribosomal proteins, for example, requires knowledge of the amino-acid sequences of the proteins, leading to an understanding of the chemical and physical basis of RNA–protein interaction in the ribosomal particle.

2. Ribosomal proteins are altered in many bacterial mutants, e.g., those resistant to antibiotics inhibiting protein biosynthesis. In *E. coli* ribosomes, ten proteins from the small subparticle (S2, S4, S5, S6, S7, S8, S12, S17, S18 and S20) and three from the large (L4, L6 and L22) have been found to be altered by mutation. In many of the altered proteins, single amino-acid replacements, clustered in "hot spots," have been found and localized. In some mutants of protein S4, more drastic differences, leading to completely different C-terminal regions as a result of frame shifts, have been determined and analyzed. These studies have been reviewed elsewhere (*19*) and are not described here.

3. The fact that ribosomes consist of numerous proteins raises the following questions. First, are the proteins present in ribosomes of a given organism, e.g., *E. coli*, completely different from each other, or do there exist regions of identical or similar amino-acid sequences among them? Second, how did the ribosomal components change during evolution, i.e., are there identical or similar regions in ribosomal proteins from organisms belonging to different classes, e.g., bacteria, plants and animals?

A preliminary answer to the question of homology among the various *E. coli* ribosomal proteins, as well as between proteins from *E. coli* and other organisms, can be given by immunological, electrophoretic and chromatographic methods (reviewed in *19, 20*). However, a more direct and detailed approach to this question is a comparison of the amino-acid sequences of the proteins. As mentioned above, approximately 2500 amino-acid positions have to date been determined in *E. coli* ribosomal proteins, and these sequences, comprising about 30% of the total number of amino acids in the *E. coli* ribosome, are distributed among 32 of the proteins. No regions longer than five amino acids common to different proteins have been found (9), with the exception of two proteins (L7 and L12) that have identical primary sequences and differ only in the presence of an acetyl group at the N terminus of protein L7 (16). These direct results are in very good agreement with the conclusions drawn from immunological studies on proteins isolated from *E. coli* ribosomes (21).

A comparison of the amino-acid sequences of ribosomal proteins from different organisms has up to now been limited to *E. coli* on the one hand and to only two different bacteria (*Bacillus stearothermophilus* and *Halobacter cutirubrum*) on the other. Yaguchi et al. (22) compared the N-terminal regions (15 amino acids) of 21 proteins isolated from *B. stearothermophilus* 30 S subparticles with the sequences of *E. coli* proteins. Similarly, Higo and Loertscher (23) made a comparison of five *E. coli* and *B. stearothermophilus* 30 S proteins (25–30 amino acids each). The first authors found, for each *E. coli* 30 S protein (with the exception of S1), a homologous protein from *B. stearothermophilus* ribosomes. The degree of homology, depending on the particular protein, varied from 25 to 90% of the regions sequenced. These sequence homologies support earlier structural and functional comparisons of the proteins isolated from these organisms (24, 25).

It remains to be seen, by comparison of protein sequences from more organisms, whether a correlation exists between the degree of conservation of the primary structure during evolution and other properties of the proteins, e.g., ability to bind to RNA, or presence in an active ribosomal site. The hypothesis, that functionally important proteins are mainly conserved, is supported by the finding that the structures of proteins L7 and L12, which have important functions during initiation, elongation and termination (reviewed in 26), have been relatively strongly conserved during evolution. This has been shown by comparing corresponding regions of the protein chains from *E. coli*, *B. stearothermophilus* and *H. cutirubrum* (27). Further, immunological and biochemical studies indicate that proteins related to *E. coli* proteins L7 and L12 are present in ribosomes from yeast and rat liver (28, 29).

B. Ribosomal RNA

1. PRIMARY STRUCTURE

The primary structures of the three ribosomal RNAs have been extensively studied during the last ten years. The sequence of 5 S RNA was completed some years ago (*30*), and that of the 16 S RNA is about 90% completed (*31–33*). Sequencing of the 23 S RNA has advanced to the stage where large sections of the 3′ and 5′ regions of the molecule are known, as well as several sections in the central region (*34*).

The current state of the 16 S RNA sequence is given in Fig. 1, which also indicates those regions whose sequences are not yet certain. The sequences of some sections were determined by Santer and Santer (*35*). They obtained results very similar to those cited above (*31–33*), and confirmed many of the connections between the sections. Moreover, Uchida *et al.* (*36*), from analyses of the T1 ribonuclease oligonucleotides, confirmed almost all the oligonucleotide sequences reported by Fellner *et al.* (*32*); very few changes were incorporated into the latest sequence [(*33*) and Fig. 1]. The 3′-terminal oligonucleotide was sequenced by Shine and Dalgarno (*37*) and confirmed by Ehresmann *et al.* (*38*) and by Noller and Herr (*39*).

A number of interesting properties of the ribosomal RNAs can be derived from the RNA base-sequence determinations. (a) The 16 S RNA contains about 1600 nucleotides, and the 23 S RNA about 3200. (b) The overall base composition of the 16 S RNA is 31.8% guanine, 24.8% adenine, 23.3% cytosine and 20.0% uracil. (c) A low level of heterogeneity of approximately 1% of the nucleotides in the 16 S RNA was found. These tend to be clustered at a few points along the sequence, and comparable levels of heterogeneity were found both in 5 S RNA [reviewed by Monier (*40*)] and 23 S RNA (*34*). (d) Although substantial duplication of sequence was found within the 5 S RNA (*30*), no such repeated sequences have yet been detected in either the 16 S RNA or the 23 S RNA. Moreover, a comparison of the T1 ribonuclease oligonucleotides of 16 S and 23 S RNAs revealed no similarities, and the methylated regions and the 5′ and 3′ termini are completely different. (e) Several palindromes were detected in the 16 S RNA sequence, i.e., the sequences are the same whether they are read from the 5′ or 3′ end. These occur in sections L, R, I′, I, K, C′$_2$ and D, and some of the longer examples are indicated in Fig. 1. Moreover, some base-paired hairpin loops contain 2-fold axes of symmetry.

2. SECONDARY STRUCTURE

The overall secondary structures of the isolated large ribosomal RNAs have been studied extensively by different spectroscopic methods and by spectrophotometric titration (see References *41–43* for review of the earlier work) 60–70% of the nucleotides, were estimated to be involved in base-pairing.

Mainly from the efforts of Cox (*44*) and Spencer and their co-workers, more detailed information has been obtained on the nature of the

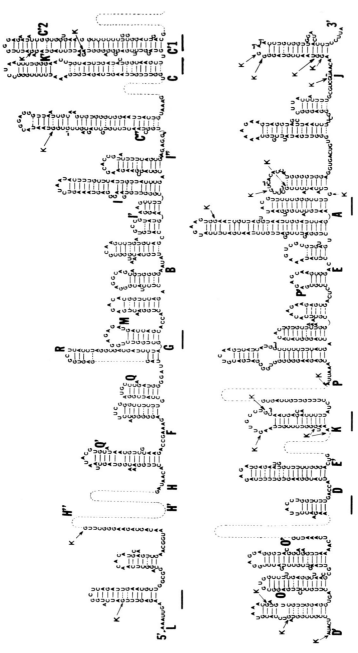

Fig. 1. The current sequence of the 16 S RNA with a theoretically evaluated base-pairing scheme [from Ehresmann et al. (33)]. -----, Regions for which the order of the oligonucleotides is still uncertain. Some "hairpins" that exhibit perfect, or near-perfect, palinfromes or symmetry about a 2-fold axis are underlined. Guanine residues that are readily modified by kethoxal in the 30 S subunit are indicated by a K.

double-helical regions. They isolated "hairpins" of RNA after mild alkaline or ribonuclease treatment and then determined their size and structure. The relative uniformity of these structures was indicated by their capacity to form microcrystals (45). The "hairpins" were shown to contain 10–15 base-pairs and to be in the nucleic acid "A" conformation, with an average of $11\frac{1}{3}$ base-pairs per helical turn.

Secondary structural maps of the intact RNAs are much more difficult to elucidate. However, by applying the rules of Tinoco *et al.* (46) to the nucleotide sequence, it has been possible to propose a secondary structure map of the 16 S RNA (33) (see Fig. 1). The structure comprises 40 "hairpins," and 63% of the nucleotides are base-paired, comprising 34% G · C, 23% A · U and 6% G · U pairs. One outstanding feature of the proposed scheme is that there are many nucleotides that loop out from the double-helical regions, and they are predominantly A residues (53% A, 19% G, and 14% C and U). The map is similar to that proposed for the coat-protein cistron of bacteriophage MS2 (47) in its general form and in the stability of the "hairpins."

A number of methods are available for checking this secondary structure, namely, preferential chemical modification of single-stranded regions with base-specific reagents, preferential cutting of single-stranded regions with nucleases, and binding to single-stranded regions by complementary oligonucleotides. To date, these methods have been applied with varying degrees of success mainly to tRNA and 5 S RNA structures (48–50).

Each of the methods has been applied to the study of yeast tRNA[Phe], and the recent determination of the three-dimensional structure of this molecule (51, 52) makes it possible to evaluate the various methods objectively. Chemical modification gives very good agreement (49), and nuclease digestion also correlates well (53). Oligonucleotide binding, on the other hand, correlates rather poorly (54, 55), which could arise from alterations in the tRNA structure induced by the oligonucleotides (56).

Application of these methods to the large ribosomal RNAs is obviously considerably more difficult and has been confined to some chemical modification and nuclease digestion studies on the 16 S RNA. The results correlate fairly well with the secondary structural map of Fig. 1. For example, Noller (57) modified the 16 S RNA in 30 S subparticles and identified the modification sites. They are indicated in Fig. 1. Although some occur in unpaired G-residues, others are in putative base-paired regions. Moreover, although many cuts produced in the 16 S RNA during partial digestion with T_1 and pancreatic ribonucleases occurred in single-stranded regions, some were also in the proposed base-paired regions (33.)

The question has often been raised whether the RNA has the same structure in the free state as in the ribosome. It has been demonstrated by a number of physicochemical methods that the total amount of base-pairing is approximately the same in both (see References 41–43 for reviews), and there is presently no evidence to suggest that gross changes occur in the base-pairing during assembly into 30 S and 50 S subparticles.

3. Tertiary Structure

There is much evidence to suggest that changes in the RNA structure occur when the Mg^{2+} concentration is varied. The sedimentation coefficients of both large ribosomal RNAs increase significantly on addition of Mg^{2+} (58, 59), suggesting the formation of a more contracted structure. The range in which this change occurs, namely 10^{-3} to 10^{-2} M Mg^{2+}, also corresponds to the range in which the proteins start to bind specifically and strongly to the RNAs (59, 60). Further, the addition of Mg^{2+} also produces a markedly increased resistance of the RNA toward ribonuclease digestion (61). It was concluded from all these results that, above a critical level of Mg^{2+} (1 mM) the RNAs fold into a more compact structure that may, or may not, result in tertiary structural interactions within the RNA molecules. Recent work on the binding site of protein S4 at the 5'-end of the 16 S RNA has provided direct evidence for such an interaction (62–64).

4. Conformational Heterogeneity

A well-known property of tRNAs and 5 S RNAs is that they can be conformationally heterogeneous. Denatured forms of both have been isolated and identified, and shown to be renaturable (65, 66). Recent evidence suggests that the large ribosomal RNAs prepared by standard methods are also conformationally heterogeneous. Sedimentation study shows a high level of heterogeneity, removable by heating in the presence of Mg^{2+} (59). A similar conclusion was drawn from electrophoretic studies on 16 S and 23 S RNAs; at low temperatures and in the absence of Mg^{2+}, 16 S and 23 S RNA can be resolved into three and two components, respectively, whereas only one component was observed on heating with Mg^{2+} (67).

III. Protein–RNA Interactions

Some twenty proteins bind specifically to the three isolated ribosomal RNAs of *E. coli*. These include six proteins that bind to 16 S RNA

[S4, S7, S8, S15, S17 and S20 (68–73)], three that bind to 5 S RNA [L5, L18 and L25 (74–77)] and eleven that bind to 23 S RNA [L1, L2, L3, L4, L6, L13, L16, L17 (possibly), L20, L23, and L24 (78–80)].

Many attempts have been made to locate, at least partially, the binding sites of these proteins on the respective RNA molecules. The success of these methods has been mainly due to the recent rapid progress in the base-sequencing of 16 S and 23 S RNA and to the known sequence of 5 S RNA (see preceding section). Two principal methods have been employed. The first of these was developed by Zimmerman and his co-workers (72, 81, 82), who showed that mild digestion of 16 S RNA produces large RNA fragments, to which certain proteins can be rebound. This permitted a partial localization of proteins along the 16 S RNA. To a more limited extent, this approach has also been applied successfully to 23 S RNA (83, 84). The second method involves the digestion by nuclease of a complex between a single protein and the RNA, with a view to isolating a region of RNA protected from digestion by the presence of the bound protein. This method has yielded the most detailed information on the binding sites of proteins on 16 S RNA (62, 72, 85–88), 5 S RNA (76, 89), and 23 S RNA (90, 91), although it has the disadvantage that it is difficult to produce protected ribonucleoprotein fragments containing those proteins that bind relatively weakly.

The results obtained for the protein binding sites on both 16 S and 23 S RNA from both these methods are discussed below (see also Fig. 2).

A. Protein Binding Sites on 16 S RNA

a. S4. Ribonucleoprotein particles containing protein S4 have been isolated by ribonuclease digestion of the protein · RNA complex (62, 85, 88). The ribonucleoproteins contain up to 500 nucleotides of RNA, arising from the 5′ region of the 16 S RNA. The RNA region has recently been characterized in detail (62–64), and the sequences it contains were shown to be noncontiguous. A number of excisions were observed, constituting mainly the upper parts of the proposed hairpin loops. Further degradation of the ribonucleoprotein complex leads to a complete breakdown of the structure, and no smaller protein-containing fragments could be isolated. Further evidence that protein S4 has a large binding site on the RNA has come from electron microscopic visualization of the S4–16 S RNA complex, in which the unbound RNA appears stretched out as a "tail" (92).

A few preliminary attempts have been made to establish the parts of protein S4 involved in RNA binding. Lysine residues at positions 119

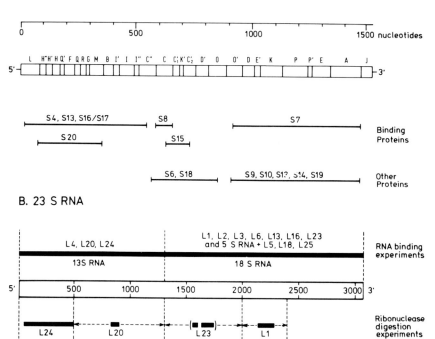

Fig. 2. (A) The location of binding sites of the 30 S subunit proteins on the 16 S RNA. The order of the lettered sections is from Ehresmann et al. (33). Protein S13 has been found at both ends of the 16 S RNA. (B) A scheme of the 23 S RNA showing the distribution of the protein binding sites.

and 147 are completely protected against reductive methylation by the 16 S RNA (93), and there is partial protection of the lysine at position 30. In another study involving chemical modifications (94) and enzyme treatment, the importance of a small number of lysine residues in binding was confirmed and the C-terminal region was also shown to be important. In addition, the cysteine and tryptophan residues at positions 31 and 167, respectively, were shown to be not involved in binding to RNA.

  b. *S20.* The binding site of protein S20 on the 16 S RNA lies within that of S4. A ribonucleoprotein particle containing S20 has been isolated electrophoretically; the RNA consists of about 250 nucleotides at the 5'-end of the corresponding S4 RNA region (see Fig. 2A) (62, 88).

  c. *S16 and S17.* A mixture of proteins S16, S17 and S20, complexed with 16 S RNA, digested with ribonuclease, yields an RNA identical with a 9 S RNA fragment constituting the S4 binding site in other experiments

(*81*). S17 is known to bind directly to the 16 S RNA (*68, 73, 81*), whereas S16 binds only in the presence of S4 and S20 (*68, 81*).

d. S8. A small region of RNA, protected by protein S8 against nuclease digestion, has been isolated chromatographically (*87*) and electrophoretically (*62*). This binding region corresponded to section C near the center of the 16 S RNA molecule. By use of the former method, a mixture of RNA fragments containing submolar amounts of most of section C was obtained. On the other hand, the latter method yielded a homogeneous fragment 40 nucleotides in length. By increasing the enzyme concentration, one half, namely the 3′ half of the fragment, could be digested away, leaving the protein complexed with the 5′ half (*62*).

A tentative secondary structure proposed for this region suggests that it lies at the base of a long hairpin loop. The unusual stability of this particular hairpinlike structure has also been demonstrated by Muto and Zimmermann (*61*), who showed that this region of the RNA is only partially denatured by exhaustive dialysis aginst water. Exmination of the base-pairs at the base of the "hairpin" (Fig. 1), taken together with the fact that S8 can preferentially attach to one half of the region, suggests that the structure may be partially opened when complexed with the protein.

e. S15. A relatively large fragment of RNA complexed with protein S15 was originally isolated on sucrose gradients; it consisted of two hairpin loops covering sections C to C′$_2$ (*72*). However, stronger digestion of the S15–16 S RNA complex yielded a smaller ribonucleoprotein particle (*62, 88*) that constituted the base of a "hairpin," adjacent to that of the S8 binding site, and the 3′ half of the S8 site. This again suggests that the RNA in the S8 binding site is partly opened up when complexed with protein. The S15 RNA site contains a total of about 120 nucleotides.

f. S8 plus S15. Zimmermann et al. (*88*) digested a complex of S8 and S15 with 16 S RNA, and obtained an RNA fragment consisting of 150 nucleotides that consisted of the two hairpin loops covering sections C to C′$_2$, and included the separate S8 and S15 binding sites described above. In addition, it included the tops of the hairpin loops that were excised in the experiments with the separate proteins.

g. Proteins S6 and S18 in the S8–S15 region. Proteins S6 and S18, which cannot independently bind to 16 S RNA, can form a complex with the RNA in the presence of S8 and S15 (*88*). Ribonuclease digestion produces a particle containing all four proteins as well as a region of RNA sedimenting at 7 S (*88*). This RNA included the region corresponding to S8–S15, but it also included an additional part of section C at the 5′ end and a large part of section O at the 3′ end. Although it cannot be ruled out that the two proteins, S6 and S18, had bound to S8 and/or

S15 and rendered these regions of RNA sterically inaccessible to the ribonuclease, the results strongly suggest that these proteins bind to sections C and O.

h. S7. Protein S7 was originally positioned in the 3′-proximal region of the 16 S RNA. Zimmermann *et al.* (*72*) isolated an 8 S fragment containing S7 from the 3′ half of the 16 S RNA, after nuclease digestion of an S7–16 S RNA complex. However, despite the large size of this RNA fragment, the protein did not rebind specifically to it.

i. Proteins S7, S9, S10, S13, S14 and S19. Evidence that this group of proteins binds within the 3′ half of the 16 S RNA has come from two sources. First, a ribonucleoprotein fragment containing these proteins was isolated from 30 S particles (see Section IV, A, 2 on topography), and the RNA protected by the proteins in this fragment extends from section O′ to the beginning of section A (see Fig. 2A) (*95, 96*). Second, Muto *et al.* obtained a similar fragment after digestion of a reconstituted intermediate particle (*81*). The fragment in this case contained proteins S7, S9, S13 and S19, and the RNA extended somewhat farther into section A. All these results are summarized in Fig. 2A.

B. Protein Binding Sites on 23 S RNA

Although a large number of proteins bind to 23 S RNA, relatively few of these have yielded specifically protected RNA fragments. However, the 23 S RNA can be split, by very mild nuclease digestion of 50 S subparticles, into two fragments of 18 S and 13 S, respectively (*83*); a partial localization of the proteins along the 23 S RNA has been possible by binding the individual proteins to these isolated RNA fragments. The binding tests showed that three proteins, namely L4, L20 and L24, can reassemble to the 13 S (5′-proximal) RNA, and that six proteins, namely L1, L2, L3, L6, L13, L16 and L23, can reassemble to the 18 S (3′-proximal) RNA (see Fig. 2B) (*84*).

Ribonuclease digestion of reconstituted protein · RNA complexes has led to the isolation of some protected RNA fragments. In the case of L24, a very large RNA fragment, similar in complexity to the corresponding S4 RNA, has been isolated. This RNA occurred at the 5′ end of the molecule (*90*), and a larger fragment could also be obtained in the absence of the protein. Ribonucleoprotein particles have also been produced for proteins L1, L20 and L23, after digestion of their respective 23 S RNA · protein complexes, and the approximate positions of their constituent RNAs within the 23 S RNA sequence are indicated in Fig. 2B (*91*). Both L1 and L23 yielded relatively small RNA fragments (about 150 nucleotides in length), with fairly contiguous sequences. L20 showed some nonspecific binding to the 23 S RNA, but a homogeneous RNA fragment to which L20 could be specifically rebound was also obtained (*91*).

IV. Topography of Ribosomal Subparticles

An understanding of the spatial arrangement of proteins and RNA within the ribosome is important for the correlation of ribosome structure with the protein synthetic function. This problem of topography can be divided into two parts, namely, the arrangement of the individual proteins relative to the RNA strand (discussed in the preceding section), and the three-dimensional arrangement of the proteins relative to each other.

Before the three-dimensional arrangement of the proteins can be discussed, a number of points must be borne in mind. First, the ribosomal proteins not only vary considerably in size, as already mentioned, but, more important in this context, it seems that they can also have very irregular shapes. Isolated protein L7/L12 for example has a very elongated conformation (*97*). If such shapes are preserved within the ribosomal particle, it is clear that this will strongly influence the way in which a model of the protein arrangement can be drawn. Second, the proteins cannot all be isolated in equimolar amounts at all stages of the protein synthetic cycle (e.g., *98–101*), and although this question of which proteins are "unit" and which are "fractional" still requires clarification, it is certain that the isolated ribosome cannot be considered a homogeneous entity. It remains to be seen whether the removal or addition of a fractional protein can seriously distort the arrangement of the other proteins. Third, considerable changes in ribosomal conformation could occur as a result of interaction between the two ribosomal subparticles, or as a result of interaction with the various components of protein synthesis (tRNA, mRNA, factors, etc). Such changes have been observed in the case of binding of initiation factor IF3 to the 30 S particle (*102*), or of streptomycin binding (*103, 104*). Again, it remains to be seen whether such changes are accompanied by radical changes in the protein arrangement within the particle. Fourth, only a small proportion of ribosomes are active in protein synthesis in *in vitro* systems (e.g., *105*), and this raises the possibility that distortions of the protein arrangement may occur as a result of the isolation procedures used. Finally, the procedures used to make the structural investigations may themselves cause distortions, and many of the chemical reactions involved in these procedures give very low yields of product.

Despite these reservations, several techniques have been used during the last few years to examine the spatial arrangement of the *E. coli* ribosomal proteins. The problem can be divided into three parts, *viz.* the topography of the 30 S particle, the topography of the 50 S particle, and the organization of the subparticle interface. These three topics are discussed in that order, and the success of the various experimental approaches may be judged by the measure of agreement among the various results.

A. The 30 S Subparticle

1. Use of Bifunctional Protein Cross-Linking Reagents

The use of bifunctional cross-linking reagents is the most direct way of determining which pairs of proteins are neighbors within the ribosome. The chief problem involved in the use of such reagents is the identification of the proteins contained in the cross-linked complex. In some early work, identification of proteins irreversibly linked together by means of a bifunctional sulfhydryl reagent (*106*) was made solely on the basis of the disappearance of particular proteins from a polyacrylamide gel of the ribosomal proteins and the appearance of a new species of the expected molecular weight. Such a method has clearly only a limited application, and later work has made use of more positive methods of identification of the proteins in the isolated cross-linked product, notably peptide analysis of the cross-linked complex (*107*), or the use of Ouchterlony double-diffusion tests with antibodies specific to the individual proteins (*108–110*). However, most attention has been given to the development of cleavable reagents, which enable the proteins in the complex to be regenerated and identified by polyacrylamide gel electrophoresis. The most widely used reagents for this purpose have been the bis-imido esters, and, of these, bis-methyl suberimidate proved in earlier work to have the most useful properties (*111*). The reagent attacks the ϵ-amino groups of lysine, and the cross-link may be broken by ammonolysis. However, some authors have found that the latter reaction does not proceed very satisfactorily, and other more susceptible reagents have been developed accordingly. The most notable of these are methyl 4-mercaptobutyrimidate (*112*) and bis-azide compounds derived from tartaric acid (*113*). Both these compounds react with amino groups, but the former does not of itself introduce a protein–protein cross-link. Instead, the mercaptobutyrimidate effectively adds an SH group to the protein, and these SH groups can subsequently be cross-linked by mild oxidation. Cleavage is later accomplished by *in situ* reduction in a diagonal gel electrophoresis system (*114*). The tartaric acid derivatives form direct cross-links between the proteins, which can later be cleaved by periodate oxidation of the vicinal OH groups.

The protein pairs in the 30 S particle so far identified by this approach are listed in Table I, which also indicates the reagent used and the method of identification of the proteins. The methods have also been applied to the cross-linking of initiation factors to the ribosome, and these results are also included in the table. The possibility remains that some of the reagents can themselves distort the protein arrangement, but the consistency of the results obtained with the different reagents argues against this. Furthermore, in one case, a cross-linked protein pair (S5-S8) has been incorporated back into a reconstituted 30 S particle, without impairment of protein synthetic activity (*120*). One other type of cross-linking reaction that should be mentioned in this section is the formation of a specific bond between the 3′ terminus of the 16 S RNA and a ribosomal protein, tentatively identified as S1 (*121*). This was accomplished by periodate oxidation of the 3′-terminal ribose of the RNA, followed by formation of a Schiff's base with a lysine amino group, which was then reduced with borohydride to stabilize the complex.

2. Isolation of Specific Ribonucleoprotein Fragments

If the 30 S ribosome is treated with nucleases under suitably mild conditions, it can be broken down into fragments of ribonucleoprotein that can be analyzed for their protein or RNA content. Those proteins found together in such fragments have been considered to be neighbors in the intact particle. This is a reasonable assumption, since it is unlikely that a ribonucleoprotein complex consisting of widely separated groups of proteins joined by an unprotected RNA strand could survive the nuclease treatment. Using this approach, the 30 S particle can be readily split into two fragments of unequal size, the one containing proteins S4, S5, S6, S8, S15, S16 (S17), S20, and possibly S13 and S18, and the other containing proteins S7, S9, S10, S19, and S13 or S14 (*122–124, 96*). These fragments have been isolated both by gel electrophoresis and by sucrose density gradient centrifugation, and their protein content has been determined by gel electrophoresis. Various smaller fragments have been isolated, containing subsets of these two groups of proteins, the smallest containing S8 and S15, and the other S7 and S19 (*123*). A similar subdivision of the 30 S proteins into three groups has been accomplished by other workers (*82, 88*), who analyzed hydrolyzates of a protein-deficient reconstitution-intermediate (RI) particle. In this case, the protein groups were: S4, S16, S17, S20, containing RNA from the 5′-proximal region of the RNA (cf. Section III, A on RNA–protein interactions already discussed); S6, S8, S15 and S18, containing RNA from the central region of the 16 S RNA; and S7, S9, S13 and S19, with RNA from the 3′-proximal region of the 16 S molecule (*82, 88*). The RNA sequences in the last of

TABLE I
30 S PROTEIN CROSS-LINKS[a]

Protein pair	Cross-linking agent	Identification of proteins	Reference
S2–S3	HS-butyrimidate	Reduction and gel EP	115
	Tartaroyl (N$_3$)$_2$	Oxidation and gel EP; antibodies	113
S2–S5	Tartaroyl(N$_3$)$_2$	Oxidation and gel EP	113
S2–S8	Tartaroyl(N$_3$)$_2$	Oxidation and gel EP	113
S3–S10	Tartaroyl(N$_3$)$_2$	Oxidation and gel EP	113
S4–S5	Tartaroyl(N$_3$)$_2$	Oxidation and gel EP; antibodies	113
	Me$_2$suberimidate	Ammonolysis and gel EP	117
S4–S8	HS-butyrimidate	Reduction and gel EP	118a
S4–S9	HS-butyrimidate	Reduction and gel EP	118a
S4–S12	HS-butyrimidate	Reduction and gel EP	118a
S4–S13	HS-butyrimidate	Reduction and gel EP	114
S5–S8	Me$_2$adipimidate	Antibodies	108
	HS-butyrimidate	Reduction and gel EP; antibodies	115
	Me$_2$suberimidate	Ammonolysis and gel EP	116
	Tartaroyl(N$_3$)$_2$	Oxidation and gel EP; antibodies	113
	Me$_2$suberimidate	Ammonolysis and gel EP	117
S5–S9	Me$_2$suberimidate	Ammonolysis and gel EP	111
	HS-butyrimidate	Reduction and gel EP	114
S6–S8	HS-butyrimidate	Reduction and gel EP	118
S6–S18	Tartaroyl(N$_3$)$_2$	Oxidation and gel EP	113
S7–S8	HS-butyrimidate	Reduction and gel EP	118a
S7–S9	Me$_2$adipimidate	Antibodies	109
	Me$_2$suberimidate	Antibodies	109
	HS-butyrimidate	Reduction and gel EP	114
	Tartaroyl(N$_3$)$_2$	Oxidation and gel EP; antibodies	113
	Me$_2$suberimidate	Ammonolysis and gel EP	117
S8–S9	HS-butyrimidate	Reduction and gel EP	118
S8–S13	HS-butyrimidate	Reduction and gel EP	118
S13–S19	Me$_2$suberimidate	Antibodies	110
	Me$_2$adipimidate	Antibodies	112
	HS-butyrimidate	Reduction and gel EP; antibodies	115
	Tartaroyl(N$_3$)$_2$	Oxidation and gel EP; antibodies	113
S14–S19	Me$_2$suberimidate	Antibodies	110
S18–S21	Ph(NMal)$_2$	Disappearance of protein	106
	Ph(NMal)$_2$	Antibodies	108
	HS-butyrimidate	Reduction and gel EP	114
S4–S5–S8	Me$_2$suberimidate	Ammonolysis and gel EP	117
S6–S14–S18	Me$_2$suberimidate	Ammonolysis and gel EP	116
S7–S13–S19	Me$_2$suberimidate	Ammonolysis and gel EP	117
S11–S18–S21	C(NO$_2$)$_4$	Peptide analysis	107
IF-3–S12	Me$_2$suberimidate	Antibodies	119
IF-3–S1, S11, S12, S13, S19	HS-butyrimidate	Reduction and gel EP	118
IF-2–S1, S2, S11, S12, S13, S14, S19	HS-butyrimidate	Reduction and gel EP	118

[a] Abbreviations: gel EP, polyacrylamide gel electrophoresis; HS-butyrimidate, mercaptobutyrimidate; tartaroyl(N$_3$)$_2$, tartaroyl diazides; Me$_2$adipimidate, bis-(methyl)adipimidate; Me$_2$suberimidate, bis(methyl)suberimidate; Ph(NMal)$_2$, N,N'-phenylenedimaleimide; C(NO$_2$)$_4$, tetranitromethane. For details of methods, see text, Section IV, A, 1.

these fragments are in good agreement with the sequences found in a similar fragment isolated from complete 30 S particles (*95, 96*). Furthermore, in the latter case (*96*), the breaks found in the RNA corresponded well with sites on the RNA found to be reactive toward kethoxal in the intact 30 S particle (*57*).

In this type of approach, the proteins are not covalently attached to the RNA, and therefore special care must be taken to ensure that the observed results are genuinely specific. The importance of this point has been enhanced by the finding that under conditions that promote unfolding of the ribosomal subparticle, such as EDTA or very low salt, the proteins lose their specific sites of attachment to the RNA and are able to exchange freely with heterologous RNA (*125*). Some ribonucleoprotein fragments reported were produced under such conditions and are very likely artifacts (*126*), though it should be noted that a ribonucleoprotein complex containing only S8 and S15 (cf. the foregoing discussion) has been isolated in the presence of EDTA (*127*).

3. Assembly Mapping

It has been known for some time that the 30 S proteins interact with one another in a very specific manner during *in vitro* reconstitution from protein and RNA. These interactions have been incorporated into an "assembly map" (*68*), which has since been expanded by the addition of some proteins (S12, S15, and separated S16 and S17) (*128*), which were not included in the original map (*68*). Some of these interactions have been confirmed by other workers (*129*). Although these interactions are not necessarily a direct reflection of the protein neighborhoods within the 30 S particle, there is good agreement between both those proteins which are found together in cross-linked pairs and in ribonucleoprotein fragments with those related in the assembly map. Therefore, it seems reasonable to conclude that most if not all of the assembly interactions are indeed direct reflections of the ribosomal topography. The question of whether all proteins in the complete 30 S particle have substantial contact with the RNA has yet to be settled, and it has been suggested (*130*) that it may be more appropriate to think of the assembly interactions as being between regions of ribonucleoprotein as opposed to simply interactions between proteins.

Assembly of the 30 S ribosome *in vitro* has a rather high activation energy (*131*), which implies that the assembly map may not be a good indicator of the *in vivo* assembly process, and in fact there is not very good agreement between the assembly map and the estimated order of addition of proteins *in vivo* (*132*). 30 S particles can be formed by reconstitution from ribosomal proteins and precursor 16 S RNA (*133, 134*), but the energy of activation is lowered only when proteins from nascent as opposed to completed ribosomes are used (*135*), suggesting that *in vivo* assembly may be helped by a recycling factor. Further information relevant in this context comes from experiments with methyl-deficient 16 S RNA (*136*); a distinct group of proteins is required for methylation of the ribosomal RNA, and these proteins are not all "early" proteins in the assembly map, but represent rather those proteins that the ribonucleoprotein fragment work cited above has shown to be clustered in the 5′-proximal or central regions of the RNA. Another group of proteins caused inhibition of the methylation (*136*).

4. Chemical Modification

Differential reactivity to chemical reagents has been used by many workers as a probe of ribosomal topography, the assumption being that those proteins most exposed on the ribosome surface will react most strongly with a particular protein reagent. Among the methods used have been digestion with trypsin (*137–140*), and, more recently, reaction with relatively small molecules such as kethoxal (*141*), various aldehydes (*142, 143*), acetic anhydride (*144*), or sulfhydryl reagents (*145–148*). However, interpretation of the results of such experiments is rather complicated, since exposure or lack of exposure of a particular reactive group is not necessarily a reflection of the protein topography in its wider sense. A reactive group on a protein could be shielded by RNA or by the tertiary structure of the protein itself. Furthermore, it is not easy to predict how far a chemical reagent can penetrate into the ribosome, and it is therefore not surprising that the degree of agreement between the various results is not very high. A good summary of the recent data has been made by Benkov and Delihas (*141*). At the present time, data from reactivity toward very large molecules are easier to interpret, and two methods are noteworthy in this context. The first of these is measurement of the accessibility to protein-specific antibodies (*149, 150*), and the second is the iodination of ribosomal proteins catalyzed by lactoperoxidase (*151, 152*). The general conclusion from both methods is that all 30 S proteins have some accessible groups on the ribosome surface.

5. Models of the Protein Arrangement

Several workers have attempted to coordinate the available data into three-dimensional models of the protein arrangement (*123, 153, 154*). In these models, the proteins have been represented as spheres, and the authors have been careful to point out that the arrangements are only schematic, serving mainly as a means of testing the self-consistency of the various approaches. The models show a good measure of agreement, although data obtained more recently on the relative distances between various pairs of proteins labeled with fluorescent markers (*155*) fit rather better with the first of these models (*123*) than with the other two.

However, a technique has recently been developed that casts doubt on the whole validity of this type of model-building, *viz.* the direct visualization by electron microscopy of the proteins in complexes formed between ribosomal subparticles and protein-specific immunoglobulins. This technique has become possible because the 30 S particle has a readily recognizable shape in the electron microscope, and, as mentioned above, each ribosomal protein has some antigenic determinants that are accessible to the corresponding protein-specific antibody (*149*). Thus, if the ribosome is treated with a single protein-specific antibody, a ribosome · antibody · ribosome complex is formed with the "Fab" arm of the antibody attached to a point on the ribosome surface. Several proteins have been localized on the surface of the 30 S particle by this method, namely S5, S13 and S14 (*156*), and S4 and S14 (*157*). Some examples are illustrated schematically in Fig. 3.

The technique already has a high resolving power, but the most important feature of the results is that whereas some proteins show a sin-

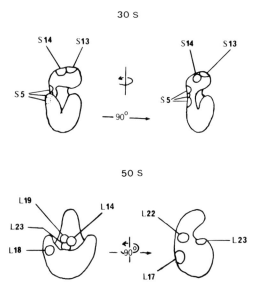

FIG. 3. Location of ribosomal proteins on the surface of the subparticles by visualization of bound antibodies by electron microscopy. The upper diagrams show the location of proteins S4, S5 and S13 on the 30 S subparticle (*156*), and the lower diagrams show proteins L14, L18, L19 and L23 on the 50 S subunit (*178*).

gle surface-binding-site for their cognate antibodies, others appear to have multiple binding sites over a wide area of the surface [e.g., S4 (*157*)]. Other proteins appear to share the latter property (*158*), which indicates that the conformation of the proteins within the ribosomal particle can be greatly extended. This, as suggested at the beginning of this section, alters the whole conception of the 30 S topography, and rather changes the interpretation that must be made of other information (such as the protein cross-linking results discussed above). However, it should be noted that the overall dimensions of the 30 S particle, as measured by electron microscopy, differ significantly from those estimated by low-angle X-ray scattering in solution (*159*). Therefore, the possibility arises that the subparticles prepared for electron microscopy may have undergone some distortion.

B. The 50 S Subparticle

1. Use of Bifunctional Protein Cross-Linking Reagents

Progress on the more complex 50 S particle has inevitably lagged behind that on the 30 S, but nevertheless the protein cross-linking approach is being applied with success. Bis-methyl suberimidate, as well as methyl 4-mercaptobutyrimidate (*118*), has been used to identify a number of protein pairs (*116, 160*). It is reasonable to expect that many more protein pairs in the 50 S subparticle will be identified by this approach in the near future.

2. Isolation of Specific Ribonucleoprotein Fragments

Isolation of specific fragments from the 50 S subparticle has proved to be much more difficult than from the 30 S. Reports of such fragments appeared some time ago in the literature (*83, 161*), but these were isolated under conditions subsequently shown to promote random exchange of the ribosomal proteins (*125*). Furthermore, the protein analysis in the early work was not reported in standard nomenclature, and is therefore difficult to relate to other data. More recently, some specific fragments from the 50 S particle have been described (*162, 163*). In one case (*162*), two small fragments were isolated, containing proteins L1 and L9, and L5, L18 and L25, respectively, and in the other (*163*), a series of fragments were isolated from various protein-deficient core particles derived from the 50 S ribosome. Many of these fragments were relatively large, and contained most of the proteins present in the corresponding core particle, but four small fragments were also identified, containing: L3, L4 and L24; L1, L3, L8/L9, L23 and L24; L1 and L23; and L5 and

L18. However, it is clear that, although the 23 S RNA can be readily broken into two very well-defined halves (83, 164–166), this is not reflected in the behavior of the ribonucleoprotein complex. The finding, by neutron scattering, that the centers of mass of protein and RNA are widely separated in the 50 S particle (167) as opposed to the 30 S (168) could explain this difference in the behavior of the two subparticles after nuclease treatment.

3. Assembly Mapping

It is perhaps the lack of an assembly map of the 50 S particle that has most significantly retarded progress on the topography of the larger subparticle. A reassembly procedure for 50 S ribosomes from *E. coli* was reported in 1971 (169), but this report was not confirmed, and although 50 S ribosomes from *B. stearothermophilus* were reconstituted some time ago from their constituent RNA and protein by taking advantage of the thermal stability of this organism (170), a successful total reconstitution of the *E. coli* 50 S particle has only recently been described (171). The method involves a two-step incubation procedure, the second step of which corresponds to the "partial reconstitution" process (172, 173), which was developed to reassemble protein-deficient core particles into active ribosomes. However, some assembly interactions have been determined using the latter process (174).

4. Chemical Modification

Chemical modification techniques have been applied to the 50 S particle in the same way as they have to be the 30 S. In addition to the literature already cited in connection with the 30 S particle, two other publications should be noted, namely, the reaction of 50 S subparticles with Celite-bound fluorescein isothiocyanate (175), and further work on the lactoperoxidase-catalyzed iodination of the proteins (176). Again, the reader is referred to the paper by Benkov and Delihas (141), or to this last paper (176) for a summary of the recent data. The general conclusion is that the 50 S proteins are not all accessible to modifying reagents, though, as with the 30 S particle, there is not very good agreement between the various sets of data. In contrast, most 50 S proteins appear to have some antigenic determinants on the subparticle surface accessible to their cognate antibodies (158).

5. Models of the Protein Arrangement

There is as yet not nearly enough information to allow the building of models of the arrangement of all the 50 S proteins. However, a number of proteins have been located on the surface of the particle using the electron-microscope technique described for the 30 S particle. Proteins L1 and L19 (*177*), and L14, L17, L18, L19, L22, and L23 (*178*) have been localized by this method. It is to be expected that progress on the 50 S particle, using this method and those described above, will be fairly rapid in the immediate future.

C. The 30–50 S Subparticle Interface

The interface between the 30 S and 50 S particles is obviously a highly important area in terms of ribosomal function. Direct evidence for the involvement of proteins in this region is rather difficult to obtain, but there are nevertheless a number of lines of evidence that implicate various proteins at the interface.

First, 70 S ribosomes containing labeled 50 S particles have been cross-linked with glutaraldehyde, and the total protein has been extracted and tested against antibodies to 30 S proteins (*179*). It would be expected that a 30 S protein cross-linked to a 50 S protein would cause coprecipitation of radioactivity in the presence of antibody, and protein S16 was identified by this process. Second, antibodies to a number of proteins have been shown to inhibit subparticle reassociation, or to cause aggregation of both the individual subparticles (suggesting the presence on the "wrong" subparticle of small amounts of the protein concerned). The proteins concerned are S9, S11, S12, S14, S20, L1, L6, L14, L15, L19, L20, L23, L26 and L27 (*180*). Of these, protein S20 is a special case, as it is identical with L26 (*1*). Further, the combined amounts of S20, originally designated a "fractional" protein (*98*), and L26, give a value of one mole per ribosome (*181*). This indicates that the protein can associate with either subparticle, and is therefore presumably located at the subparticle interface. Third, proteins S11 and S12 appear to be able to bind preferentially to the 23 S RNA (*180*), again suggesting that they are interface proteins.

Other more indirect lines of evidence have implicated other proteins. For example, specific proteins can restore the ability of inactive protein-deficient cores to form 70 S particles (e.g., *182*), and antibodies to several proteins have been shown to inhibit interface-related functions such as tRNA binding (*183*), (see also Section V, A, 2 on function). Chemical modification has also been used to probe the interface question, with some strange results. Although a number of 50 S proteins (L2, L26, L28 and L18) showed a reduced reactivity in the 70 S particle as opposed to the 50 S (*176*), several 30 S proteins already implicated as being at or near the interface (e.g., S9) actually showed an enhanced reactivity in the 70 S particle (*184*). These results were obtained with the enzymic iodination technique already discussed and were interpreted as indicating a pronounced conformational change of the 30 S particle upon 70 S formation. In any event, it is clear that much remains to be done before the nature of the subparticle interface is fully understood.

V. Ribosomal Function

The description of bacterial ribosomal function presented below is based on a recently published review (185), and is mainly concerned with reports that have appeared subsequently. As in that review, we adopt the classical two-site "donor–acceptor" model of ribosome function, and the discussion is divided into the three established steps of initiation, elongation, and termination of protein synthesis.

A. Initiation

1. Steps in the Initiation Process

"Initiation" covers a series of steps, starting with the native 30 S subparticle and ending with the "70 S initiation complex," which consists of 70 S ribosomes, fMet-tRNA^{fMet} (bound in the P-site, or puromycin-sensitive site), and mRNA. These steps are presented schematically in Fig. 4, which also shows tentatively the stages where the various initiation factors (IFs) enter and leave the complex.

The first stage of the reaction is the binding of IF-3 (MW 22,500) to the native 30 S subparticle. This induces a conformational change in the 30 S particle (186, 102), which may then no longer be able to associate with 50 S subparticles to form a 70 S ribosome. In log-phase cells, there is one molecule of IF-3 per twenty 30 S subparticles (187). The next stage of the process is the binding of initiator tRNA and mRNA, but the precise sequence of events involved here is still under discussion. Evidence is accumulating that the binding of fMet-tRNA precedes the binding of mRNA (188–190), and even deacylated tRNA can be bound to 30 S subparticles in 1:1 stoichiometry in the absence of mRNA (191). Thus, it appears that the sequence of events resembles the situation in eukaryotic systems (192), but it must be emphasized that this view is not universally accepted. It was reported that, prior to binding of fMet-tRNA, the E. coli 30 S subparticle forms a stoichiometric complex with mRNA and IF-3 (193). Further, 30 S subparticles from E. coli are capable of binding MS2 RNA even in the absence of initiation factors (194).

The binding of the initiator tRNA (fMet-tRNA) depends on IF-2

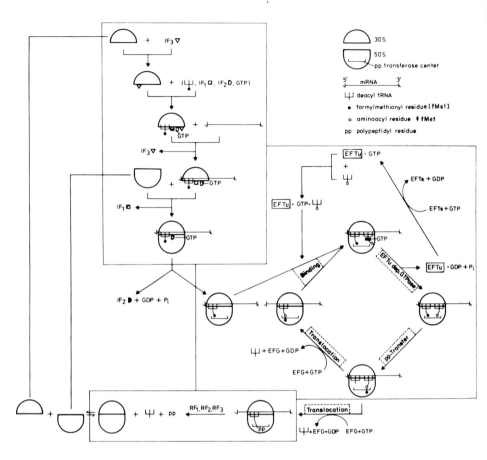

Fig. 4. A schematic representation of ribosomal function during protein biosynthesis in prokaryotes. The boxes indicate (from top to bottom) initiation, elongation and termination, respectively.

(MW 90,000) and GTP, and the binding is stimulated both by IF-1 (MW 9400) and IF-3 (*195, 196*). The most important requirements for the initiator tRNA seem to be the specific tRNA molecule (tRNA^{fMet}), and the blocked NH$_2$ group of the amino acid, whereas the nature of the amino acid itself seems to be less important. This is indicated by the following observations: Mischarged and formylated tRNA^{fMet} (i.e., fVal-tRNA^{fMet} or fPhe-tRNA^{fMet}) works as well in initiation as fMet-tRNA^{fMet} itself (*197*). IF-3 can stimulate the binding of Ac · Phe-tRNA^{Phe} to 30 S subparticles in the presence of poly(U) (*198*), and, further, fMet-tRNA^{fMet} can "chase" the Ac · Phe-tRNA from the 30 S initiation complex containing the latter [i.e., 30S · (Ac-Phe-tRNA) · poly(U)] (*199*). This last observation can be taken as a hint that codon–anticodon interaction does not play an important part in the binding of initiator tRNA to the P-site. The precise steps at which IF-1 enters and leaves the initiation complex are not known at all, and no specific function has as yet been assigned to this factor, other than its ability to stimulate the IF-2 and IF-3-dependent functions. It should be noted that in eukaryotic systems, in contrast to prokaryotic systems, a free amino group on the aminoacylated initiator tRNA (Met-tRNA^{Met}) is necessary for activity in initiation.

Formation of the 30 S initiation complex, 30S · (fMet-tRNA) · mRNA plus GTP and factors, is completed by the binding of mRNA, and IF-3 appears to leave the complex after association of the 50 S subparticle, or alternatively after both mRNA and fMet-tRNA have been bound (*193, 200–202*). Association of the 50 S subparticle to give the final 70 S initiation complex has no factor requirement. As mentioned previously (see Section IV), this last step may induce a second conformational change in the 30 S subparticle, as some 30 S proteins (S9 and S18) can be enzymically iodinated significantly more strongly in the 70 S ribosome than in the isolated 30 S particle (*184*).

Dissociation of IF-2 from the complex is accompanied (or probably caused) by the IF-2-dependent GTP hydrolysis (GTP to GDP plus P_i). The fMet-tRNA^{fMet} is now in such a position that the peptidyltransferase can connect the formylmethionyl residue via a peptide bond to a substrate bound in the A-site (e.g., puromycin or aminoacyl-tRNA), and with this step the initiation complex enters the elongation cycle.

2. Ribosomal Components Involved in Initiation

a. Binding of IF-3 to the 30 S Particle. It has been reported that IF-3 binds to the 16 S RNA rather than to the ribosomal proteins (*203*). However, this finding was interpreted differently by other workers (*204*), who claimed that the binding to both 16 S and 23 S RNA is nonspecific, whereas the stoichiometric (1:1) binding of IF-3 to the 30 S particle is protein-dependent. Recently it was reported that IF-3 could be crosslinked to protein S12 (*119*) and in addition to S1, S11, S13 and S14 (*154*) (see also Section IV, A, 1). This agrees with the finding that both IF-3 and S12 are functionally related in the process of mRNA recognition; two subspecies of IF-3 have been described with a high selectivity for different mRNAs (*205*), whereas the different cistron specificity between *E. coli* and *B. stearothermophilus* depends mainly on S12, and to a lesser extent on S10 and 16 S RNA (*206*). The fact that the specificity was affected at all by the 16 S RNA suggests that the 16 S RNA may interact with mRNA during initiation (see below). According to another recent report, S1 is important for this cistron specificity (207); *B. stearothermophilus* ribosomes incubated with protein S1 from *E. coli* showed *E. coli* specificity with respect to phage RNA cistrons.

b. Binding of IF-1, IF-2, GTP, and Initiator tRNA. The sequence of binding of these components is not yet known. A ternary complex between IF-2, GTP, and fMet-tRNA, analogous to the elongation complex between EF-Tu, GTP, and aminoacyl-tRNA, could not be detected. IF-2 was cross-linked to S13, S19, and to a lesser extent to S1, S11 and S12 (154). Various other techniques have been applied to the study of the ribosomal components important for initiation factor-dependent binding of fMet-tRNA, such as single-protein-omission experiments (128), addition of fractional proteins (208), and inhibition by specific antibody fragments (183). Proteins S3, S14 and S21 were shown by at least two of these techniques to be involved in fMet-tRNA binding.

In addition, the 16 S RNA may play a part in the recognition of the initiator tRNA. Binding of fMet-tRNA to the 30 S particle can be prevented by kasugamycin (209, 210), and resistance to this drug in mutants may be due to an alteration in either the protein or the RNA moiety of the small subparticle. Lack of methylation of two adjacent adenine residues 23 nucleotides from the 3' end of the 16 S RNA can confer resistance to the drug (211), again suggesting that the ribosomal RNA may be involved in recognition of the initiator tRNA. On the other hand, alteration of proteins S4 (212) or S2 (213) also confer drug resistance.

A detailed and extended discussion of various aspects of initiation will be found in a future volume (see chapter by M. Grunberg-Manago. and F. Gros). The GTPase activity dependent on IF-2 is discussed below, together with other factor-dependent GTP hydrolyses (Section V, B, 3).

c. Binding of mRNA. Besides the importance of IF-3, S12 and S1 for cistron selectivity already mentioned, protein S1 is also directly involved in mRNA binding (214, 215), and this dependence on S1 seems to be independent of the functional state of the ribosome. The binding of artificial or natural mRNA (216) is observed only when S1 is present on the ribosome. Further, although S1 is easily lost during isolation of the small subparticle, it is found with a 1:1 stoichiometry on ribosomes active in poly(Phe) synthesis (217). S1 seems to be located near the 3' end of the 16 S RNA (121) (see Section IV, A, 1 on topography of ribosomal subparticles). Proteins S4 and S18 also appear to be close to the location where codon–anticodon interaction takes place; after an affinity-label analog of the trinucleotide AUG was allowed to react with 70 S ribosomes, labeled 30 S subparticles programmed for initiation factor-dependent binding of fMet-tRNA could be isolated. Analysis revealed that S4 and S18 were the primary targets of the affinity label (*218*).

The 16 S RNA may play a crucial role in the recognition process of the initiation region on the mRNA. Since the triplet AUG codes for the internal methionyl-tRNA as well as for the initiator fMet-tRNA^{fMet}, it is clear that codon–anticodon interaction between AUG and a tRNA is not sufficient for recognition of the initiation signal. Sequence studies of the 3′ end of 16 S RNA (*37*) and of the initiator regions of different mRNAs [see Shine and Dalgarno (*37*) for review] demonstrate that the sequence G-A-U-C-A-C-C-U-C-C-U-U-A$_{OH}$ at the 3′ end of 16 S RNA is complementary to a sequence of 4–6 nucleotides (e.g., G-G-A-G-G-U) found 12–18 nucleotides away from the initiation AUG triplet on the mRNA, in the direction of the 5′ end.

The 5′ end of the 16 S RNA may also be important for recognition of the initiator region of mRNA. A comparison of the known sequence of the 5′-terminal region of 16 S RNA with some initiator regions from RNA phages revealed a striking homology of sequence in hairpinlike structures containing the initiator AUG triplet at the top of the hairpin loop. A model for mRNA recognition by means of rRNA · mRNA hybrid structure formation has been proposed (*219*).

d. Association with 50 S Subparticle. Some information is available on the ribosomal components necessary for subparticle association. Partial reconstitution experiments with protein-deficient cores derived from the 30 S particle show that proteins S2, S5 and S9 are important for coupling of the subparticles. Since a 70 S ribosome is required in the EF-G dependent GTPase activity, it is not surprising that this same set of proteins can stimulate the latter reaction (*220*). However, single-protein-omission tests (i.e., reconstitution of the 30 S particle from protein and RNA, but with one protein absent) demonstrate that the omission of S2 or S5 does not influence the stimulatory effect of the reconstituted particle, whereas the omission of proteins S4, S7, S8, S9, S11 and S15 has a substantial effect (*221*). As the author pointed out, S4, S7, S8 and to a lesser extent S9 are important structural proteins, and therefore their effect may be simply due to their fundamental role in 30 S assembly. Thus the most interesting proteins from the point of view of subparticle association are more probably S11, S15 and possibly S9. The different results from partial, as opposed to total, reconstitution experiments may indicate that there are several attachment points at the subparticle interface. Several proteins have been implicated to be at this interface (see Section IV, C), but it remains to be seen how many of these are necessary for formation of the 70 S ribosome.

B. Elongation

1. Steps in the Elongation Cycle

After the IF-2-dependent GTP hydrolysis and concomitant release of IF-2, the 70 S initiation complex is free of all initiation factors. The complex at this stage consists of the 70 S ribosome, bound to the initiator region of the mRNA and carrying the initiator tRNA in the P-site (puromycin-sensitive site). The next step is the binding of a ternary complex between EF-Tu (MW 47,000), GTP and aminoacyl-tRNA to the adjacent A-site (puromycin-insensitive site). The anticodon of this tRNA molecule binds to the complementary mRNA codon in the A-site. GTP hydrolysis then takes place, and the resulting GDP · EF-Tu complex leaves the ribosome, to be later regenerated by the factor EF-Ts (MW 34,000) to GTP · EF-Tu, which in turn can associate with a further aminoacyl-tRNA molecule, thus re-forming the ternary complex of EF-Tu, GTP and aminoacyl-tRNA. When EF-Tu has left the ribosome, the aminoacyl-tRNA in the A-site is ready to accept the peptidyl (or formylmethionyl) residue from the tRNA in the P-site, and peptide bond formation can occur [see Lucas-Lenard and Lipmann (222) for review]. The aminoacyl residue in the tRNA can isomerize from the 2'- to the 3'-position of the terminal nucleotide (223), and vice versa, with the short half-reaction time of 0.3 msec (224). The actual position of the aminoacyl residue may be different in the various steps of the elongation process (225–230); the evidence can be summarized as follows. (a) The synthetase (or ligase) acylates the tRNA in the 2'-position, which position is favored in the ternary complex between EF-Tu, GTP and aminoacyl-tRNA. (b) Once it is bound to the A-site, the aminoacyl-tRNA must isomerize to the 3'-position, as only this configuration shows significant acceptor activity. (c) After the subsequent translocation to the P-site, the situation is not clear, but there are hints that, regardless of the position of the aminoacyl residue, the adjacent hydroxyl group must be free.

The peptidyltransferase reaction has been studied by affinity labeling experiments using a puromycin analog, which could be covalently bound to the 23 S RNA in the 50 S particle. By extrapolation, it was shown that two molecules of the puromycin analog must react with the RNA in order to block the peptidyltransferase activity, and the reaction with the RNA can be completely inhibited by chloramphenicol (231), which binds specifically to the A-site moiety of the peptidyltransferase center. Two interesting conclusions were drawn by the authors. (a) The two reaction sites of the puromycin analog were interpreted as being the

A and P regions of the peptidyltransferase center, supporting the classical "two-site" model of protein synthesis. (b) Although chloramphenicol binds only to the A region of the peptidyltransferase center, it prevents reaction with both sites. Thus, the reaction of the affinity label seems to be cooperative, in that reaction with the A region of the peptidyltransferase center enhances the reactivity of the P region. The cooperative effect between both sites is supported by several other observations. Puromycin has a much higher affinity for polysomes carrying peptidyl-tRNA than for washed ribosomes carrying acetylphenylalanyl-tRNA or polylysyl-tRNA (232). Initiation-factor-dependent binding of Ac-Phe-tRNA to the P-site strongly enhances the subsequent EF-Tu-dependent binding of Phe-tRNA to the A-site (233). Further, a transient binding of some drugs (e.g., chloramphenicol) stimulates peptidyltransferase activity (234), indicating that a substrate bound to the A region optimizes the conformation of the peptidyltransferase center. Comparative studies with a series of substrates in the P region of the peptidyltransferase center demonstrate that the hydrophobicity of the first two amino acids plays an important part with respect to the donor activity of the substrate; the more pronounced the hydrophobicity, the better is the donor activity (235). This is in good agreement with the fact that the loss of one positive charge on the aminoacyl residue as a result of N-acetylation changes a molecule that is purely a substrate for the A region of the peptidyltransferase center [viz. C-A-C-C-A(Leu)] to one that is purely a substrate for the P region [C-A-C-C-A(Leu-Ac)].

After peptide bond formation (peptidyl transfer), the peptidyl-tRNA is now located in the A-site, prolonged by one amino-acid residue. The subsequent translocation is directed by EF-G (MW 83,000), and EF-G-dependent cleavage of GTP occurs. The term translocation means the displacement of the ribosome by one codon length (ca 10 Å) toward the 3′ end of the mRNA. This movement is characterized by three events: (a) the deacylated tRNA leaves the ribosome; (b) the peptidyl tRNA moves from the A-site to the P-site; and (c) a new codon enters the A-site.

The movement along the mRNA during translocation seems to be phased by the codon–anticodon interaction in the A-site. Evidence for this comes from the analysis of a frameshift suppressor tRNA whose anticodon interacts with four instead of three nucleotides on the mRNA (236). After this suppressor tRNA binds to the A-site, the ribosome moves a distance of four (instead of three) nucleotides toward the 3′ end of the mRNA. Thus, the translation becomes again in phase with the codon frame, if the original frameshift mutation was caused by addition of one extra nucleotide.

After translocation, EF-G and GDP leave the ribosome, which is then ready to enter a new cycle of elongation.

2. Ribosomal Components Involved in the Elongation Process

This topic has been discussed in detail in a previous review (*185*), and therefore the following discussion is primarily concerned with the more recent work.

The ribosomal environment of the peptidyl residue of peptidyl tRNA bound to the P-site has been examined by affinity labeling experiments (*237, 238*), and the various authors found similar groups of proteins to be involved, namely, L27, L2, L13, L14, L15, L16, L32 or L33. By using homologs of the affinity label Br-Ac-(Gly)$_n$-Phe-tRNA^{Phe} of various chain lengths (*239*), some of these proteins could be arranged in order. With increasing numbers of glycine residues (n), the following proteins were labeled: L2 and L16 (n = 2), L26 or 27 (n = 4 to 6), L32 or L33 (n = 9 to 10), and L24 (n = 10 to 16). These proteins are therefore located in the vicinity of the peptidyl residue. Affinity labeling experiments with analogs of fMet-tRNA (*240, 241*) resulted in approximately the same protein pattern, indicating that the puromycin-reactive site of fMet-tRNA in the 70 S initiation complex is similar or identical to that of peptidyl tRNA, at least with respect to the orientation of the tRNA on the 50 S subparticle. The ribosomal RNA is also involved in this region, since a photoaffinity label analog of peptidyl-tRNA attached to 70 S ribosomes reacted to the extent of 65% to 70% with RNA as opposed to protein in the 50 S particle (*242*). The binding to 23 S RNA of a puromycin analog (*231*), already mentioned above, should also be borne in mind in this context.

The environment of the aminoacyl residue of aminoacyl-tRNA bound to the A-site was identified by chloramphenicol binding (*243*), by affinity labeling (*244*), and by binding studies with aminoacylated fragments of tRNA (*245*). Proteins L6 and L16 were identified from these experiments, and L6 was further found to be the primary target of N-iodoacetylpuromycin (*246*). Proteins L11 and L16 seem to be neighbors at the peptidyltransferase center, as concluded from the observations that there is only one binding site for chloramphenicol per 50 S subparticle (*243, 247*), but protein L11 appears to be involved in chloramphenicol binding in addition to the main drug-binding protein L16. Thus, both proteins may be part of the one binding site (*245*).

It has been suggested that the tRNA molecule itself interacts with the 5 S RNA (*248*). All tRNA molecules, with the exception of eukaryotic initiator tRNAs (*249, 250*) and tRNAs needed in cell wall synthesis (*251, 252*), contain the sequence T-ψ-C-G in loop IV, and the partially complementary sequence G-A-A-C is found in all the prokaryotic 5 S RNA species that have to date been sequenced. Experimental support for this idea has come from studies on the binding of T-ψ-C-G to 5 S RNA (*253*), and competition experiments with this oligonucleotide and aminoacyl-tRNA (*254*). Also, in "magic spot" formation, T-ψ-C-G could be used as a substitute for the mRNA-directed binding of deacylated tRNA to the A-site (*255*). However, the T-ψ-C-G region of tRNA in solution is not accessible. This can be demonstrated by considering the tertiary structure of crystallized tRNA^{Phe} and comparing the sequence of this RNA molecule with the conserved regions of 56 other tRNA species of known primary structure. All the nucleotides hydrogen-bonded to T-ψ-C in tRNA^{Phe} belong to the conserved regions (*256*). This inaccessibility was also shown by chemical modification studies (*257*). Further, the postulated complementary binding sequence (G-A-A-C) on the 5 S RNA is not accessible in the 50 S subparticle, since the G-specific reagent kethoxal does not react with a G-residue in this specific region of the 5 S RNA (*258*), although other G-residues in 5 S RNA are attacked.

Therefore, if the tRNA · 5 S RNA binding hypothesis is to be tenable, one must postulate a conformational change in both tRNA and 5 S RNA upon binding to the A-site. In fact, such a conformational change of aminoacyl-tRNA was reported in the ternary complex between EF-Tu, GTP and aminoacyl-tRNA (*259*), since the oligonucleotide G-A-A-C could well be bound to the ternary complex, as opposed to the free aminoacyl-tRNA. On the other hand, no change in the secondary structure of aminoacyl-tRNA could be detected by nuclear magnetic resonance, either after formation of the ternary complex with EF-Tu, or after binding to the 70 S particle (*260*).

It is also possible that the T-ψ-C-G loop of the tRNA could interact with a complementary sequence on 16 S or 23 S RNA, and the fully complementary sequence (C-G-A-A) occurs 4 or 5 times in the 16 S RNA sequence alone (*33*). Reaction of kethoxal with 16 S RNA in the 30 S subparticle leads to a loss of the ability to bind tRNA, whereas mRNA binding is not impaired (*261*). A detailed discussion of the possible binding of tRNA to 5 S RNA is presented elsewhere in this volume (see chapter by V. A. Erdmann).

The enzymic activity causing formation of the peptide bond is termed peptidyltransferase and is an intrinsic property of the 50 S subparticle (*262, 263*). Partial reconstitution experiments demonstrate that the activity is dependent on the presence of protein L11 (*245, 264*), and this finding was confirmed by an elegant affinity-labeling experiment (*265*). A nonradioactive photoaffinity label analog of the peptidyl-tRNA was bound to the P-site and was allowed to react with tritiated puromycin by peptide-bond formation. The sample was irradiated, and the ribosomal proteins were analyzed for radioactivity; most of the counts were associated with L11 and, to a lesser extent, with L18. In contrast, a protein-deficient core particle derived from the 50 S subparticle has been described that is fully active in the peptidyltransferase assay, and yet contains no L11 as judged by two-dimensional electrophoresis of the proteins (*266, 267*); the authors concluded that L11 is not the peptidyltransferase protein.

More recently, a detailed study of these core particles in the "fragment assay" has revealed that the latter conclusion is not necessarily valid (*268*). This is because less than 1% of a normal population of 50 S particles is active in the "fragment assay," whereas the lower limit of detection of a protein by two-dimensional electrophoresis is 5–10% of its normal content. Further, inactive particles tend to lose proteins (including L11) more easily than active particles. It follows that undetectable amounts of L11 could remain attached to the core particle, without impairment of activity. A recent report concludes, from partial reconstitution experiments, that L16 is essential for the peptidyltransferase activity (*269*). However, the authors describe reconstituted particles apparently free from L16 that still show low but significant activity. This implies that L16 is not an absolutely essential protein for the peptidyltransferase activity, and it is already known that the L11-dependent activity can be markedly stimulated by L16 (*245*). L11 is therefore still a good candidate for the peptidyltransferase protein.

3. Factor-Dependent GTPase Activity

The three largest ribosomal factors, namely IF-2, EF-Tu and EF-G, induce GTPase activity when bound to ribosomes, and this subject has been recently reviewed in detail (*270*). The GTPase activity can be studied by two methods: "uncoupled GTPase activity," which involves only 70 S ribosomes, GTP and the appropriate factor, or "coupled GTPase activity" in which mRNA and aminoacyl-tRNA are also present. All three factors bind to overlapping ribosomal sites, in which the most important proteins are L7 and L12. Core particles lacking L7 and L12 show a drastic decrease in EF-Tu and EF-G binding and consequently a decrease in their induced GTPase activity [see review by Möller (*270*)], although these cores were able to associate with 30 S particles (*271*).

A similar, though less drastic, effect has been observed in the case of IF-2 GTPase activity in the absence of L7 and L12 (*272–274*). Thus, binding of IF-2 is not so strongly dependent on L7 and L12 as that of

EF-Tu and EF-G, reflecting the fact that IF-2 binds to the 30 S subparticle (see Section V, A, 1). Nevertheless, the binding sites of the three factors are overlapping, since EF-G competes with the IF-2-dependent AUG-directed binding of fMet-tRNA (275). In addition to IF-2, EF-Tu and EF-G, the release factors (RF) also bind to the ribosome via L7 and L12, although the RF-dependent hydrolysis of the nascent peptidyl-tRNA does not require hydrolysis of GTP (276).

It is clear that proteins L7 and L12 enable or facilitate the binding of a variety of factors directing different ribosomal functions. To avoid competition between the factors, the factor-binding must be a precisely controlled process in each step of protein synthesis, and this seems to require a highly defined structure of L7 and L12. In contrast to other proteins, L7 and L12 have been conserved during evolution from proto eukaryotes (see Section II, A), as demonstrated by studies with protein-specific antibodies (277). Further, eukaryotic "L7" and "L12" proteins can be extracted from the cytoplasmic ribosomes of yeast or rat liver, and are functionally interchangeable with L7 and L12 from *E. coli* ribosomes (270, 277). In contrast, the factors involved in eukaryotic cytoplasmic protein synthesis are not interchangeable with prokaryotic factors, with the exception of the prokaryotic EF-Tu, which can replace the eukaryotic T1, but not vice versa [see Lucas-Lenard and Lipmann (222) for review].

As already mentioned (Section II, A), proteins L7 and L12 have identical amino-acid sequences and differ only by an acetyl group on the N-terminal serine of L7. The physical assembly of these proteins into the 50 S particle depends on L6 and L10, as shown by reconstitution experiments (278), and also on L18, as demonstrated by antibody studies (279). L7 and L12 do not themselves constitute the site of GTPase activity, since ribosomes depleted of these proteins can exert EF-Tu- and EF-G-dependent GTP hydrolysis in the presence of methanol (280).

A complex between protein and 5 S RNA exerting factor-independent GTPase and ATPase activity has been described; the complex contained proteins L5, L18 and L25 (281). This finding is in good agreement with the result of a photoaffinity label experiment with a GDP analog, which reacted with L11, L5, L30 and L18 (282). Further, a series of ribosomal fragments was isolated from the 50 S particle after ribonuclease digestion (see Section IV, A, 2), and several of the smaller fragments were tested for factor-independent GTPase activity; these experiments point to L18 as being the protein responsible for the GTPase activity (163).

A recent report claims that factor EF-Tu carries its own GTPase center. Kirromycin, an antibiotic not related to any known class of antibiotic, binds to EF-Tu and prevents dissociation of the factor from the ribosome. When kirromycin was bound to EF-Tu, the (EF-Tu) · antibiotic complex showed a small but significant GTPase activity (283).

As mentioned above, L11, the protein intimately involved in the peptidyltransferase activity (264), can be labeled by the photoaffinity GDP analog, which suggests that the peptidyltransferase and GTPase centers are close to one another. Proteins L11 and L18 were further the main targets of a photoaffinity label probe of the peptidyltransferase center [(265) and see Section V, B, 2].

The question remains, what is the function of the GTP hydrolysis? Poly(U)-dependent synthesis of poly(Phe) can be demonstrated in a system containing poly(U), Phe-tRNA and ribosomes, but without factors, GTP or S_{100} enzymes (284, 285), and the existence of this "nonenzymic translation system" shows that GTP hydrolysis is not necessary for the basic ribosomal functions such as tRNA binding, peptide bond formation or translocation. GTP is needed for the dissociation of EF-G from the complex between ribosome, factor and GTP, since the energy of the GTP cleavage was used neither for the translocation nor for removal of the deacylated tRNA (286). Thus, the general function of the factor-dependent GTPase activities may be simply to promote the dissociation of the ribosome-bound factor.

This interpretation is supported by studies with thiostrepton, a drug that binds to protein L11 (287). Thiostrepton blocks the coupled GTPase activity of the factors IF-2, EF-Tu and EF-G and prevents dissociation of these factors, whereas the uncoupled GTPase activities and translocation are not affected. The overall effect of the drug is therefore an inhibition of the binding of aminoacyl-tRNA to the A-site [(288) and Vasquez (289) for review]. The mode of action of the drug implies that: (a) the three factors IF-2, EF-Tu and EF-G bind to overlapping sites; (b) translocation does not require GTP hydrolysis; (c) the GTPase and peptidyltransferase centers are in close proximity (as already suggested above), since thiostrepton binds to L11 as well as inhibiting the coupled GTPase activities; (d) a second GTPase center may exist, since thiostrepton inhibits the coupled GTPase activity in contrast to the uncoupled GTPase activity. (In the case of EF-Tu, this second center may be located on the factor itself, as suggested by the kirromycin experiment discussed above.)

In addition to its function as an energy supply for the release of factors, GTP may also have a second steric function, not involving cleavage, at least in the case of EF-Tu. EF-Tu requires GTP to form its ternary complex with aminoacyl-tRNA, prior to binding to the ribosome. This implies that the GTP may be acting allosterically, changing the conformation of EF-Tu in such a way as to allow it to bind the aminoacyl-tRNA and associate with the ribosome. After cleavage to GDP, the EF-Tu could then "refold" and leave the ribosome. This idea is supported by spin-label experiments (*290*), and by tritium exchange (*291*) and fluorescence studies (*292*), which suggest that the conformation of (EF-Tu) · GTP differs from that of (EF-Tu) · GDP.

C. Termination

When a termination codon enters the ribosomal A-site after some rounds of the elongation cycle, protein synthesis stops. A release factor, either RF-1 (MW about 44,000), which is specific for the codons UAG and UAA, or RF-2 (MW about 47,000), which is specific for UAA and UGA, promotes the release of the polypeptide. The release factor activity is enhanced in the presence of a third factor, RF-3, and GTP or GDP (*293*). In this case, the role of GTP and GDP is not clear; the GTP is not cleaved in prokaryotic systems, but in contrast does appear to be hydrolyzed in eukaryotic systems (*294*).

There are some indications that the termination signal consists of more than one nonsense codon. Strains containing efficient nonsense suppressors grow at the same rate as those without suppressors, indicating that the nonsense mutation is suppressed more efficiently than the termination signal. In support of this idea, two adjacent terminator codons were found at the end of the coat protein genes from the RNA phages f2 and R17 (*295*), and in *E. coli* the frequency of such tandem codons was estimated to be 13% of all termination signals (*296*). The function of this second terminator codon can probably be fulfilled, though less effectively, by a codon similar to a terminator triplet. This was demonstrated by the finding that a nonsense mutation followed by the triplet CAA (which resembles UAA) was not suppressed as efficiently as the same nonsense codon followed by other triplets (*297*).

It has been suggested that the 16 S RNA from the 30 S particle plays a part in the recognition of the termination signal. The evidence came from sequence studies of the 3'-terminus of RNA from the small subparticle of ribosomes from *E. coli*, yeast, *Drosophila*, and rabbit reticulocytes, which showed that in all cases the last three nucleotides were U-U-A$_{\text{OH}}$, complementary to the nonsense codon UAA (*37, 298*).

Antibiotics that inhibit peptide-bond formation show the same specificity against the peptidyl-tRNA hydrolysis reaction required for the termination process [see Vasquez (*289*) for review]. Thus it is most likely that this hydrolysis is made by the peptidyltransferase itself, and this concept was confirmed by antibody studies on a model termination system, where it was found that only anti-L11 and anti-L16 inhibit the termination process (*299*). L11 and L16 are neighbors at the peptidyltransferase center, and L11 is intimately involved in the peptidyltransferase activity, as already discussed (see Section V, B, 2).

In addition to hydrolytic release of the polypeptide chain, the termination process also involves release of the deacylated tRNA, release of mRNA and dissociation of the ribosome. The precise sequence of these steps is not yet clear, but it has been reported that the mRNA is released, leaving the ribosome as a 70 S particle (*300*), which is then free to equilibrate with the 30 S and 50 S subparticle pools. The 30 S and 50 S subparticles can subsequently reenter a new cycle of protein synthesis.

VI. Conclusion

The amino acid and nucleotide sequences of a substantial number of the ribosomal proteins and nucleic acids have now been established, and it can be expected that the primary structures of all the ribosomal components will soon be determined. On the other hand, the problem of how these components are arranged within the ribosomal particles remains one of the most challenging questions in ribosome research. A final solution to this topography problem is too complex even to contemplate at the present time, but the techniques that are summarized here for the investigation both of RNA–protein interactions and the spatial organization of the proteins have already led to a crude understanding of many features of the ribosomal structure. These techniques are by no means exhausted, and it is to be expected that their further application, combined with new methods, will lead to fairly rapid progress in this direction.

On the functional side, research over the past two or three years has led to a detailed knowledge of most of the individual reactions involved in protein synthesis. In many cases, this has reached the point where functional studies have become intimately connected with structural questions, involving the elucidation of those ribosomal components specifically involved in the function under consideration. [For Addendum to article, see page 323.]

References

1. H. G. Wittmann, *in* "Ribosomes" (M. Nomura, P. Lengyel and A. Tissières, eds.), p. 93. Cold Spring Harbor Press, Cold Spring Harbor, New York, 1974.
2. E. Kaltschmidt and H. G. Wittmann, *PNAS* **67**, 1276 (1970).
3. G. R. Craven, P. Voynow, S. J. Hardy and C. G. Kurland, *Bchem* **8**, 2906 (1969).
4. E. Kaltschmidt, M. Dzionara and H. G. Wittman, *Mol. Gen. Genet.* **109**, 292 (1970).

5. G. Mora, T. Palaih, L. Lutter, G. R. Craven, D. Donner and C. G. Kurland, *Mol. Gen. Genet.* **112**, 229 (1971).
6. E. Kaltschmidt, *Anal. Biochem.* **43**, 25 (1971).
7. B. Wittmann-Liebold, *Z. Physiol. Chem.* **352**, 1705 (1971).
8. B. Wittmann-Liebold, *Z. Physiol. Chem.* **354**, 1415 (1973).
9. B. Wittmann-Liebold, A. W. Geissler, and E. Marzinzig, *J. Supramol Struct.* **3**, 426 (1975).
10. J. Reinbolt and E. Schiltz, *FEBS Lett.* **36**, 250 (1973).
11. B. Wittmann-Liebold and B. Greuer, in preparation.
12. H. Hitz, D. Schäfer and B. Wittmann-Liebold, *FEBS Lett.* **56**, 259 (1975).
13. H. Stadler, *FEBS Lett.* **48**, 114 (1975).
14. R. Chen and B. Wittmann-Liebold, *FEBS Lett.* **52**, 139 (1975).
15. M. Yaguchi, *FEBS Lett.* **59**, 217 (1975).
16. C. Terhorst, W. Möller, R. A. Laursen and B. Wittmann-Liebold, *EJB* **34**, 138 (1973).
17. K. Bitar and B. Wittmann-Liebold, *Z. Physiol. Chem.* **356**, 1343 (1975).
17a. N. V. Dovgas, L. F. Markova, T. A. Mednikova, L. M. Vinokurov, Y. B. Alakov and Y. A. Ovchinnikov, *FEBS Lett.* **53**, 351 (1975).
18. K. Bitar, *BBA* **386**, 99 (1975).
19. H. G. Wittmann and B. Wittmann-Liebold, in "Ribosomes" (M. Nomura, P. Lengyel and A. Tissières, eds.), p. 115. Cold Spring Harbor Press, Cold Spring Harbor, New York, 1974.
20. G. Stöffler, in "Ribosomes" (M. Nomura, P. Lengyel and A. Tissières, ed.), p. 615. Cold Spring Harbor Press, Cold Spring Harbor, New York, 1974.
21. G. Stöffler and H. G. Wittmann, *JMB* **114**, 122 (1971).
22. M. Yaguchi, A. T. Matheson and L. P. Visentin, *FEBS Lett.* **46**, 296 (1974).
23. K. I. Higo and K. Loertscher, *J. Bact.* **118**, 180 (1974).
24. K. Isono, S. Isono, G. Stöffler, L. P. Visentin, M. Yaguchi and A. T. Matheson, *Mol. Gen. Genet.* **127**, 191 (1973).
25. K. Higo, W. Held, L. Kahan and M. Nomura, *PNAS* **70**, 944 (1973).
26. W. Möller, in "Ribosomes" (M. Nomura, P. Lengyel and A. Tissières, eds.), p. 711. Cold Spring Harbor Press, Cold Spring Harbor, New York, 1974.
27. G. Oda, A. R. Strøm, L. P. Visentin and M. Yaguchi, *FEBS Lett.* **43**, 127 (1974).
28. G. Stöffler, I. G. Wool, A. Lin and K. H. Rak, *PNAS* **71**, 4723 (1974).
29. D. Richter and W. Möller, in "Lipmann Symposium" (D. Richter, ed.), p. 524. de Gruyter, Berlin, 1974.
30. G. G. Brownlee, F. Sanger and B. G. Barrell, *Nature (London)* **215**, 735 (1967).
31. P. Fellner, C. Ehresmann and J. P. Ebel, *Biochimie* **54**, 853 (1972).
32. C. Ehresmann, P. Stiegler, P. Fellner and J. P. Ebel, *Biochimie* **54**, 901 (1972).
33. C. Ehresmann, P. Stiegler, G. A. Mackie, R. A. Zimmermann, J. P. Ebel and P. Fellner, *NARes.* **2**, 265 (1975).
34. C. Branlant, J. Sriwidada, A. Krol, P. Fellner and J. P. Ebel, *Biochimie* **57**, 175 (1975).
35. M. Santer and U. V. Santer. *J. Bact.* **116**, 1304 (1973).
36. T. Uchida, L. Bonen, H. W. Schaup, B. J. Lewis, L. Zablen and C. Woese, *J. Mol. Evol.* **3**, 63 (1974).
37. J. Shine and L. Dalgarno, *PNAS* **71**, 1342 (1974).
38. C. Ehresmann, P. Stiegler and J. P. Ebel, *FEBS Lett.* **49**, 47 (1974).

39. H. F. Noller and W. Herr, *Mol. Biol. Rep.* **1**, 437 (1974).
40. R. Monier, in "The Mechanism of Protein Synthesis and Its Regulation" (L. Bosch, ed.), p. 353. North-Holland Publ., Amsterdam, 1972.
41. W. Möller and R. A. Garrett, in "Protein Synthesis" (E. H. McConkey, ed.), p. 229. Dekker, New York, 1971.
42. R. A. Garrett and H. G. Wittmann, *Advan. Prot. Chem.* **27**, 277 (1973).
43. P. Fellner, in "Ribosomes" (P. Lengyel, M. Nomura and A. Tissières, eds.), p. 169. Cold Spring Harbor Press, Cold Spring Harbor, 1974.
44. R. A. Cox, in "Proceedings of the International Symposium on Biochemistry of Ribosomes and mRNA, Schloss Reinhardsbrunn" (R. Lindigkeit, P. Langen and J. Richter, eds.), p. 23. Deut. Akad. Wissenschaften, Berlin, 1967.
45. M. Spencer and F. Poole, *JMB* **11**, 314 (1965).
46. I. Tinoco, O. C. Uhlenbeck and M. D. Levine, *Nature (London)* **230**, 362 (1971).
47. W. Min Jon, G. Haegeman, M. Ysebaert and W. Fiers, *Nature (London)* **237**, 82 (1972).
48. F. Cramer, *Progr. Nucl. Acid. Res.* **11**, 391 (1971).
49. J. D. Robertus, J. E. Ladner, J. T. Finch, D. Rhodes, R. S. Brown, B. F. C. Clark and A. Klug, *NARes.* **1**, 927 (1974).
50. R. Monier, in "Ribosomes" (P. Lengyel, M. Nomura and A. Tissières, eds.), p. 141. Cold Spring Harbor Press, Cold Spring Harbor, New York 1974.
51. J. D. Robertus, J. E. Ladner, J. T. Finch, D. Rhodes, R. S. Brown, B. F. C. Clark and A. Klug, *Nature (London)* **250**, 546 (1974).
52. S. H. Kimm, F. L. Suddath, G. J. Quigley, A. McPherson, J. L. Sussman, A. H. J. Wang, N. C. Seeman and A. Rich, *Science* **185**, 435 (1974).
53. G. von Ehrenstein, in "Aspects of Protein Biosynthesis" (C. B. Anfinsen, Jr., ed.), p. 139. Academic Press, New York, 1970.
54. O. C. Uhlenbeck, J. Baller and P. Doty, *Nature (London)* **225**, 508 (1970).
55. O. Pongs, R. Bald and E. Reinwald, *EJB* **32**, 117 (1973).
56. O. C. Uhlenbeck, J. G. Chirikjian and J. R. Fresco, *JMB* **89**, 495 (1974).
57. H. F. Noller, *Bchem* **13**, 4694 (1974).
58. K. Cammack, D. S. Miller and K. H. Grinstead, *BJ* **117**, 745 (1970).
59. C. Schulte, C. A. Morrison and R. A. Garrett, *Bchem* **13**, 1032 (1974).
60. C. Schulte and R. A. Garrett, *Mol. Gen. Genet.* **119**, 345 (1972).
61. A. Muto and R. A. Zimmermann, *JMB*, submitted for publication.
62. E. Ungewickell, R. A. Garrett, C. Ehresmann, P. Stiegler and P. Fellner, *EJB* **51**, 165 (1975).
63. E. Ungewickell, P. Stiegler, C. Ehresmann and R. A. Garrett, *NARes* **2**, 1867 (1975).
64. G. Mackie and R. A. Zimmermann, *JBC* **250**, 4700 (1975).
65. T. Lindahl, A. Adams and J. R. Fresco, *PNAS* **55**, 941 (1966).
66. M. Aubert, G. Bellemare and R. Monier, *Biochemie* **55**, 135 (1973).
67. D. R. Morris, J. E. Dahlberg and A. E. Dahlberg, *NARes.* **1**, 1249 (1974).
68. S. Mizushima and M. Nomura, *Nature (London)* **226**, 1214 (1970).
69. H. W. Schaup, M. Green and C. G. Kurland, *Mol. Gen. Genet.* **109**, 193 (1970).
70. H. W. Schaup, M. Green and C. G. Kurland, *Mol. Gen. Genet.* **112**, 1 (1971).
71. R. A. Garrett, K. H. Rak, L. Daya and G. Stöffler, *Mol. Gen. Genet.* **114**, 112 (1971).
72. R. A. Zimmermann, A. Muto, P. Fellner, C. Ehresmann and C. Branlant, *PNAS* **69**, 1282 (1972).
73. W. A. Held, B. Ballou, S. Mizushima and M. Nomura, *JBC* **249**, 3103 (1974).

74. P. N. Gray, R. A. Garrett, G. Stöffler and R. Monier, *EJB* **28**, 412 (1972).
75. J. Horne and V. A. Erdmann, *Mol. Gen. Genet.* **119**, 337 (1972).
76. P. N. Gray, G. Bellemare, R. Monier, R. A. Garrett and G. Stöffler, *JMB* **77**, 133 (1973).
77. R. S. T. Yu and H. G. Wittmann, *BBA* **324**, 375 (1973).
78. G. Stöffler, L. Daya, K. H. Rak and R. A. Garrett, *JMB* **62**, 411 (1971).
79. G. Stöffler, L. Daya, K. H. Rak and R. A. Garrett, *Mol. Gen. Genet.* **114**, 125 (1971).
80. R. A. Garrett, S. Müller, P. Spierer and R. A. Zimmermann, *JMB* **88**, 553 (1974).
81. A. Muto, C. Ehresmann, P. Fellner and R. A. Zimmermann, *JMB* **86**, 411 (1974).
82. R. A. Zimmermann, A. Muto and G. A. Mackie, *JMB* **86**, 433 (1974).
83. B. Allet and P. F. Spahr, *EJB* **19**, 250 (1971).
84. P. Spierer, R. A. Zimmermann and G. Mackie, *EJB* **52**, 459 (1975).
85. H. W. Schaup and C. G. Kurland, *Mol. Gen. Genet.* **114**, 350 (1972).
86. H. W. Schaup, M. L. Sogin, C. Woese and C. G. Kurland, *Mol. Gen. Genet.* **114**, 1 (1971).
87. H. W. Schaup, M. L. Sogin, C. G. Kurland and C. Woese, *J. Bact.* **115**, 82 (1973).
88. R. A. Zimmermann, G. A. Mackie, A. Muto, R. A. Garrett, E. Ungewickell, C. Ehresmann, P. Stiegler, J. P. Ebel and P. Fellner, *NARes.* **2**, 279 (1975).
89. P. Gray, G. Bellemare and R. Monier, *FEBS Lett.* **24**, 156 (1972).
90. C. Branlant, A. Krol, J. Sriwidada, P. Fellner and R. R. Crichton, *FEBS Lett.* **35**, 265 (1973).
91. C. Branlant, A. Krol, J. Sriwidada, J. P. Ebel, P. Sloof and R. A. Garrett, *FEBS Lett.* **52**, 195 (1975).
92. N. Nanninga, R. A. Garrett, G. Stöffler and G. Klotz, *Mol. Gen. Genet.* **119**, 175 (1972).
93. R. Amons, W. Möller, E. Schiltz and J. Reinbolt, *FEBS Lett.* **41**, 135 (1974).
94. L. Daya-Grosjean, J. Reinbolt, O. Pongs and R. A. Garrett, *FEBS Lett.* **44**, 253 (1974).
95. M. Székely, R. Brimacombe and J. Morgan, *EJB* **35**, 574 (1973).
96. A. Yuki and R. Brimacombe, *EJB* **56**, 23 (1975).
97. K. P. Wong and H. H. Paradies, *BBRC* **61**, 178 (1974).
98. P. Voynow and C. G. Kurland, *Bchem* **10**, 517 (1971).
99. T. A. Bickle, G. A. Howard and R. R. Traut, *JBC* **248**, 4862 (1973).
100. P. Thammana, C. G. Kurland, E. Deusser, J. Weber, R. Maschler, G. Stöffler and H. G. Wittmann, *Nature NB* **242**, 47 (1973).
101. E. Deusser, H. J. Weber and A. R. Subramanian, *JMB* **84**, 249 (1974).
102. H. H. Paradies, A. Franz, C. C. Pon and C. Gualerzi, *BBRC* **59**, 600 (1974).
103. L. Brakier-Gingras, L. Provost and H. Dugas, *BBRC* **60**, 1238 (1974).
104. M. I. Sherman and M. V. Simpson, *PNAS* **64**, 1388 (1969).
105. W. Gilbert, *JMB* **6**, 374 (1963).
106. F. N. Chang and J. G. Flaks, *JMB* **68**, 177 (1972).
107. C. T. Shih and G. R. Craven, *JMB* **78**, 651 (1973).
108. L. C. Lutter, H. Zeichhardt, C. G. Kurland and G. Stöffler, *Mol. Gen. Genet.* **119**, 357 (1972).
109. L. C. Lutter, U. Bode, C. G. Kurland and G. Stöffler, *Mol. Gen. Genet.* **129**, 167 (1974).

110. U. Bode, L. C. Lutter and G. Stöffler, *FEBS Lett.* **45**, 232 (1974).
111. T. A. Bickle, J. W. B. Hershey and R. R. Traut, *PNAS* **69**, 1327 (1972).
112. R. R. Traut, A. Bollen, T. T. Sun, J. W. B. Hershey, J. Sundberg and L. R. Pierce, *Bchem* **12**, 3266 (1973).
113. L. C. Lutter, F. Ortandel and H. Fasold, *FEBS Lett.* **48**, 288 (1974).
114. A. Sommer and R. R. Traut, *PNAS* **71**, 3946 (1974).
115. T. T. Sun, A. Bollen, L. Kahan and R. R. Traut, *Bchem* **13**, 2334 (1974).
116. C. Clegg and D. Hayes, *EJB* **42**, 21 (1974).
117. D. Barritault, A. Expert-Bezançon, M. Milet and D. H. Hayes, *FEBS Lett.* **50**, 114 (1975).
118. R. R. Traut, personal communication.
118a. A. Sommer and R. R. Trout, *JMB* **97**, 471 (1975).
119. D. A. Hawley, L. I. Slobin and A. J. Wahba, *BBRC* **61**, 544 (1974).
120. L. C. Lutter and C. G. Kurland, *Nature NB* **243**, 15 (1973).
121. R. A. Kenner, *BBRC* **51**, 932 (1973).
122. J. Morgan and R. Brimacombe, *EJB* **29**, 542 (1972).
123. J. Morgan and R. Brimacombe, *EJB* **37**, 472 (1973).
124. H. E. Roth and K. H. Nierhaus, *FEBS Lett.* **31**, 35 (1973).
125. I. Newton, J. Rinke and R. Brimacombe, *FEBS Lett.* **51**, 215 (1975).
126. P. Schendel, P. Maeba and G. R. Craven, *PNAS* **69**, 544 (1972).
127. C. T. Chow, L. P. Visentin, A. T. Matheson and M. Yaguchi, *BBA* **287**, 270 (1972).
128. W. A. Held, S. Mizushima and M. Nomura, *JBC* **248**, 5720 (1974).
129. M. Green and C. G. Kurland, *Mol. Biol. Rep.* **1**, 105 (1973).
130. C. G. Kurland, *J. Supramol. Struct.* **2**, 178 (1974).
131. P. Traub and M. Nomura, *JMB* **40**, 391 (1969).
132. J. Marvaldi, J. Pichon and G. Marchis-Mouren, *BBA* **269**, 173 (1972).
133. G. Mangiarotti, E. Terco, A. Ponzetto and F. Altruda, *Nature (London)* **247**, 147 (1974).
134. J. W. Wireman and P. S. Sypherd, *Nature (London)* **247**, 552 (1974).
135. G. Mangiarotti, E. Terco, C. Perlo and F. Altruda, *Nature (London)* **253**, 569 (1975).
136. P. Thammana and W. A. Held, *Nature (London)* **251**, 682 (1974).
137. G. R. Craven and V. Gupta, *PNAS* **67**, 1329 (1970).
138. F. N. Chang and J. G. Flaks, *PNAS* **67**, 1321 (1970).
139. R. R. Crichton and H. G. Wittmann, *Mol. Gen. Genet.* **114**, 95 (1971).
140. P. Spitnik-Elson and A. Breiman, *BBA* **254**, 457 (1971).
141. K. Benkov and N. Delihas, *BBRC* **60**, 901 (1974).
142. L. Kahan and E. Kaltschmidt, *Bchem* **11**, 2691 (1972).
143. G. Moore and R. R. Crichton, *BJ* **143**, 607 (1974).
144. L. P. Visentin, M. Yaguchi and H. Kaplan, *Can. J. Biochem.* **51**, 1487 (1973).
145. A. S. Acharya and P. B. Moore, *JMB* **76**, 207 (1973).
146. J. Ginzburg, R. Miskin, and A. Zamir, *JMB* **79**, 481 (1973).
147. F. N. Chang, *JMB* **78**, 563 (1973).
148. A. Bakardjieva and R. R. Crichton, *BJ* **143**, 599 (1974).
149. G. Stöffler, R. Hasenbank, M. Lütgehaus, R. Maschler, C. A. Morrison, H. Zeichhardt and R. A. Garrett, *Mol. Gen. Genet.* **127**, 89 (1973).
150. D. A. Hawley and L. I. Slobin, *BBRC* **55**, 162 (1973).
151. R. V. Miller and P. S. Sypherd, *JMB* **78**, 527 (1973).
152. C. Michalski and B. H. Sells, *EJB* **49**, 361 (1974).
153. A. Bollen, R. J. Cedergren, D. Sankoff and G. Lapalme, *BBRC* **59**, 1069 (1974).

154. R. R. Traut, R. Heimark, T. T. Sun, J. W. B. Hershey and A. Bollen, in "Ribosomes" (M. Nomura, P. Lengyel and A. Tissières, eds.), p. 271. Cold Spring Harbor Press, Cold Spring Harbor, New York, 1974.
155. K. K. Huang, R. H. Fairclough, and C. R. Cantor, *JMB* **97**, 443 (1975).
156. G. W. Tischendorf, H. Zeichhardt and G. Stöffler, *Mol. Gen Genet.* **134**, 209 (1974).
157. J. A. Lake, M. Pendergast, L. Kahan and M. Nomura, *PNAS* **71**, 4688 (1974).
158. G. Stöffler and G. W. Tischendorf, in "Topics in Infectious Diseases," p. 117. Springer-Verlag, Berlin and New York, 1975.
159. W. E. Hill and R. J. Fessenden, *JMB* **90**, 719 (1974).
160. D. Hayes, personal communication.
161. H. Kagawa, L. Jishuken and H. Tokimatsu, *Nature NB* **237**, 74 (1972).
162. I. Newton and R. Brimacombe, *EJB* **48**, 513 (1974).
163. H. E. Roth and K. H. Nierhaus, *JMB* **94**, 111 (1975).
164. C. Ehresmann and J. P. Ebel, *EJB* **13**, 577 (1970).
165. B. K. Baha, *BBA* **353**, 292 (1974).
166. K. A. Hartmann and N. W. Clayton, *BBA* **335**, 201 (1974).
167. P. B. Moore, D. M. Engelman and B. P. Schoenborn, *PNAS* **71**, 172 (1974).
168. P. B. Moore, D. M. Engelman and B. P. Schoenborn, *JMB* **91**, 101 (1975).
169. H. Maruta, T. Tsuchiya and D. Mizuno, *JMB* **61**, 123 (1971).
170. M. Nomura and V. A. Erdmann, *Nature (London)* **228**, 744 (1970).
171. K. H. Nierhaus and F. Dohme, *PNAS* **71**, 4713 (1974).
172. T. Staehelin and M. Meselson, *JMB* **16**, 245 (1966).
173. K. Hosokawa, R. K. Fujimura and M. Nomura, *PNAS* **55**, 198 (1966).
174. J. H. Highland and G. A. Howard, *JBC* **250**, 831 (1975).
175. N. Hsuing and C. Cantor, *ABB* **157**, 125 (1973).
176. D. J. Litman and C. R. Cantor, *Bchem* **13**, 512 (1974).
177. M. R. Wabl, *JMB* **84**, 241 (1974).
178. G. W. Tischendorf, H. Zeichhardt and G. Stöffler, *Mol. Gen. Genet.* **134**, 187 (1974).
179. T. T. Sun, R. R. Traut and L. Kahan, *JMB* **87**, 509 (1974).
180. C. A. Morrison, R. A. Garrett, H. Zeichhardt and G. Stöffler, *Mol. Gen. Genet.* **127**, 359 (1973).
181. H. J. Weber, *Mol. Gen. Genet.* **119**, 233 (1972).
182. R. C. Marsh and A. Parmeggiani, *PNAS* **70**, 151 (1973).
183. J. C. Lelong, D. Gros, F. Gros, A. Bollen, R. Maschler and G. Stöffler, *PNAS* **71**, 248 (1974).
184. D. J. Litman, C. C. Lee and C. R. Cantor, *FEBS Lett.* **47**, 268 (1974).
185. O. Pongs, K. H. Nierhaus, V. A. Erdmann and H. G. Wittmann, *FEBS Lett.* **40**, S28 (1974).
186. C. L. Pon and C. Gualerzi, *PNAS* **71**, 4950 (1974).
187. P. H. van Knippenberg, J. van Duin and H. Lentz, *FEBS Lett.* **34**, 95 (1973).
188. M. Noll and H. Noll, *JMB* **89**, 477 (1974).
189. S. D. Bernal, B. M. Blumberg and T. Nakamoto, *PNAS* **71**, 774 (1974).
190. G. Jay and R. Kämpfer, *PNAS* **71**, 3199 (1974).
191. R. A. Grajevskaja, V. B. Odinzov, E. M. Saminsky and S. E. Bresler, *FEBS Lett.* **33**, 11 (1973).
192. M. H. Schreier and T. Staehelin, *Nature NB* **242**, 35 (1973).
193. C. Vermeer, W. van Alphen, P. van Knippenberg and L. Bosch, *EJB* **40**, 295 (1973).
194. W. Szer and S. Leffler, *PNAS* **71**, 3611 (1974).

195. J. Dondon, T. Godefroy-Colburn, M. Graffe and M. Grunberg-Manago, *FEBS Lett.* **45**, 82 (1974).
196. S. D. Bernal, B. M. Blumberg, J. J. Wang and T. Nakamoto, *BBRC* **60**, 1127 (1974).
197. R. Giegé and J. P. Ebel, *FEBS Lett.* **37**, 166 (1973).
198. S. D. Bernal, B. M. Blumberg and T. Nakamoto, *PNAS* **71**, 774 (1974).
199. B. M. Blumberg, S. D. Bernal and T. Nakamoto, *Bchem.* **13**, 3307 (1974).
200. S. Sabol and S. Ochoa, *Nature NB* **234**, 233 (1971).
201. C. L. Pon, S. M. Friedman and C. Gualerzi, *Mol. Gen. Genet.* **116**, 192 (1972).
202. J. Thibault, A. Chestier, D. Vidal and F. Gros, *Biochimie* **54**, 829 (1972).
203. C. Gualerzi and C. L. Pon, *BBRC* **52**, 792 (1973).
204. S. Sabol, D. Meier and S. Ochoa, *EJB* **33**, 332 (1973).
205. S. L. Huang and S. Ochoa, *ABB* **156**, 84 (1973).
206. W. A. Held, W. R. Gette and M. Nomura, *Bchem* **13**, 2115 (1974).
207. K. Isono and S. Isono, *EJB* **56**, 15 (1975).
208. J. van Duin, P. van Knippenberg, M. Dieben and C. G. Kurland, *Mol. Gen. Genet.* **116**, 181 (1972).
209. T. L. Helser, J. E. Davies and J. E. Dahlberg, *Nature NB* **233**, 12 (1971).
210. A. Okuyama, N. Machiyama, T. Kinoshita and N. Tanaka, *BBRC* **43**, 196 (1971).
211. T. L. Helser, J. E. Davies and J. E. Dahlberg, *Nature NB* **235**, 6 (1972).
212. R. A. Zimmermann, Y. Ikeya and P. F. Sparling, *PNAS* **70**, 71 (1973).
213. A. Okuyama, M. Yoshikawa and N. Tanaka, *BBRC* **60**, 1163 (1974).
214. J. van Duin and C. G. Kurland, *Mol. Gen. Genet.* **109**, 169 (1970).
215. M. Tal, M. Aviram, A. Kanarek and A. Weiss, *BBA* **281**, 381 (1972).
216. W. Szer and S. Leffler, *PNAS* **71**, 3611 (1974).
217. J. van Duin and P. H. van Knippenberg, *JMB* **84**, 185 (1974).
218. O. Pongs and E. Lanka, *PNAS* **72**, 1505 (1975).
219. P. H. van Knippenberg, *NARes.* **2**, 79 (1975).
220. G. Sander, R. C. Marsh and A. Parmeggiani, *FEBS Lett.* **33**, 132 (1973).
221. J. A. Cohlberg, *BBRC* **57**, 225 (1974).
222. J. Lucas-Lenard and F. Lipmann, *Annu. Rev. Biochem.* **40**, 409 (1971).
223. B. E. Griffin, M. Jarman, C. B. Reese, J. E. Sulston and D. R. Trentham, *Bchem* **5**, 3638 (1966).
224. R. Wolfenden, D. H. Rammler and F. Lipmann, *Bchem* **3**, 329 (1964).
225. S. Chládek, D. Ringer and J. Žemlička, *Bchem* **12**, 5135 (1973).
226. M. Sprinzl and F. Cramer, *Nature NB* **245**, 3 (1973).
227. T. H. Fraser and A. Rich, *PNAS* **70**, 2671 (1973).
228. G. Chinali, M. Sprinzl, A. Parmeggiani and F. Cramer, *Bchem* **13**, 3001 (1974).
229. D. Ringer and S. Chládek, *FEBS Lett.* **39**, 75 (1974).
230. S. M. Hecht, J. W. Kozarich and F. J. Schmidt, *PNAS* **71**, 4317 (1974).
231. P. Greenwell, R. J. Harris and R. H. Symons, *EJB* **49**, 539 (1974).
232. S. Pestka, *JBC* **247**, 4669 (1973).
233. M. Springer and M. Grunberg-Manago, *BBRC* **47**, 477 (1972).
234. R. Miskin and A. Zamir, *JMB* **87**, 121 (1974).
235. J. C. H. Mao, *BBRC* **52**, 595 (1973).
236. D. L. Riddle and J. Carbon, *Nature NB* **242**, 230 (1973).
237. H. Oen, M. Pellegrini, D. Eilat and C. R. Cantor, *PNAS* **70**, 2799 (1973).
238. A. P. Czernilofsky, E. E. Collatz, G. Stöffler and E. Kuechler, *PNAS* **71**, 230 (1974).

239. D. Eilat, M. Pellegrini, H. Oen, Y. Lapidot and C. R. Cantor, *JMB* **88**, 831 (1974).
240. R. Hauptmann, A. P. Czernilofsky, H. O. Vorma, G. Stöffler and E. Kuechler, *BBRC* **56**, 331 (1974).
241. M. Sapori, M. Pellegrini, P. Lengyel and C. R. Cantor, *Bchem* **13**, 5432 (1974).
242. A. S. Girshovich, E. S. Bochkareva, V. M. Kramarov and Y. A. Dochimikov, *FEBS Lett.* **45**, 213 (1974).
243. D. Nierhaus and K. H. Nierhaus, *PNAS* **70**, 2224 (1973).
244. O. Pongs, R. Bald and V. A. Erdmann, *PNAS* **70**, 2229 (1973).
245. S. Dietrich, I. Schrandt and K. H. Nierhaus, *FEBS Lett.* **47**, 136 (1974).
246. O. Pongs, *Acta Biol. Med. Germ.* **33**, 629 (1974).
247. R. Fernandez-Muñoz, R. E. Monro and D. Vasquez, in "Methods in Enzymology," Vol. 20: Nucleic Acids and Protein Synthesis, Part C (K. Moldave and L. Grossman, eds.), p. 481. Academic Press, New York, 1971.
248. B. G. Forget and S. M. Weissman, *Science* **158**, 1695 (1967).
249. M. Simsek, G. Petrissant and U. L. Rajbhandary, *PNAS* **70**, 2600 (1973).
250. G. Petrissant, *PNAS* **70**, 1046 (1973).
251. G. G. Lorringer and R. J. Roberts, *FP* **30**, 1217 (1971).
252. R. J. Roberts, *Nature NB* **237**, 44 (1972).
253. V. A. Erdmann, M. Sprinzl and O. Pongs, *BBRC* **54**, 942 (1973).
254. V. A. Erdmann, M. Sprinzl, D. Richter and S. Lorenz, *Acta Biol. Med. Germ.* **33**, 605 (1974).
255. D. Richter, V. A. Erdmann and M. Sprinzl, *PNAS* **71**, 3226 (1974).
256. S. H. Kim, J. L. Sussman, F. L. Suddath, G. J. Quigley, A. McPherson, A. H. J. Wang, N. C. Seeman and A. Rich, *PNAS* **71**, 4970 (1974).
257. M. Yoshida, Y. Kaziro and T. Ukita, *BBA* **166**, 646 (1968).
258. H. F. Noller and W. Herr, *JMB* **90**, 181 (1974).
259. U. Schwarz, R. Lührmann and H. G. Gassen, *BBRC* **56**, 807 (1974).
260. R. G. Shulman, C. W. Hilbers and D. L. Miller, *JMB* **90**, 601 (1974).
261. H. F. Noller and J. B. Chaires, *PNAS*, **69**, 3115 (1972).
262. R. E. Monro, *JMB* **26**, 147 (1967).
263. B. E. H. Maden, R. R. Traut and R. E. Munro, *JMB* **35**, 333 (1968).
264. K. H. Nierhaus and V. Montejo, *PNAS* **70**, 1931 (1973).
265. N. Hsiung, S. A. Reines and C. R. Cantor, *JMB* **88**, 841 (1974).
266. J. P. G. Ballesta and D. Vasquez, *FEBS Lett.* **48**, 266 (1974).
267. G. A. Howard and J. Gordon, *FEBS Lett.* **48**, 271 (1974).
268. H. Hampl, G. Wystup and K. H. Nierhaus, in preparation.
269. V. G. Moore, R. E. Achison, G. Thomas, M. Moran and H. F. Noller, *PNAS* **72**, 844 (1975).
270. W. Möller, in "Ribosomes" (M. Nomura, P. Lengyel and A. Tissières, eds.), p. 711. Cold Spring Harbor Press, Cold Spring Harbor, New York 1974.
271. G. Sander, R. C. Marsh and A. Parmeggiani, *BBRC* **47**, 866 (1972).
272. A. Kay, G. Sander and M. Grunberg-Manago, *BBRC* **51**, 979 (1973).
273. A. H. Lockwood, U. Maitra, N. Brot and H. Weissbach, *JBC* **249**, 1213 (1974).
274. R. Mazumder, *PNAS* **70**, 1931 (1973).
275. S. Lee-Huang, H. Lee and S. Ochoa, *PNAS* **71**, 2928 (1974).
276. N. Brot, W. P. Tate, C. T. Caskey and H. Weissbach, *PNAS* **71**, 89 (1974).
277. G. Stöffler, I. G. Wool, A. Lin and K. H. Rak, *PNAS* **71**, 4723 (1974).
278. P. J. Schrier, J. M. Maassen and W. Möller, *BBRC* **53**, 90 (1973).

279. G. Stöffler, R. Hasenbank, J. W. Bodley and J. M. Highland, *JMB* **86**, 171 (1974).
280. J. P. G. Ballesta and D. Vasquez, *FEBS Lett.* **28**, 337 (1972).
281. M. Gaunt-Klöpfer, and V. A. Erdman, *BBA* **390**, 226 (1975).
282. J. A. Maassen and W. Möller, *PNAS* **71**, 1277 (1974).
283. M. Wolf, G. Chinali and A. Parmeggiani, *PNAS* **71**, 4910 (1974).
284. S. Pestka, *JBC* **243**, 2810 (1968).
285. L. P. Gavrilova and A. S. Spirin, in "Methods in Enzymology," Vol. 30: Nucleic Acids and Protein Synthesis, Part F (L. Grossman and K. Moldave, eds.), p. 452. Academic Press, New York, 1974.
286. N. Inone-Yokosawa, C. Ishikawa and Y. Kaziro, *JBC* **249**, 4321 (1974).
287. J. H. Highland, G. A. Howard, E. Ochsner, G. Stöffler, R. Hasenbank and J. Gordon, *JBC* **250**, 1141 (1975).
288. A. H. Lockwood, P. Sarkar, U. Maitra, N. Brot and H. Weissbach, *JBC* **249**, 5831 (1974).
289. D. Vasquez, *FEBS Lett.* **40**, S63 (1974).
290. H. Yokosawa, N. Inone-Yokosawa, K. I. Arai, M. Kawakita and Y. Kaziro, *JBC* **248**, 375 (1973).
291. M. P. Printz and D. L. Miller, *BBRC* **53**, 149 (1973).
292. L. J. Crane and D. L. Miller, *Bchem* **13**, 933 (1974).
293. A. L. Beaudet and C. T. Caskey, in "The Mechanisms of Protein Synthesis and Its Regulation" (L. Bosch, ed.), p. 133. North Holland Publ., Amsterdam, 1973.
294. W. P. Tate, A. L. Beaudet and C. T. Caskey, *PNAS* **70**, 2350 (1973).
295. J. L. Nichols, *Nature (London)* **225**, 147 (1970).
296. P. Lu and A. Rich, *JMB* **58**, 513 (1971).
297. H. Yahato, Y. Ocada and A. Tsugita, *Mol. Gen. Genet.* **106**, 208 (1969).
298. L. Dalgarno and J. Shine, *Nature NB* **245**, 26 (1973).
299. W. P. Tate, C. T. Caskey and G. Stöffler, *JMB* **93**, 375 (1975).
300. A. Subramanian and B. D. Davies, *JMB* **74**, 45 (1973).

Structure and Function of 5 S and 5.8 S RNA

VOLKER A. ERDMANN

Max-Planck-Institut für
Molekulare Genetik
Abt. Wittmann
Berlin-Dahlem, Germany

I. Introduction	45
II. Primary Sequences of 5 S and 5.8 S Ribosomal RNA	46
A. Prokaryotic 5 S RNA	47
B. Eukaryotic 5 S RNA	51
C. Eukaryotic 5.8 S RNA	54
D. Precursors of 5 S and 5.8 S RNA	56
III. Secondary and Tertiary Structures of 5 S and 5.8 S RNA	58
A. Physical Characterization of Native and Denatured 5 S RNAs and 5.8 S RNA	58
B. Enzymic Hydrolysis	62
C. Chemical Modification	65
D. Oligonucleotide Binding	68
E. Proposed Models for 5 S and 5.8 S RNA	69
IV. Complexes between 5 S RNA and Protein	71
A. Proteins Binding 5 S RNA	71
B. Ribosomal Proteins in the Neighborhood of 5 S RNA in Ribosomes	74
C. Regions of *E. coli* and *B. stearothermophilus* 5 S RNA Interacting with Proteins	75
D. Binding of Oligonucleotides to 5 S RNA-Protein Complexes	78
V. Function of 5 S and 5.8 S RNA	79
A. Activity of 50 S Ribosomal Subunits Lacking 5 S RNA	79
B. Reconstitution of 50 S Ribosomal Subunits with Modified 5 S RNAs	79
C. Functional Activities of 50 S Ribosomal Subunits with Modified 5 S RNAs	80
D. Importance of 5 S RNA in Binding tRNA to Ribosomes	82
E. Enzymic Activities Associated with 5 S RNA · Protein Complexes	83
F. Function of 5.8 S RNA	84
VI. Outlook	85
References	85
Note Added in Proof	90

I. Introduction

The principles of protein synthesis, as it occurs on the ribosome, are now understood, and therefore recent studies concentrate upon the detailed structure and function of the numerous components involved. The aim of these investigations has been to understand the synthesis of proteins at the molecular level. One component that has been intensively investigated is the 5 S RNA, and the accumulated data concerning this ribosomal RNA are discussed here.

5 S RNA is a small ribonucleic acid associated with the large ribosomal subunits of eukaryotic (60 S) and prokaryotic (50 S) ribosomes. It first came to attention in 1963 (1), and it was soon determined that its sequence (2) did not contain modified nucleotides, such as are found in tRNAs (3). The discovery of 5 S RNA was followed by investigations concerning its sequence and physical properties, to obtain information concerning its structure and function. It turned out that the primary structures of 5 S RNA cannot be folded in a single pattern, such as the cloverleaf structure of tRNA (4, 5). After a method for the total reconstitution of bacterial 50 S ribosomal subunits was developed (6), it became possible to add a new approach to the research concerning the structure (7, 8) and function (9) of 5 S RNA.

II. Primary Sequences of 5 S and 5.8 S Ribosomal RNA

After the initial report (1) that *Escherichia coli* 50 S ribosomal subunits contain 5 S RNA in addition to 23 S RNA, similar 5 S RNAs were noted in other prokaryotic ribosomes and in those of eukaryotic cells (10–12, 17), plant cells and plant chloroplasts (13). More recently, one laboratory reported the presence of 5 S RNA in plant mitochondrial ribosomes (14), although other reports claimed that animal (15) and ascomycete (16) mitochondrial ribosomes do not contain 5 S RNA. In contrast to 5 S RNA, 5.8 S RNA has been found only in eukaryotic ribosomes (18, 19) and not at all in prokaryotic ribosomes. This RNA had previously been termed 7 S or 5.5 S RNA, but was renamed 5.8 S RNA after its sequence was determined (19).

5 S and 5.8 S RNAs are separately discussed here in Sections II, A–C: Prokaryotic 5 S RNA, Eukaryotic 5 S RNA and Eukaryotic 5.8 S RNA. There are three reasons for dealing with the prokaryotic and eukaryotic 5 S RNAs separately: their sequences differ significantly from each other; eukaryotic 5 S RNAs seem to be primary transcription products (20) whereas prokaryotic 5 S RNAs are derived from precursor molecules (see below); reconstitution experiments show that different prokaryotic 5 S RNAs can be incorporated into *Bacillus stearothermophilus* 50 S subunits to yield active ribosomes, whereas eukaryotic 5 S RNAs cannot (21), indicating that they may have different functions. 5.8 S RNA deserves separate consideration because of its different size and because it is found only in eukaryotic ribosomes.

A. Prokaryotic 5 S RNA

The first prokaryotic 5 S RNA to be sequenced was that of *E. coli* (*2*); its primary structure is shown in Fig. 1. The now well known ^{32}P-"fingerprinting" (chromatographic) technique was developed especially for the determination of the sequence of this small ribosomal RNA (*22, 23*). To simplify the comparison of the different prokaryotic 5 S RNAs, I have divided its structure into regions A–G, of which regions A and E are drawn as double-stranded helical structures (Figs. 1 and 2).

Figure 1 accommodates the sequences of *E. coli* (*2*), *Pseudomonas fluorescens* (*24*), *Bacillus stearothermophilus* (*25*), *B. megaterium* (*26*), *Photobacterium* 8265 (*26a*) and *Anacystis nidulans* (*27*) 5 S RNAs. *A. nidulans* is a blue-green alga (prokaryote), and its 5 S RNA sequence is included in this figure, as it shows great similarities to bacterial 5 S RNAs (*28*). Chromatograms of 5 S RNA from *Proteus vulgaris, B. subtilis,* and *Salmonella typhimurium* are similar to those for 5 S RNA from *E. coli* and their primary sequences may assumed to be nearly identical (*24*). Comprehensive studies involving oligonucleotide cataloguing of 5 S RNAs from the four prokaryotic families of the Enterobacteriaceae, the Bacillaceae, the Achromobacteraceae, and the Pseudomonadaceae suggest that in general the primary structures of all these bacterial 5 S RNAs are very similar, but that within a family the sequences may differ in composition by 2–10% (*29*).

Comparison of the different bacterial 5 S RNA sequences (Fig. 1) with that of *E. coli* shows that these RNA molecules are of a similar length of 120 nucleotides and contain no modified nucleotides. It is also always possible to base-pair the 5′ and 3′ ends with each other. This region of 5 S RNA, which consists of 9–12 base pairs, I have designated as region A (Figs. 1 and 2). Within region A there is only 24% homology between *B. stearothermophilus, B. megaterium, A. nidulans,* and *P. fluorescens* when compared to *E. coli*. Therefore, region A of 5 S RNA is constant only in the sense that it is base-paired; it is variable with respect to the number of base-pairs and the nucleotides involved. The homology estimate of 24% suggests that this sequence is not conserved.

Sequence B connects the double-stranded region A with region C, which starts with the sequence C-A-C-C in all bacterial 5 S RNAs (Figs. 1 and 2). Region B is also heterologous with respect to length (15–19 nucleotides) and base composition. The homology among bacteria is 22%, which indicates that this region too is not conserved but varies greatly between the different 5 S RNAs.

Fragment C is strongly conserved in all prokaryotic 5 S RNAs and connects the constant-sequence C-A-C-C-Y-G with C-C-G-A-A-C defined

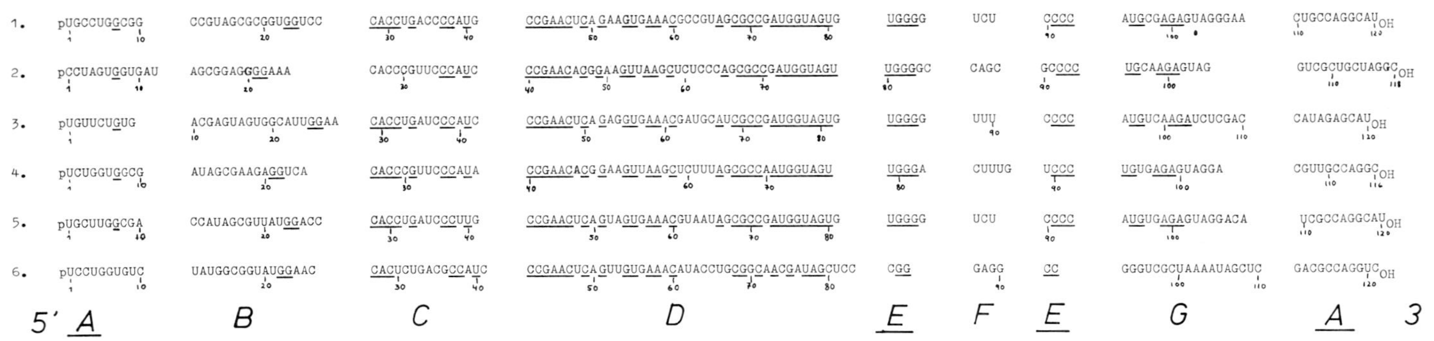

1. pUGCCUGGCGG CCGUAGCGCGGUGGUC CAGCGUGACGCCAUG CCGAACUCA GAAGUGAAACGCGUAGCGCCGAUGGUAGUG UGGGG UCU CCCC AUGGCGAGCGUAGGAA CUGCCAGGCAU_OH
   1   10   20   30   40   50   60   70   80   90   100   110   120

2. pCCUAGUGGUGAU AGCGGAGGGAAA CACCGGUUCDCAUC CCGAACAGGGAAGUUAAGCUUCUCCCAGCGCCGAUGGUAGUG UGGGGC CAGG GCCCC UGCAAGCAGUAG GUCCGCUAGG_OH
   1   10   20   30   50   60   70   80   90   100   115

3. pUGUUCUGUG ACGAGUAGUGGCAUUGGAA CAGCCGUUCDCAUC CCGAACUCA GAGUGAAACAGUAGUCGCUGGCGGUAGUG UGGGG UUU CCCC AUGCAAGAAUCUCGAC CAUAGACAU_OH
   1   10   20   30   40   50   60   70   80   90   100   110

4. pCCUGUGGCG AUAGCGAAGAGGUCA CAGCCGUUCDCAUA CCGAACAGG GAAGUAAGCUCUUUUAGCGGCCAAUGGUAGU UGGGA CUUUG UCCC UGUGAGAGUAGGA CGUUGCCAGGCG_OH
   1   10   20   30   40   50   60   70   80   90   100   110

5. pUGCUUGCCA CCCAUGGGUUAUGGACC CAGCUGAUCCGUUG CCGAACAGU AGUGAAACGUAAUAGCGGCCGAUGGUAGUG UGGGG UCU CCCC AUGUGAGAGUAGGACA UGCCCAGGAU_OH
   1   10   20   30   40   50   60   70   80   90   100   110   120

6. pCCUGGUGC UAUGGCGGUAUGGAAC CAGUCUGACGGCAUC CCGAAGUCA GUGUGAAACAUACUCGGGCGAAGGAUGUUCC CGG GAGG CC GGUCGCUAAAAUAGCUC GAGCCAGGUC_OH
   1   10   20   30   40   50   60   70   80   90   100 110   120

5' A   B   C   D   E   F   G   A   3

Fig. 1. Primary structures of prokaryotic 5 S RNAs. 1, *Escherichia coli* (2); 2, *Bacillus stearothermophilus* (25, including small modifications according to References 25a and 169); 3, *Pseudomonas fluorescens* (24); 4, *Bacillus megaterium* (26); 5, *Photobacterium* 8265 (26a); and 6, *Anacystis nidulans* (27) 5 S RNA. Underlining nucleotides in sequences 1–5 indicates identical nucleotides at similar positions in all prokaryotic class I 5 S RNAs. The underlined nucleotides in sequence 6 (prokaryotic class II 5 S RNA) indicates similarities to class I 5 S RNAs.

as the start of region D (Figs. 1 and 2). Fragment C always contains 14 nucleotides, and the sequence is strongly conserved. The homology of 77% between all bacterial 5 S RNAs indicates that this part of the molecule may be important in the interaction with ribosomal proteins (see Section IV).

Region D of 5 S RNA is another conserved region. It includes the sequence C-C-G-A-A-C and extends to the U-G-G-G sequence of region E (Figs. 1 and 2). This part of 5 S RNA is again conserved with respect to base sequence and number of nucleotides. The middle or center of this sequence shows the largest deviation. Evidence is presented below (Section IV, C) that suggests that parts of region D are important for ribosomal protein interaction and for the binding of tRNA to the ribosome (Sections IV and V). Analysis of region D reveals that, besides the sequence C-C-G-A-A-C, two other regions contained in it are strongly conserved. One of these sequences is A-A-R-C which is always 10 nucleotides away from the last adenine of the C-C-G-A-A-C sequence (Figs. 1 and 2). The third strongly conserved part in region D occurs just prior to the start of region E, and consists of the following nucleotides: C-G-C-C-R-A-U-G-G-U-A-G-U. This section includes the sequence U-G-G, common to all pro- and eukaryotic 5 S RNAs as well as to eukaryotic 5.8 S RNA at similar locations.

Another strongly conserved region (75% homology) is the double-stranded region E, which consists of four G·C base-pairs, except in A. nidulans, which has an additional A·U base-pair (Figs. 1 and 2). This part of 5 S RNA may also be important in protein binding. At the start of this region, we again find the sequence U-G-G.

Region F is the loop region connecting the two strands involved in base-pairing of region E. This area of the molecule is not very strongly conserved, shows 39% homology, and consists of 3 to 4 nucleotides (Figs. 1 and 2).

The last part of the 5 S RNA to be discussed is region G, which, with 47% homology, can be classified as intermediately conserved. This section links the two double-stranded regions E and A (Figs. 1 and 2). It seems interesting that this area of the molecule can vary significantly in length, for B. stearothermophilus 5 S RNA has only 9 nucleotides in this area, whereas E. coli and P. fluorescens have 12 (Fig. 1). This difference in length is probably compensated for by the double-stranded region A, since B. stearothermophilus, which has a short region G, has a long double-stranded region A.

Figure 2 shows all the constant elements in bacterial 5 S RNAs (class I). It is noteworthy that one finds conservation not only of nucleotide sequences, but also of the number of nucleotides that separate them.

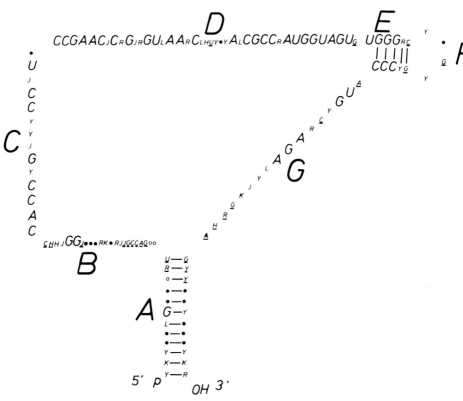

FIG. 2. General structure for prokaryotic 5 S RNAs (class I). Underlined letter (e.g., A) indicates that this base appears sometimes at this position. ●, Position occupied by either A, C, G or U; ○, possible additional position for either A, C, G or U; H indicates either A or C; J indicates either A or U; K indicates either G or C; L indicates either G or U; R = purine; Y = pyrimidine.

Regions A, B, F and G of *A. nidulans* 5 S RNA show 29, 18, 25 and 44% homologies, respectively, to *E. coli* 5 S RNA (Fig. 1). This divergence is similar to the one observed between other bacterial 5 S RNAs. Comparison with the conserved regions C and D of bacterial 5 S RNA shows that the corresponding sequences of *A. nidulans* are 79 and 61% identical. Further, region E also consists of at least four G · C base-pairs.

The following partial sequences of *A. nidulans* are identical to (and at identical positions of) bacterial 5 S RNAs: C-A-C (start of region C), C-C-G-A-A-C (start of region D) and A-A-R-C (10 nucleotides away from the last adenine of the C-C-G-A-A-C part of region D). Since all 5 S RNAs contain in region D and/or E the sequence U-G-G, it seems surprising that *A. nidulans* does not have this sequence, although it has a similar sequence (C-G-G) at this position (Fig. 1).

Oligonucleotide separations show that the 5 S RNA from another blue-green alga *Oscillatoria tenuis* is about 75% homologous to the one from *A. nidulans* (*28*). Therefore it seems permissible to consider this RNA as similar to bacterial 5 S RNAs. Since *A. nidulans* and most likely *O. tenuis* 5 S RNAs have some significant sequence differences from bacterial 5 S RNAs, I have classified them as prokaryotic class II, in contrast to the bacterial 5 S RNAs (prokaryotic class I).

A degree of heterogeneity among *E. coli* 5 S RNAs themselves may exist. For example, at positions 3 and 92 we find either C or U, at position 12 either C or A, and at positions 13 and 116 either G or U (*2, 30, 31*). Similarly, G or U may occur in position 7 of *P. fluorescens* 5 S RNA (*24*).

B. Eukaryotic 5 S RNA

All eukaryotic 5 S RNAs so far sequenced contain more than one phosphate at their 5' end, suggesting that they are primary transcription products. The eukaryotic 5 S RNAs may also be divided into two classes: class I contains the sequence C-Y-G-A-U at the start of region D, which is complementary to parts of loop IV in eukaryotic initiator tRNAs (*3*); and class II contains the sequence C-A-G-A-A-C at the same region, which is similar to the one found at this position in prokaryotes. Part of this sequence, namely G-A-A-C, is complementary to parts of loop IV in all tRNAs (*3*) used for protein chain elongation.

The following 5 S RNAs of the class I group have been sequenced: human KB cell (*32*), *Xenopus laevis* kidney cell (*33–35*), *X. laevis* ovary (*34, 35*), chicken (as cited in reference *36*), *Saccharomyces carlsbergensis* (*37*), and *Torulopis utilis* (*38, 39*). From chromatograms, it can further be concluded that the following 5 S RNAs are either identical or very similar to that from KB cells: HeLa cell (*40*), marsupial (*41*), several mouse cell lines (*42, 43*), rat pituitary (*44*), rabbit reticulocyte (*44*), *Drosophila* (*45*), and maize (*45*). In Fig. 3 the primary structures of some eukaryotic 5 S RNAs are shown and, in a way similar to that just used for prokaryotic 5 S RNAs; a generalized structure for these RNAs is shown in Fig. 4. Dividing the eukaryotic 5 S RNA structures into regions A–G illustrates that they are structurally related to the prokaryotic RNAs and, in addition, the conserved regions are more easily visualized.

Comparison of regions A–G of the different eukaryotic 5 S RNAs with those of KB cell 5 S RNA shows that these molecules are more conserved than their prokaryotic counterparts. As can be seen from Fig. 3, the two yeast 5 S RNAs differ the most from the other eukaryotic

1. (pp)pGUCUACGGC CAUACCACCCUGAACG CGGCCGAUCUGU CUGAUCUCGGAAGCUAAGCAGGGUCGGGCGGUUAGUACUUUGGA UGG GAGA CC GCCUGGGAAUACCGGGU GCUGUAGGCU(U)$_{OH}$
                                     10   20   30   40   50   60   70   80   90   100   110   120

2. (pp)pGCCUACGG CACACCACCCUGAAAG UGCCCGAUCUUGU CUGAUCUCGGAAGCUAAGCAGGGUCGGGCCUUGUUAGUACUUUGGA UGGG AGAC CCC UGGGAAUACCAGGU GUCGUACGCUU(U)$_{OH}$
                               10   20   30   40   50   60   70   80   90   100   110   120

3. (pp)pGGUUGCGGC CAUACCAUCUAGAAAG CACGGUUUCCGU CCGAUUAACCUGUAGUUAAGCUGGUAAGAGCCUGACCGAGUAGUGUAG UGG GUGA CC UAACGCGAAACCUAGG GUGCAAUCU_{OH}
                               10   20   30   40   50   60   70   80   90   100   110   120

4. (pp)pGGUUGCGGC CAUAUCUAGCAGAAAG CACGGUUUCGGU CCGAUCAACUGUAGUUAAGCUGCUAAGAGCCUGAUCCGAGUAGUGUAG UGG GUGA CC AUACGGAAACCAGGU GCUGAAUCU_{OH}
                               10   20   30   40   50   60   70   80   90   100   110   121

5. (pp)pGCCUAGGCAUCC CACCCUGUAACG CCCGAUCUGGU CUGAUCUCGGAAGCUAAGCAGGGUCGGGCCUGGUUAGUACUUGGA UGG GAGA CC GCCUGGGAUACC GGGUGCUGUAGGCUU_{OH}
                                 10   20   30   40   50   60   70   80   90   100   110   120

6. (pp)pAUGCUACGGU CAUACACGAAAG CACGGAUCCAU CAGAACUUCGGAAGCUAAGCGGUUGGGCUUACAACAGGUAGUACUUGGGU UGG AGGAUUA CCU GAGUGGAAGCCC GACGCAGGUU_{OH}
                                10   20   30   40   50   60   70   80   90   100   110   120

5' _A_   _B_   _C_   _D_   _E_ _F_ _E_   _G_   _A_   3'

FIG. 3. Primary structures of eukaryotic 5 S RNAs. 1, KB cell (32); 2, Xenopus laevis (33–35); 3, Saccharomyces carlsbergensis (37); 4, Torulopsis utilis (38, 39); 5, chicken (as cited in Reference 36); and 6, Chlorella pyrenoidosa (46, 47) 5 S RNA. Underlining of nucleotides in sequences 1–5 indicates identical nucleotides at similar positions in all eukaryotic class I 5 S RNAs. The underlined nucleotides in sequence 6 (eukaryotic class II 5 S RNA) indicates similarities to class I 5 S RNAs.

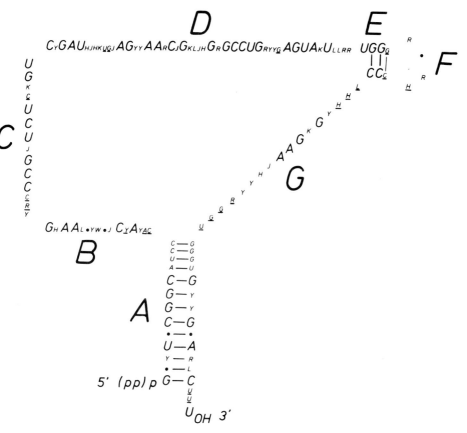

FIG. 4. General structure for eukaryotic 5 S RNAs (class I). Underlined letter (e.g., A) indicates that this base appears sometimes at this position. ●, Position occupied by either A, C, G or U; ○, possible additional position for either A, C, G or U; H indicates either A or C; J indicates either A or U; K indicates either G or C; L indicates either G or U; R = purine; Y = pyrimidine.

class I species. Nevertheless, there exist several sequences common to all 5 S RNAs in this group.

Figure 4 shows the constant elements of eukaryotic class I 5 S RNAs. In region A one finds at identical positions the nulceotide sequences C-G-G-C (at the 5' end), which can be base-paired with the sequence G-Y-Y-G found at the 3' end. The total number of possible base-pairs is nine, except in the case of chicken 5 S RNA, which may have an additional set of four base-pairs (see also Fig. 3).

Region B (Figs. 3 and 4), which is made up of 16 nucleotides (except chicken 5 S RNA, which has 13), has at its start the constant sequence Y-A-Y-C and ends with A-A-H-G (H here stands for either C or A).

The number of nucleotides varies from 11 to 14 in region C, although it always starts with a pyrimidine and ends with G-U. Between these two markers lies the constant sequence C-C-G-J-U-C-U (J here represents either A or U).

The next region (D) of the molecule is made up of 45–47 nucleotides; it always starts with the sequence C-Y-G-A-U and reaches to the U-G-G sequence of region E (Figs. 3 and 4). At 9–11 nucleotides away from the 3′ end of C-Y-G-A-U, we find another strongly conserved segment, namely, A-A-R-C. Prokaryotic 5 S RNAs have the same sequence at the identical position. Six nucleotides separate A-A-R-C from the conserved section G-R-G-C-C-U-G-R. The last identical sequence in region D is A-G-U-A.

Next we find the second base-paired region (E), which starts always with U-G-G and consists of at least two base-pairs. This U-G-G sequence is separated by 40 nucleotides (except in yeast, where it is 42 nucleotides) from C-Y-G-A-U, which is at the start of region D. Region F contains four nucleotides and links the two strands base-paired in region E (Figs. 3 and 4).

Similar to prokaryotic 5 S RNAs, the two double-stranded regions A and E are connected by region G which varies considerably in base composition and length (13–16 nucleotides). This part of the molecule contains always a G-A-A sequence (Figs. 3 and 4).

Class II of eukaryotic 5 S RNAs consists so far of one species, namely, cytoplasmic *Chlorella pyrenoidosa* 5 S RNA. Its primary sequence (*46, 47*) is given in Fig. 3. There are two reasons for setting this 5 S RNA apart from the class I eukaryotic 5 S RNAs. (a) Regions A, F and G show only 50% homology to KB cell 5 S RNA. (b) It is the only eukaryotic 5 S RNA that has, at the beginning of region D, the sequence C-A-G-A-A-C, and is therefore very similar to that of bacterial 5 S RNAs, and different from C-Y-G-A-U of the species found in class I. Besides the differences pointed out, *Chlorella* 5 S RNA contains almost all the other constant sequences, at similar positions, found in class I eukaryotic 5 S RNAs.

C. Eukaryotic 5.8 S RNA

The large eukaryotic ribosomal subunit contains 5.8 S RNA as well as 5 S RNA (*18, 19*). For its isolation from 28 S RNA, it is required to use conditions that denature hydrogen bonds. Unlike 5 S RNA, 5.8 S RNA contains several modified nucleotides. For example, 5.8 S RNA from yeast contains four modified bases (*19*), that from HeLa cells four and one 2'-O-methylated G (48), and that from Novikoff ascites hepatoma four and one 2'-O-methylated U and G (49). The molecule is over 150 nucleotides long, and heterogeneity is found at its 5' end (49, 50). Little is yet known about its structure and function. Nevertheless 5.8 S RNA is included in this review to allow comparisons between it and the 5 S RNAs.

The primary sequence of *Saccharomyces cerevisiae* 5.8 S RNA (19, 50) is shown in Fig. 5. It is the only 5.8 S RNA whose sequence has been determined. To a certain degree, the structure also can be divided into regions A–G, as has been done above for the 5 S RNAs (Fig. 5). The 5.8 S RNA structure indicated in Fig. 5 resembles, to a certain extent, eukaryotic and prokaryotic 5 S RNAs, except that it contains a tail of 37

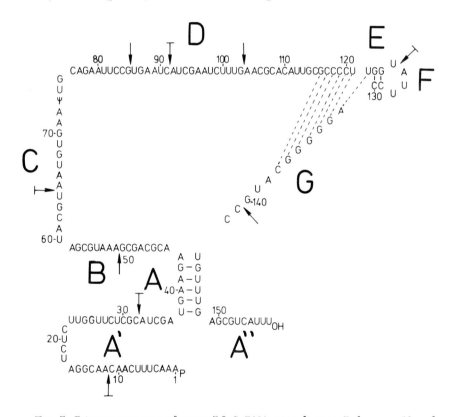

FIG. 5. Primary sequence of yeast 5.8 S RNA according to References 19 and 50. Since this RNA has some sequence similarities to eukaryotic and prokaryotic 5 S RNAs, its structure is drawn in this fashion to allow easier comparison. Points of pancrease (↦) and T1 ribonuclease (→) hydrolysis are taken from References 19 and 50.

nucleotides at the 5' end (region A') and another of 10 nucleotides at the 3' end (region A").

Surprisingly, 5.8 S RNA exhibits greater similarities to *E. coli* than to KB cell 5 S RNAs. The following regions of 5.8 S RNA are most related to *E. coli* 5 S RNA: B (47%), C (44%), and F (50%). Region D starts with the sequence C-A-G-A-A-U, which is similar to that of prokaryotic 5 S RNAs at a similar location. Regions E and G of 5.8 S RNA are to 71 and 50% homologous to KB cell 5 S RNA. The remaining two regions, A and D, are 30% or less homologous to similar regions of either pro- or eukaryotic 5 S RNAs. It seems interesting that part of (or near to) region E, seven G·C base-pairs are possible. In prokaryotic 5 S RNAs, we find at least four, and in eukaryotic 5 S RNA two such base-pairs. 5.8 S RNA has also a U-G-G sequence that is 46 nucleotides away from the start of region D. This distance consists of 45–47 nucleotides in eukaryotic, and 39–42 bases in prokaryotic, 5 S RNAs. Region D of 5.8 S RNA contains, besides the C-A-G-A-A-U sequence, three other sequences of the type Y-G-A-A-Y, all of which could theoretically interact with parts of loop IV in eukaryotic tRNAs (except initiator tRNA).

D. Precursor 5 S and 5.8 S RNA

1. PROKARYOTIC 5 S RNAs

Precursor 5 S RNAs are difficult to isolate and generally are found only upon inhibition of protein synthesis by antibiotics (51–60). Earlier studies of *E. coli* precursor 5 S RNA show that it contains at its 5' end the additional sequence pA-U-U-U-G (53, 56) and at the 3' end a stretch of oligo (A) (57). Part of the maturation of this RNA may take place when attached to the 50 S ribosomal subunit (56), and the extended 5' end does not influence its biological activity (56). More recent findings show clearly that *E. coli* precursor 5 S RNA is part of a larger precursor RNA molecule that contains, in addition, 23 S and 16 S RNA (61). This system is certainly the most promising one so far isolated, for it should permit a detailed study of how all three ribosomal RNAs are processed during maturation.

5 S RNA precursors have also been isolated from *B. licheniformis* (59), *B. subtilis* (58, 60) and *B. megaterium* (62). The *B. licheniformis* precursor 5 S RNA may contain 240 and 220 nucleotides (59), while that of *B. subtilis* contains 178 and 152 (62). From chromatographic analysis and detailed characterization of enzymatic products obtained by incubation of *B. subtilis* precursor 5 S RNA with a new specific RNase M5 (60), it was concluded that the enzyme recognizes the sequence U-G-A-G-A-G and then hydrolyzes a phosphodiester bond 16 nucleotides away from

this sequence (*60*). Since the specific precursor 5 S RNA contains additional sequences at both the 5' and 3' end of the mature 5 S RNA, RNase M5 must recognize the pU-G-A-G-A-G at the beginning of the 5' end as well as the U-G-A-G-A-G sequence at an internal position near the 3' end of the precursor, which is actually part of mature *B. subtilis* 5 S RNA (*60*). Mature *B. megaterium* 5 S RNA has also the U-G-A-G-A-G, and mature *B. stearothermophilus* 5 S RNA a similar (C-A-A-G-A-G) sequence 16 nucleotides away from the 3' end (Fig. 1). This suggests that these two organisms have a maturation enzyme very similar or identical to RNase M5 of *B. subtilis*. Further studies on 5 S RNA as well as tRNA (*63*) maturation could yield fundamental information concerning the factors involved in RNA–protein interaction.

2. Eukaryotic 5 S RNAs

There is no experimental evidence suggesting the existence of precursor 5 S RNAs in eukaryotic organisms, nor that the 5 S RNA genes are closely linked to those of 28 S, 18 S and 5.8 S ribosomal RNA. The following observations support instead the idea that eukaryotic 5 S RNAs are primary transcription products. (a) Tri- or diphosphate groups are always found at the 5' end of 5 S RNA, indicating that the transcription starts at this position on the 5 S RNA gene (*20, 64–66*) and does not include parts of the spacer regions between successive genes (*67, 68*). (b) A search for precursor 5 S RNAs of HeLa cells by gel electrophoresis was negative (*69*). Hence, the *in vivo* synthesis of eukaryotic 5 S RNA differs distinctly from that of prokaryotic 5 S RNA or eukaryotic 5.8 S RNA, the last two being secondary transcription products.

3. Eukaryotic 5.8 S RNA

The results of experiments with *X. laevis* suggest that the 5.8 S RNA cistron is an integral part of the 28 S and 18 S ribosomal RNA cistrons (*70*). Based on hybridization studies, a model has been proposed that places the 5.8 S RNA cistron between the two larger ribosomal RNA cistrons (*70*). These observations are supported by the finding that in HeLa cells the 5.8 S RNA is most likely covalently bound to 45 S and 32 S precursor ribosomal RNAs (*71*). Apparently, in a step between the conversion of 32 S to 28 S RNA, the 5.8 S RNA is cleaved from the larger precursor (*71*). Similar conclusions have also been drawn from studies with yeast (*72*).

Two reports describe heterogeneity at the 5' end of 5.8 S RNA, which is interpreted to be due to the existence of precursors. The first report describes the observation that 5.8 S RNA of *Saccharomyces cerevisiae* has two minor forms, one starting at the 5' end with pU-A-U-U-A-A and the other with pA-U-A-U-U-A-A (50). The second experiment indicates that Novikoff ascites hepatoma 5.8 S RNA starts at the 5′ end 51% of the time with pG, 41% with pC, and 4% each with pA and pU (49).

III. Secondary and Tertiary Structures of 5 S and 5.8 S RNA

After the discovery of 5 S (1) and 5.8 S RNA (18), experiments were carried out to analyze their primary, secondary and tertiary structures. The information obtained to date is based upon studies that can be grouped into sequential studies (see Section II), physical characterization, limited ribonuclease digestion, chemical modication, and the binding of oligonucleotides. In this section, the results from each of these methods are separately discussed and then compared with previously proposed structural models for 5 S and 5.8 S RNA.

A. Physical Characterization of Native and Denatured 5 S RNAs and 5.8 S RNA

1. Chromatography and Gel Electrophoresis

5 S RNA is easily separated from the larger ribosomal RNAs and tRNAs by Sephadex G-100 (73–78) and G-200 (77) chromatography. Separation can also be accomplished on columns of methylated serum albumin kieselguhr (MAK), and methylated serum albumin on silicic acid (MASA) (17, 75, 79). In addition, 1 M NaCl (80) and 5 M LiCl (81) have been used to separate 5 S RNA from larger ribosomal RNAs, since only the latter are precipitated under these conditions.

Incubation of native *E. coli* 5 S RNA with urea yields denatured forms (75). The denaturation is reversible in the presence of Mg^{2+} and requires an energy of activation of 62 kcal/mol (82). The denatured 5 S RNA must have a different conformation from the native 5 S RNA since they can be separated by column chromatography (17, 75) and gel electrophoresis (82–84). The denatured *E. coli* 5 S RNA consists of two distinct species, which can be resolved on polyacrylamide gels (85, 86). The denatured 5 S RNAs are structurally different since they cannot be incorporated into partially reconstituted 50 S subunits (74, 87).

Analogous to the case of *E. coli* 5 S RNA, KB cell and HeLa cell 5 S RNAs may also exist in thermodynamically stable denatured forms, for in gel electrophoresis these 5 S RNA species migrate as two distinct bands (17, 88). The 5 S RNA from *Chlorella* cytoplasmic ribosomes migrates faster in gel electrophoresis than that of the chloroplasts, while HeLa cell 5 S RNA migrates in an intermediate position (89). Since the

Chlorella cytoplasmic and chloroplastic and HeLa cell 5 S RNA most likely all have 120 nucleotides, the difference in mobility suggests conformational differences.

The isolation of radioactive eukaryotic 5.8 S RNA is usually accomplished by gel electrophoresis (*48–50, 88*). To isolate larger amounts of this RNA, Sephadex G-100 chromatography has been employed successfully and has permitted the separation of yeast 28 S, 5.8 S, and 5 S RNAs into distinct peaks (*90*). Denatured forms of 5.8 S RNA have not been reported.

2. Analytical Centrifugation and X-ray Scattering

In analytical centrifugation, *E. coli* 5 S RNA sediments with an S value of 4.4–4.8 (*77, 91, 92*). The 5 S RNA from *Halobacterium cutirubrum*, whose ribosomes are stable only in the presence of 4 M KCl, sediments at 5 S (*93*). That the sedimentation of *E. coli* 5 S RNA is not greatly influenced by variation of salt is considered evidence for a rigid structure of this RNA (*91, 92*). Nevertheless, since the S value of 5 S RNA depends more upon concentration than does that of tRNA, a larger asymmetry than that of tRNA is suggested (*92*). Small-angle X-ray scattering data obtained from yeast and *E. coli* 5 S RNA support this conclusion (*94*); both RNAs have a radius of gyration (R_G) equal to 34.5 ± 1.5 Å. The overall shape seems to be prolate with an axial ratio of 5:1. The tentative interpretation is that the 5 S RNA molecule consists of a long, rigid double-helix of 40 base-pairs. The remainder of the 5 S RNA structure is located near one end of the large double-helix and most likely exists as small helical regions parallel to the major part of the molecule (*94*). This interpretation may not be entirely correct, because the same authors predicted a similar situation for tRNA (*94*). Recent X-ray crystallographic work has confirmed the asymmetry of tRNA, but not the existence of parallel helical regions at one end of the molecule (*95–97*). Instead, helical segments meeting in nearly a right angle were actually observed (*95–97*). Nevertheless, it may be concluded that prokaryotic and eukaryotic 5 S RNAs have a large degree of asymmetry and that their axial ratio is approximately 5:1.

3. Optical Studies

The optical methods most frequently employed in probing the structure of 5 S RNAs are measurements of UV absorption, optical rotatory dispersion (ORD) and circular dichroism (CD). The major information obtained concerns the amount of A·U and G·C base pairs. Since these measurements require comparison with model compounds, such as synthetic polynucleotides, and a certain flexibility of interpretation, it is not surprising that significant variations among the results have been reported. Nevertheless, the data are useful in giving the lower and upper limits of possible base-pairing in 5 S RNA.

Thermal melting of *E. coli* 5 S RNA under defined conditions exhibits a biphasic hypochromicity profile (*74, 76, 98, 99*). The early melting step is believed to be due to melting of short helical segments, which may be involved in teritary structure. Denatured *E. coli* 5 S RNA does not show such two-step melting (*99*). Sea urchin (*100*) and *B. subtilis* (*101*) 5 S RNAs also exhibit biphasic melting, whereas that of *B. stearothermophilus* does not (*78*). The total amount of hyperchromicity is similar for the different 5 S RNAs (*74, 76, 78, 98–101*), suggesting similar extents of base-pairing.

From the total amount of hypochromicity, it has been calculated that 62–64% of *E. coli* 5 S RNA is in helical regions (*92, 99*). The denatured B form of 5 S RNA apparently has only 51% of its bases in double-stranded regions. The total number of base-pairs estimated is therefore 37 (*92*) or 38–39 (*99*) for native 5 S RNA and 31 (*99*) for the denatured form. From detailed kinetic studies of native and denatured *E. coli* 5 S RNA, it has been concluded that about nine base-pairs are re-formed when the RNA structure is changed from native to denatured (*82, 102*).

A combination of ORD and UV measurements suggest that *E. coli* 5 S RNA contains 40–49 base-pairs (*103, 104*) and that the secondary structure consists of short helical segments similar to those of tRNA. UV, ORD, CD and infrared absorption studies suggest that the secondary structure of *E. coli* 5 S RNA includes 28 ± 4 G·C and 13 ± 4 A·U base-pairs (*105*).

The conclusions from these studies on *E. coli* 5 S RNA may be compared with those on other prokaryotic 5 S RNAs, from *B. subtilis* and *B. stearothermophilus* (*101, 106*). Based upon thermal denaturation curves and UV difference spectra at 23° and 85°C, it was concluded that the overall structures of these 5 S RNAs are similar and that *E. coli* 5 S RNA, with 36 base-pairs, has a slightly higher helical content than either of the other two (34 base-pairs). Thermal melting and ORD studies indicate a concentration of G·C base-pairs in a long helical segment of *E. coli* 5 S RNA. In contrast, *B. stearothermophilus* 5 S RNA consists of helical regions in which G·C and A·U base-pairs are more evenly distributed (*106*).

In experiments with ethidium bromide, which is generally believed to intercalate between adjacent base-pairs of nucleic acids (*107, 108*), *E. coli* 5 S RNA bound 12–13 and *B. stearothermophilus* 9–10 ethidium bromide molecules per 5 S RNA (*106*). The difference suggests that

E. coli 5 S RNA contains a longer helical region than B. stearothermophilus. Other experiments indicate that only 5–6 molecules of ethidium bromide bind to native and denatured E. coli 5 S RNA (109). This discrepancy may be due to differences of experimental conditions, but it calls nevertheless for clarification, since the number of ethidium bromide molecules bound per 5 S RNA has been used to propose significant differences between the structures of E. coli and B. stearothermophilus 5 S RNA (106).

Another interesting comparison is that between the structures of prokaryotic and eukaryotic 5 S RNAs. The physical characteristics of sea urchin egg and E. coli 5 S RNA were compared (100), leading to the following conclusions. (a) The sequence of sea urchin 5 S RNA differs drastically from that of E. coli, Pseudomonas fluorescens and KB cells. (b) Sea urchin 5 S RNA is not only more compact, but also more asymmetric, than E. coli 5 S RNA. (c) E. coli 5 S RNA contains several G·C base-pairs, which are involved only in three-dimensional structure. (d) Sea urchin 5 S RNA has two distinct regions, one being rich in A·U base-pairs and the other in G·C base-pairs. Since significant differences between these 5 S RNAs exist, it would be of great interest to know the sequence of sea urchin 5 S RNA.

4. NMR STUDIES

High resolution nuclear magnetic resonance (NMR) spectroscopy has recently been extensively employed to study the total number, types and location of base-pairs in RNA. The results obtained are in good agreement with X-ray crystallographic studies on tRNA (110). At 35°C and in the absence of Mg^{2+}, yeast 5 S RNA contains 21 ± 2 base-pairs (111). Addition of Mg^{2+} increases the number to 24 ± 2, which, after heating to 60°C for 5 minutes, increases to 28 ± 3. Clearly, Mg^{2+} significantly influences the total number of base-pairs. The data suggest that 85% of the 28 ± 3 base-pairs are G·C. Similar studies with E. coli 5 S RNA also indicate the presence of 28 ± 2 base-pairs, of which 24 are G·C and 4 are A·U (112). Preliminary studies show that B. stearothermophilus 5 S RNA contains 30 base-pairs, of which only a few are A·U (D. R. Kearns and V. A. Erdmann, unpublished results).

In summary, physical measurements on 5 S RNA suggest that it is more asymmetric than tRNA. The sedimentation value lies between 4.4 and 5 S. The overall structures of eukaryotic and prokaryotic 5 S RNAs are similar in that G·C base-pairs are dominant and approximate 70% of the total. Some 5 S RNAs, such as that from E. coli, are relatively easy to denature, whereupon they assume a configuration that is thermodynamically stable and distinct from that of the native form. The information obtained with different methods concerning the number of base-pairs within 5 S RNA leaves something to be desired, since the values, even for *E. coli* 5 S RNA, range from 28 to 49. Unfortunately, there are few data available on eukaryotic 5.8 S RNA.

B. Enzymic Hydrolysis

Partial enzymic hydrolysis has also been used to investigate the structures of 5 S RNA. The basic rationale for this approach is that ribonucleases will first hydrolyze unpaired regions of the RNA that are at "exposed" positions of the molecule. Again, *E. coli* 5 S RNA has been studied in most detail.

1. Pro- and Eukaryotic 5 S RNAs Free in Solution

Treatment of *E. coli* 5 S RNA with limiting amounts of T1 ribonuclease shows that the residues around G_{41} are well exposed, for it is usually at this position that the first hydrolysis occurs (*113–115*) (see also Fig. 6). The next points of cleavage occur at G_{13}, G_{56} and G_{69} (*114*). Studies with RNase IV and sheep kidney nuclease confirm that the region around position 41 in *E. coli* 5 S RNA is the most accessible for enzymic hydrolysis (*115*). However, the major point of hydrolysis with RNase IV is not directly at G_{41} but most likely at G_{44}, and sheep kidney nuclease produces cleavages first at positions 34 and 50 (*115*). It is quite evident from these results that even the primary site of RNase hydrolysis is somewhat dependent upon the ribonuclease used.

One contradictory report indicates that G_{61}, and not G_{41}, is the primary target of T1 ribonuclease, with other breaks at G_{16}, G_{18}, G_{56}, G_{100} and G_{102} (*116*). Among these, only that at G_{56} is in agreement with the studies cited above. However, since the other cleavage sites are significantly different from those of denatured 5 S RNA, in which primarily positions G_{16}, G_{24}, G_{44}, G_{61}, G_{79}, G_{98} and G_{106} are attacked (*113*), it was concluded that the 5 S RNA studied was a stable form different from both the native and denatured 5 S RNAs (*116*) (Fig. 6). The latter interpretation is supported by the fact that the studies were carried out under conditions that should not denature 5 S RNA.

Comparison of primary cleavage sites in prokaryotic and eukaryotic 5 S RNA shows that the overall structures must be similar. For example, partial hydrolysis of cytoplasmic (eukaryotic) and chloroplastic (prokaryotic) 5 S RNAs from the unicellular green alga *Chlorella* with T1 ribonuclease yields two similar fragments of 32 and 88 nucleotides (*89*). It was assumed that the cleavage occurred at similar positions, although detailed sequence data have been published only for the cytoplasmic 5 S RNA (*47*). Nucleotides around position 40 seemed unavailable for RNase

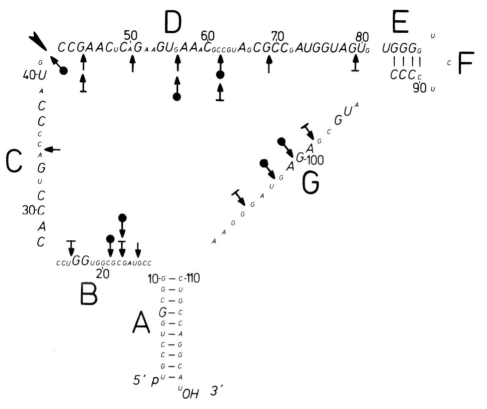

Fig. 6. Points of ribonuclease hydrolysis for native [→; (114, 115)], stable [●→; (116)] and denatured [↦; (113)] forms of *Escherichia coli* 5 S RNA. Heavy arrow indicates first point of hydrolysis. For other details, see Section III, B, 1.

cleavage, in contrast to other eukaryotic 5 S RNAs in which this region of the molecule is hydrolyzed at the same rate at the vicinity of nucleotide 90 (117).

Partial hydrolysis studies of *P. fluorescens* 5 S RNA with T2 ribonuclease show that the first cleavage occurs at C_{42} and U_{49} (117). These sites of hydrolysis are similar to the ones observed with different ribonucleases and *E. coli* 5 S RNA (Fig. 6), suggesting that both 5 S RNAs have similar overall structures.

The eukaryotic 5 S RNA analyzed in most detail by partial ribonuclease digestion is that of yeast (Fig. 7). The primary points of hydrolysis are positions 37 (117), 41 (39) and 91 (39, 117). Secondary cleavage sites are at positions 11, 54, 73 and 107 (39).

From these studies with limited ribonuclease digestion of both

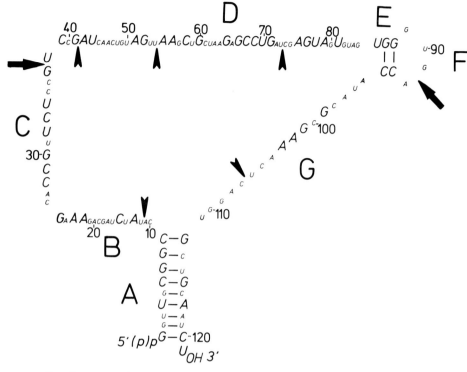

Fig. 7. Points of ribonuclease hydrolysis of *Torulopsis utilis* (yeast) 5 S RNA (39). Heavy arrow indicates sites of primary hits and are also observed at similar positions in other eukaryotic 5 S RNAs (117). Small arrow indicates cleavage sites as observed with yeast 5 S RNA. For other details see Section III, B, 1.

prokaryotic and eukaryotic 5 S RNAs it seems clear that the regions around position 40 are the most exposed. In eukaryotic 5 S RNAs, there is a second region around nucleotide 90 that is as readily hydrolyzed as that near position 40. Regions A, C and E in eukaryotic and prokaryotic 5 S RNAs are resistant to RNase attack and do not contain sites of secondary hydrolysis (Figs. 6 and 7). The resistance of regions A and E may be due to the existence of base-pairing as indicated. Region C cannot be base-paired with itself and may actually be protected from RNase hydrolysis through secondary interaction with other parts of the molecule.

Regions B, D and G of both prokaryotic and eukaryotic 5 S RNase show secondary sites of hydrolysis (Figs. 6 and 7); this suggests that these areas do not contain long base-paired regions. An alternative explanation is that after the primary hydrolysis (around nucleotide 40 and 90) has taken place, the structures of regions B, D and G are changed so that the ribonucleases can hydrolyze parts of these regions.

In eukaryotic 5 S RNAs, region F is readily attacked by ribonuclease, while in prokaryotic 5 S RNAs it is not. The reason for this behavior is not known. It may reflect significant differences in the structures of eukaryotic and prokaryotic 5 S RNAs in this region.

2. 5 S RNA IN THE RIBOSOME

Since free 5 S RNA is readily attacked at definite positions by ribonucleases, it was of interest to see whether the same positions are available when the RNA is part of the ribosome. Detailed studies with partially reconstituted and native *E. coli* 50 S ribosomal subunits show clearly that 5 S RNA is protected when part of the ribosome (*118, 119*). In native ribosomal subunits, 5 S RNA is even protected after extensive trypsin digestion of the ribosome (*119*), indicating that long stretches of the 5 S RNA are not exposed on the ribosomal surface. However, this does not prove that short sequences of resistant 5 S RNA are not exposed. In fact, chemical modification (Section III, C) shows that certain nucleotides of 5 S RNA are indeed available for chemical reaction in the ribosome.

3. EUKARYOTIC 5.8 S RNA

The partial hydrolysis of yeast 5.8 S RNA with pancreatic or T1 ribonuclease yields a large number of fragments from which the primary structure can be deduced (*19*). Assuming that the largest fragments produced reflect the first points of RNase cleavage, the primary hydrolysis points are at nucleotides 11, 32, 64, 91 and 125 for pancreatic ribonuclease and at nucleotides 51, 85, 103 and 140 for T1 ribonuclease. As can be seen from Fig. 5, the primary attacks occur mostly in region D of 5.8 S RNA, whereas regions A, C and E seem protected. [A very similar RNase hydrolysis pattern is observed with pro- and eukaryotic 5 S RNAs (Figs. 6 and 7).] However, this must be interpreted with caution, as the primary points of hydrolysis were not pointed out explicitly by the worker who determined its sequence (*19*), and sequence data from other 5.8 S RNAs are lacking.

One interesting observation made during the process of determining the sequence of yeast 5.8 S RNA deserves special mention: partial hydrolysis with T1 ribonuclease yields a fragment that includes nucleotides 115 through 140 (Fig. 5); it is rich in G·C base pairs and seems to remain base-paired even in the presene of 7 M urea (*19*).

C. Chemical Modification

Chemical modification can be a useful method for obtaining information concerning the structure and function of RNA. The ideal reagent should be specific for one of the four common bases and preferentially react with nucleotides whether base-paired or not. The ideal reagent would also allow the introduction of radioactive isotopes into the modified RNA without disturbing its structure. Unfortunately, no chemical reagent is available that meets all these criteria.

1. 5 S RNA FREE IN SOLUTION

Monoperphthalic acid reacts under defined conditions preferentially with unpaired adenines in ribonucleic acids to form the corresponding 1-N-oxides, which cannot engage in the Watson-Crick type of base-pairing (76, 120, 121). Reaction of E. coli 5 S RNA with monoperphthalic acid at room temperature shows that 10 out of 23 adenosines are oxidizable and therefore not base-paired. Oxidations carried out at 47°C, the t_m of the first melting phase, modifies six additional adenines that are believed to be involved in three-dimensional structure (76, 98). From these results, the lower and upper limits of the total number of base-pairs in E. coli 5 S RNA were predicted to be between 41 and 65% (76, 98), which corresponds to 24–39 base-pairs. Since the reaction with monoperphthalic acid does not allow the introduction of radioactive label, the adenines in 5 S RNA modified by it have not yet been determined.

The reaction of a water-soluble carbodiimide with E. coli 5 S RNA allowed the modification of two unpaired uridines (122). One of these is in position 40; the other could not be definitely identified, but could possibly be in position 14, 65, 77 or 103 (see Fig. 8).

Glyoxal and kethoxal, two guanine-specific reagents, react preferentially with G_{41} and G_{13} in E. coli 5 S RNA (123–125). Prolonged reaction time or increase in reagent concentration causes additional modifications at positions 75 and 100 and/or 102. Other reagents tested in this study were methoxyamine (reagent for cytosine) and deamination with nitrous acid. Although these two reagents did not permit the precise identification of the bases modified, the results suggest that the reacting nucleotides are in positions 34–41, 44–51 and most likely 88 (123). Denatured 5 S RNA reacts differently: glyoxal does not modify guanine at position 41 but does modify G_{61} and G_{100} and/or G_{102} (125).

2. 5 S RNA WITHIN THE RIBOSOME

Several reagents have been tested for their ability to modify the 5 S RNA in E. coli ribosomes. Reaction of 70 S ribosomes with kethoxal showed that all three ribosomal RNAs (23 S, 16 S and 5 S) have modifiable guanines (126, 127) and that kethoxal can also react with proteins. G_{41} in 5 S RNA is the major base modified in the 70 S ribosome (128). In another study, 50 S subunits were treated with kethoxal; again, G_{41} was modified (129). In addition, G_{13} also reacted.

Treatment of E. coli 50 S ribosomal subunits with monoperphthalic

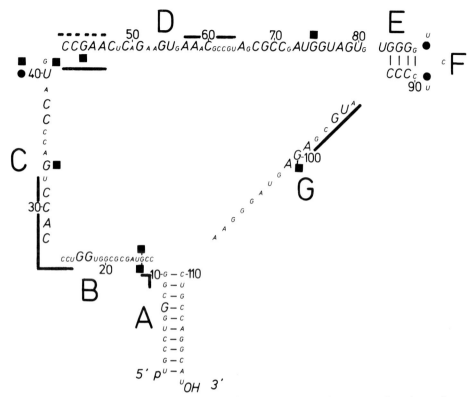

FIG. 8. Chemical modification and oligonucleotide binding to *Escherichia coli* 5 S RNA. Primary sites of modification with carbodiimide (●) (122, 130) and glyoxal or kethoxal (■) (123–130). Oligonucleotide binding (9, 132, 133): strong (——) and weak (----). Symbols outside the structure represent experiments with free 5 S RNA, and those inside the structure represent experiments with 5 S RNA· protein complexes or ribosomes. For other details see text.

acid allows the modification of two adenines (9) that are crucial for ribosome activity and for binding of the tRNA fragment T-Ψ-C-G (see Section V and below). The modified adenines have so far not been identified, but from the T-Ψ-C-G binding studies it is concluded that they are either at 45 and 46 or 57 and 58.

Other studies have dealt with the accessibility of 5 S RNA in a 23 S·5 S·protein complex, the protein being unidentified, probably containing, together with other possibilities, proteins L18 and L25 (130). In such complexes, kethoxal reacts with G_{33}, G_{41} and G_{44}, and possibly G_{51}. Carbodiimide modification suggests that U_{87} and U_{89} of 5 S RNA are exposed.

As can be seen from Fig. 8, chemical modification of 5 S RNA, whether free in solution or within the ribosome, indicates similar chemically reactive sites on the molecule. It is of interest to note that nucleotides at the end of the conserved region C and the beginning of conserved region D (nucleotides 40–55) are more reactive when in the ribosome or in the 5 S RNA·23 S RNA·protein complex than they are in the free 5 S RNA (Fig. 8).

D. Oligonucleotide Binding

1. Prokaryotic 5 S RNAs

The binding of oligonucleotides is a very sensitive method for the detection of single-stranded stacked regions in nucleic acids, for it depends upon the formation of base-pairs between the oligonucleotide and a complementary sequence in the macromolecule. This novel method had originally been introduced to analyze the anticodon loop in tRNA^{fMet} (131) and subsequently for the determination of unpaired regions in E. coli 5 S RNA (132).

The following sequences of E. coli 5 S RNA bind their complementary oligonucleotides: 9–13, 25–32, 58–65 and 94–99 (Fig. 8). In addition, weak binding of U-C-G is observed, which suggests that it may interact with positions 43–45 (132). Subsequent studies with U-U-C-G and T-Ψ-C-G support the observation that the region around positions 45 may be available for weak binding of oligonucleotides (9).

Studies (133) comparing the oligonucleotide binding regions of 5 S RNA in E. coli with those from other sources confirmed the earlier observation (132) that positions 9–13 and 28–32 are available for oligonucleotide binding in E. coli 5 S RNA, whereas B. stearothermophilus 5 S RNA binds oligonucleotides at positions 27–32 and 68–73. The first region corresponds to one in E. coli 5 S RNA, but the second does not.

2. Eukaryotic 5 S RNA

The only eukaryotic 5 S RNA analyzed for oligonucleotide binding is that from rat liver, which bound U-G-G-U, G-G-U-U, G-G-G-C, C-C-C-U, C-C-C-A and A-C-C-G (134). The rat liver 5 S RNA sequence is identical to the one from KB cells (42), suggesting single-strandedness around positions 14, 28, 59, 70 and 95. These results imply similar overall structures in pro- and eukaryotic 5 S RNAs. It is striking that in both 5 S RNAs, sequences around position 28 (region C), 60 (region D) and 96 (region G) can bind oligonucleotides, although their sequences are entirely different.

Since E. coli 5 S RNA has been most widely studied, it is possible to compare enzymic hydrolysis, chemical modification and oligonucleotide binding data for this species (Figs. 7 and 8). In general, the results of partial RNase hydrolysis and chemical modification agree much better with each other than do either of them with those from oligonucleotide binding. This seems surprising, but may reflect the fact that chemical modification and RNase hydrolysis are more easily carried out at regions that are single-stranded but not necessarily ideally stacked, while oligonucleotide binding requires good stacking as well as single-strandedness.

E. Proposed Models for 5 S and 5.8 S RNA

Based upon the primary structure and various investigations including physical, enzymic, chemical, oligonucleotide binding, and computer studies, a series of models for prokaryotic (*2, 24, 36, 92, 104, 112, 134a–c*) and eukaryotic (*24, 33, 39, 47, 134a, 134c*) 5 S RNAs as well as for yeast 5.8 S RNA (*19*) have been proposed. A number of these models are shown in Figs. 9 and 10. In addition to these models, detailed computer studies show that even for *E. coli* 5 S RNA a number of other models are still possible (Richards, personal communications). The large number of possible 5 S RNA structures indicates the difficulties one will encounter in determining the true structure of this molecule.

There is only one aspect common to all 5 S RNA models: the 5' and 3' end (Section II) are base-paired with each other (Figs. 9 and 10). It should be noted that such important factors as function, protein recognition and interaction are generally not considered in constructing these models, since such information has only recently become available. One structural *E. coli* 5 S RNA model that takes into consideration RNA–protein interaction is shown in Fig. 9b. However, none of the models proposed agrees with all the chemical, enzymic hydrolysis, and oligonucleotide binding findings, which suggests that 5 S RNA has a flexible structure, perhaps required for its biological function.

Only one model has been proposed for 5.8 S RNA (*19*); its structure is shown in Fig. 10g. Since the "hairpin" loop from nucleotide 115 to 137 is unusually stable toward urea and temperature, it was concluded that it exists within the 5.8 S RNA structure (*19*).

It may be concluded that, in building 5 S RNA structural models, it should be kept in mind that 5 S RNA is an asymmetric molecule in which less base-pairing exists than was generally anticipated. Since the conserved regions C and D are accessible not only in the free 5 S RNA, but also in the ribosome or 5 S · 23 S · protein complex, it is suggested that they are important in protein synthesis. In addition, it seems worthwhile to point out that building models for 5 S RNA is justified on the basis that they stimulate new ideas and experiments to obtain informa-

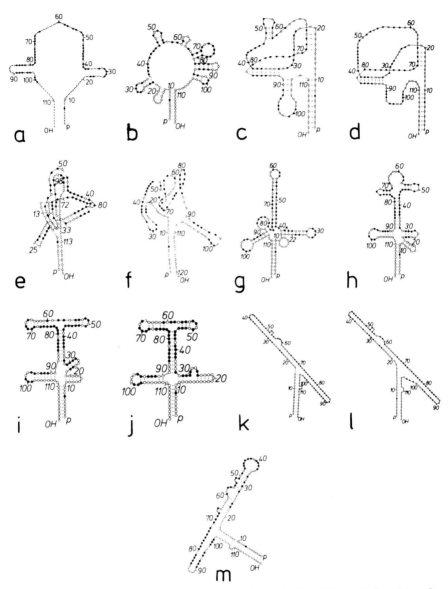

FIG. 9. Proposed structural models for prokaryotic 5 S RNAs. *Escherichia coli* 5 S RNA models: a (*2*), b (*36*), c at room temperature (*112*), d above 50°C (*112*), e (*104*), f (*134b*), g (*134c*), h (*92*), i (*24*), k (*24*) and m (*134a*). *Pseudomonas fluorescens* 5 S RNA models: j (*24*) and l (*24*). Filled circles represent the conserved bases in prokaryotic 5 S RNAs. For actual sequence of each 5 S RNA, see Fig. 1.

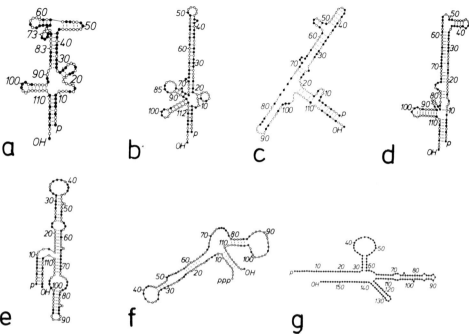

FIG. 10. Proposed structural models for eukaryotic 5 S RNAs and yeast 5.8 S RNA. KB cell 5 S RNA models: a (24), b (134c), c (134a). Xenopus laevis 5 S RNA model d (33), yeast 5 S RNA model e (39) and Chlorella cytoplasmic 5 S RNA model f (47). A 5.8 S model is represented in g (19). Filled circles represent the conserved bases in eukaryotic 5 S RNAs (a–f). For 5 S RNA sequences, see Fig. 3, and for 5.8 S RNA, Fig. 5.

tion on structure and function. The value of model-building has been demonstrated in the recent past with tRNA (135), even though none of the proposed three-dimensional models could finally withstand comparison with the X-ray structure analysis (95–97). [See article by Kim in Volume 17 of this series.]

IV. Complexes between 5 S RNA and Protein

A. Proteins Binding 5 S RNA

Two primary approaches have been used to determine the proteins interacting with 5 S RNA in the ribosome. Earlier experiments concentrated upon partial unfolding of the ribosomal subunit to determine at which step 5 S RNA and ribosomal proteins were released. The more recent approach became possible after conditions were developed for the total reconstitution of *B. stearothermophilus* 50 S ribosomal subunits (6) and for the separation of 5 S RNA from an RNA–protein mixture (136).

1. RIBOSOME UNFOLDING

Escherichia coli ribosomes can be unfolded by EDTA treatment (137) with the release of 5 S RNA (138–141). Other conditions for 5 S RNA dissociation from the 50 S ribosome are treatment with 2 M LiCl (142–145), with 50 mM phosphate (146), or with 0.5–1.0 M NH$_4$Cl at 0.1 mM Mg^{2+} (147). Ribosomes in the presence of sodium citrate and low magnesium will not release 5 S RNA, but will make it available for exchange with externally added 5 S RNA (148).

Besides the release of 5 S RNA, a number of ribosomal proteins are dissociated during unfolding (7, 141, 142, 146). The release of 5 S RNA by 2 M LiCl coincides with release of more than 15 of the 34 50 S proteins (144, 149). These released proteins are potential candidates for binding to 5 S RNA. The definite answer to this question was obtained from reconstitution experiments.

2. PROKARYOTIC 5 S RNA · PROTEIN COMPLEXES

Reconstitution experiments including *E. coli* 23 S RNA, 5 S RNA and individually purified proteins released by 2 M LiCl revealed that E-L18,[1] in conjunction with either E-L6 or E-L25, permitted 5 S · 23 S RNA complex formation (144). In different reconstitution experiments, all the *E. coli* 50 S proteins were incubated with 5 S RNA and analyzed on sucrose gradients (8). The isolated 5 S RNA·protein complex consisted of two major proteins, E-L18 and E-L25. Besides these, E-L5 was identified as a protein binding weakly to 5 S RNA. The presence of traces of proteins E-L20 and E-L30 indicated that they do not bind directly to the 5 S RNA but perhaps bind to the 5 S RNA · protein complex on the ribosome (8). The apparent discrepancy between one observation that E-L5 is a 5 S RNA-binding protein (8) and a second (144) that E-L6 binds to 5 S RNA has been resolved in favor of E-L5 (109).

There are other experiments supporting the earlier finding that E-L5, E-L18 and E-L25 are 5 S RNA-binding proteins (8). One of these included Millipore-filter binding of 5 S RNA · protein complexes (150), and the other affinity-binding of *E. coli* ribosomal proteins to immobilized 5 S RNA (151). The results from EDTA unfolding studies of 50 S subunits (141) lend support to the observation that E-L30 may be im-

[1] E and B stand for *E. coli* and *B. stearothermophilus*, respectively (8); L indicates that the protein comes from the large ribosomal subunit, in this case, the 50 S ribosome (152).

portant for interaction with the 5 S RNA · protein complex on the ribosome (8).

If 5 S RNA is incubated with early log-phase *E. coli* 50 S ribosomal proteins, several other proteins (E-L7, E-L12 and E-L10 in addition to the ones mentioned above) may associate weakly with the 5 S RNA · protein complex (J. R. Horne, K. Paehlke and V. A. Erdmann, unpublished observations).

The *B. stearothermophilus* 5 S · RNA protein complex has also been isolated and its proteins identified (8). This complex contains B-L5 and B-L22 as major proteins and traces of B-L6, B-L10, B-L26 and B-L27. Sometimes traces of B-L13 are also found (J. R. Horne and V. A. Erdmann, unpublished results); it migrates in two-dimensional electrophoresis at the same position as *E. coli* L7/L12 (*152*) and therefore most likely corresponds to it.

Hybrid 5 S RNA · protein complexes have been reconstituted from *E. coli* and *B. stearothermophilus* 50 S subunit components (8). The *B. stearothermophilus* proteins B-L5 and B-L22 bind to *E. coli* 5 S RNA, and *E. coli* proteins E-L5, E-L18 and E-L25 bind to *B. stearothermophilus* 5 S RNA. This comparative study of 5 S RNA · protein complexes permitted the first correlation of ribosomal proteins from two different bacterial species, as follows: B-L5 = E-L5; B-L22 = E-L18 (8); B-L13 = E-L7/L12; and B-L10 = E-11 (J. R. Horne and V. A. Erdmann, unpublished results). In a recent independent study, it was also concluded that B-L10 is functionally equivalent to E-L11 (*167*). Table I lists those

TABLE I

BINDING OF *Escherichia coli* AND *Bacillus stearothermophilus* RIBOSOMAL PROTEINS TO 5 S RNA[a]

Source of 5 S RNA	Source of proteins	Proteins bound
E. coli	*E. coli*	(E-L5),[b] E-L18,[b] E-L25,[b] [E-L7,[c] E-L12,[c] E-L10,[c] E-L20,[b] E-L30][b]
B. stearothermophilus	*B. stearothermophilus*	B-L5,[b] 5-L22,[b] [B-L6,[b] B-L10,[b] B-L13,[d] B-L26,[b] B-L27][b]
E. coli	*B. stearothermophilus*	B-L5,[b] B-L22[b]
B. stearothermophilus	*E. coli*	E-L5,[b] E-L18,[b] E-L25[b]

[a] Proteins enclosed in parentheses bind in less than stoichiometric amounts, and by square brackets in traces, to the corresponding 5 S RNAs. Data taken from the sources listed in footnotes b–d.

[b] From Horne and Erdmann (*8*).

[c] From J. R. Horne, K. Paehlke and V. A. Erdmann, unpublished observations.

[d] From J. R. Horne and V. A. Erdmann, unpublished observations.

proteins that bind to *E. coli* and *B. stearothermophilus* 5 S RNA or that are believed to interact with the 5 S RNA · protein complexes.

Attempts to construct 5 S RNA · protein complexes from yeast and rat liver 5 S RNA and *E. coli* proteins have failed (P. Wrede and V. A. Erdmann, unpublished results), agreeing with earlier experiments showing that eukaryotic 5 S RNAs are not incorporated during reconstitution of *B. stearothermophilus* 50 S subunits (*21*).

3. Eukaryotic 5 S RNA · Protein Complexes

As with prokaryotic ribosomes, treatment of eukaryotic ribosomes with EDTA causes the release of 5 S RNA (*153–155*). More careful analysis of the released 5 S RNA showed that it exists as a 5 S RNA · protein complex (*156–158*). Subsequently, it was found that treatment with high concentrations of monovalent ions (*158*), formamide (*159*) or urea (*160*) also releases the 5 S RNA as an RNA · protein complex.

Although the eukaryotic 5 S RNA · protein complexes were the first of this type described, they are not as well characterized as those from prokaryotes. Apparently, the complexes from rat liver (*158, 160, 161*) and rabbit reticulocyte (*158*) ribosomes contain one protein with a molecular weight of 31,000–41,000. Unfortunately, this protein has not yet been characterized by two-dimensional gel electrophoresis, so that it is not known which one of the 60 S ribosomal proteins (*162, 163*) interacts with 5 S RNA.

From sedimentation experiments, it is concluded that the eukaryotic 5 S RNA · protein complex sediments at approximately 7 S (*157, 159, 161*) and is 48.5% RNA (*159*).

B. Ribosomal Proteins in the Neighborhood of 5 S RNA in Ribosomes

Since there is only incomplete information available concerning the proteins in neighborhood of either 5 S or 5.8 S RNA in eukaryotic ribosomes, this section deals exclusively with those of *E. coli* and *B. stearothermophilus*.

1. *E. coli* 50 S Subunit

The results presented in Section IV, A, 2 and summarized in Table I show that E-L5, E-L18 and E-L25 bind directly to 5 S RNA and that proteins E-L7, E-L12, E-L10, E-L20 and E-L30 interact loosely with the 5 S RNA · protein complex. The results suggest that these proteins are neighbors in the ribosome. In addition, functional and affinity-labeling experiments, as summarized in recent reviews (*164–166*), support the assumption that the above-mentioned proteins are in the neighborhood of 5 S RNA, and that proteins E-L2, E-L6, E-L11 and E-L16 must also be at this region of the ribosome.

The observation that proteins E-L6, E-L10 and E-L7/L12 are important for binding of elongation factors E-FG and E-FTu to the ribosome suggests further that these factors bind in the vicinity of 5 S RNA on the ribosome (*164–166*). Therefore, the results of the protein topography studies suggest that 5 S RNA is at or near the ribosomal A-site, which is further supported by functional investigations concerning 5 S RNA (see Section V).

2. *B. stearothermophilus* 50 S SUBUNIT

The *B. stearothermophilus* 50 S ribosomal subunit is less thoroughly investigated than that of *E. coli*. Nevertheless, a number of proteins besides B-L5 and B-L22 are known to interact with the 5 S RNA, namely, B-L6, B-L10, B-L13, B-L26 and B-L27 (see Section IV, A, 2 and Table I). More recent affinity-labeling experiments involving EF-G dependent binding of GDP imply that proteins B-L2, B-L10 and B-L22 are in the neighborhood of the site at which this elongation factor binds (see Section V, E. and Reference *167*).

C. Regions of *E. coli* and *B. stearothermophilus* 5 S RNA Interacting with Proteins

Several techniques may be applied to study the sites of RNA that interact with proteins. Each of them requires the formation of specific RNA · protein complexes, which may then be analyzed by partial ribonuclease digestion, chemical modification, or binding of oligonucleotides. Protective effects seen by the presence of proteins may then be attributed to their interaction with the specific regions of the RNA. Such studies have been carried out with prokaryotic 5 S RNAs and are described here.

1. *E. coli* 23 S RNA · 5 S RNA · PROTEIN COMPLEX

The *E. coli* 23 S RNA · protein complex was the first analyzed to determine which regions of the 5 S RNA interact with ribosomal proteins (*7, 130, 168*). Besides the two ribosomal RNAs, the complex contained *E. coli* proteins E-L6, E-L18 and E-L25 (*130*). It is important to note that only two of the three 5 S RNA-binding proteins, namely E-L18 and E-L25, were present in this complex, but that the third, E-L5 (*8*),

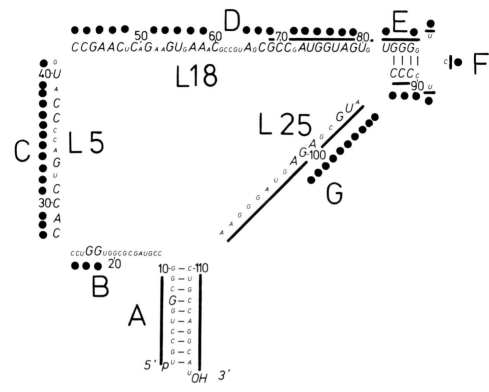

FIG. 11. 5 S RNA sequences protected from ribonuclease digestion by bound ribosomal proteins. *Escherichia coli* 5 S RNA·23 S RNA·E-L6·E-L18·E-L25 complex (———) (*130*). Summary of a comparative study involving 5 S RNA and protein components from *E. coli* and *Bacillus stearothermophilus* (●●●●) (*169, 169a*). Approximate regions at which E-L5, E-L18 and E-L25 are believed to interact with the 5 S RNAs from *E. coli* and *B. stearothermophilus* are taken from References *169* and *169a*. Note that the 5 S RNA structure shown is that of *E. coli*, in which the large letters represent nucleotides found in similar positions in all prokaryotic class I 5 S RNAs.

was not. Partial ribonuclease digestion of such a complex showed that the proteins were primarily able to protect the sequences 1–11 and 69–120 (Fig. 11 and Reference *130*).

Since partial hydrolysis studies with pancreatic ribonuclease show that within this protected fragment positions 88 and 89 are most sensitive to hydrolysis, it is concluded that this region of the complexed 5 S RNA is not interacting with ribosomal proteins. The results of chemical modification of the RNA in this complex support those of the enzymic hydrolysis experiments, showing that the most accessible G residues are

STRUCTURE AND FUNCTION OF 5 S AND 5.8 S RNA   77

G_{33}, G_{41}, G_{44} and G_{51}. In addition, U_{87} and U_{89} of 5 S RNA, which belong to the sequence within the 23 S RNA · 5 S RNA · protein complex at which secondary RNase hydrolysis occurs, are readily modified (*130*). From this comprehensive study, it was concluded that the 5 S RNA-binding proteins interact and therefore protect the two double-stranded regions, 1–10 base-paired with 110–119, and 82–86 base-paired with 90–94 (*130*). Since these regions are also most likely base-paired in free 5 S RNA (*23*), the results suggest that the structure of 5 S RNA in this region does not change significantly when proteins interact with it.

2. E. coli AND B. stearothermophilus 5 S RNA · PROTEIN COMPLEXES

A comparative study has been carried out with 5 S RNA · protein complexes involving components from *E. coli* and *B. stearothermophilus* (*169, 169a*). The four different 5 S RNA · protein complexes similar to the ones shown in Table I were treated with pancreatic and T1-ribonuclease, and the RNA fragments protected by the proteins were identified. All combinations contained protein E-L18 (or its equivalent B-L22), but only some contained E-L5 (or B-L5) and/or E-L25. (A summary of the protected fragments of these different complexes is shown in Fig. 11.) The results indicate that the 5 S RNA-binding proteins interact primarily with regions of the 5 S RNA reaching from the conserved G-G sequence of region B to the conserved A-G-A sequence of region G (Fig. 11). This part of the 5 S RNA encompasses the most conserved regions of prokaryotic 5 S RNAs (Fig. 1). These results also support earlier 50 S reconstitution experiments, which showed that a number of different prokaryotic, but not eukaryotic, 5 S RNAs can be incorporated into the subunits of *B. stearothermophilus* (*21*) and *E. coli* (*170, 171*). These previous studies had already suggested that certain regions of the 5 S RNA important for protein interaction and ribosomal function have been conserved during evolution.

Another interesting result is that in none of the 5 S RNA fragments is the complete 3′ and 5′ base-paired stem (region A) protected (Fig. 11 and References *169* and *169a*). Again, comparison of region A from different 5 S RNAs reveals that the only constant element in this stem is a G in position 7 (Fig. 1). Since all other nucleotides are variable between different 5 S RNAs, it would be difficult to envision specific RNA–protein interactions at this part of the molecule.

The comparative study has also yielded valuable information on the possible regions of 5 S RNA with which the different proteins interact (*169, 169a*). Apparently protein E-L5 binds primarily to the sequence 30–50, E-L18 to 55–98 and E-L25 around position 100. Since the protective effect of E-L25 was not as pronounced as the others, it seems possible that E-L25 binding is itself dependent upon interaction with another protein, most likely E-L18. This partial binding dependence had also previously been observed in the 23 S RNA · 5 S RNA · protein complex of *E. coli* (*130*).

Ethidium bromide binding studies to *E. coli* 5 S RNA and 5 S RNA · protein complexes indicate that protein E-L18 competes with three out of five ethidium bromide molecules for 5 S RNA binding sites (*109*). Since the addition of E-L18 to 5 S RNA did not induce a significant hyperchromic effect, it was concluded that E-L18 binds to base-paired regions and by doing so prevents the binding of ethidium bromide.

It is difficult at this time to compare fully the partial RNase hydrolysis studies of the four 5 S RNA · protein complexes from *E. coli* and *B. stearothermophilus* (Fig. 11) with that of the *E. coli* 23 S RNA · 5 S RNA · protein complex (Fig. 11), as 23 S RNA was present and protein E-L5 was absent in the latter study. The importance of E-L5 has been inferred from the observation that it significantly stabilizes the binding of E-L18 (*109*).

The larger fragments of all 5 S RNA · protein complexes discussed above include the conserved nucleotide sequence around position 80, which has been suggested to be base-paired with 23 S RNA (*172*).

D. Binding of Oligonucleotides to 5 S RNA-Protein Complexes

Although the 5 S RNA · protein complexes have not been as thoroughly mapped by oligonucleotide binding as has free 5 S RNA, several interesting observations have been made. The *E. coli* 5 S RNA · protein complex binds the tRNA fragment T-Ψ-C-G and the synthetic oligonucleotide U-U-C-G, whereas free 5 S RNA binds these oligonucleotides less strongly (*9*). N-oxidation of 5 S RNA within the ribosome allowed the modification of two adenosines that, in subsequent reconstitution experiments, were shown to be important for ribosomal function. 5 S RNA · protein complexes containing such modified RNA lose their ability to bind either T-Ψ-C-G or U-U-C-G (*9*).

Experiments with *B. stearothermophilus* 5 S RNA, 5 S RNA · protein complexes and U-U-C-G (*90*) support earlier findings with the *E. coli* system (*9*). Comparison of *E. coli* and *B. stearothermophilus* 5 S RNA sequences suggest strongly that these oligonucleotides bind to the C-G-A-A sequence at the beginning of region D (Fig. 1), which is common to all prokaryotic 5 S RNAs.

The binding of T-Ψ-C-G and U-U-C-G to 5 S RNA · protein complexes and not to free 5 S RNA were the first indications that ribosomal RNAs may undergo conformational changes upon interaction with protein.

V. Function of 5 S and 5.8 S RNA

The availability of reconstitution techniques for 50 S ribosomal subunits (6, 173–174a) has allowed a detailed analysis of the possible function of prokaryotic 5 S RNA. Since the reconstitution of active 60 S ribosomal subunits has not been achieved, information on the function of eukaryotic 5 S and 5.8 S RNA is sparse.

A. Activity of 50 S Ribosomal Subunits Lacking 5 S RNA

Reconstitution experiments with *B. stearothermophilus* 50 S ribosomal subunits established for the first time that 5 S RNA plays an important role in ribosomal function (*173*). Reconstituted particles lacking 5 S RNA exhibit greatly reduced activities in (a) polypeptide synthesis, (b) peptidyltransferase activity, (c) peptide-chain-termination-factor-R1-dependent binding of U-A-A, (d) EF-G-dependent binding of GTP and (e) nonenzymic binding of poly(U)-directed Phe-tRNA. Since the particles lacking 5 S RNA were also devoid of several proteins, the reconstitution experiments suggest that at least one function of 5 S RNA is the binding of several 50 S proteins. Subsequently, these 5 S RNA-binding proteins were identified as B-L5 and B-L22 (*8*).

E. coli reconstituted 50 S particles also show a dependence upon 5 S RNA (*174, 174a*), and the structural importance of this small RNA within the 50 S ribosomal subunit is now generally accepted.

It is important to point out that the particles lacking 5 S RNA give little information concerning possible specific functions of 5 S RNA. Therefore other experimental procedures are necessary.

B. Reconstitution of 50 S Ribosomal Subunits with Heterologous 5 S RNAs

5 S RNAs from different organisms in which "modification" has been carried by evolution present a promising approach to study 5 S RNA function. Such studies have been carried out by total reconstitution of *B. stearothermophilus* (*6, 21*) and partial reconstitution of *E. coli* 50 S ribosomal subunits (*170, 171*) with different pro- and eukaryotic 5 S RNAs. Although the partially reconstituted particles from *E. coli* were not biologically active (*170, 171*), the results otherwise agree with those of *B. stearothermophilus* 50 S reconstitutions.

As summarized in Table II, the reconstitution experiments established several major points concerning the structure and function of 5 S RNA: (a) all prokaryotic 5 S RNAs tested can be incorporated into prokaryotic 50 S ribosomal subunits (*6, 21, 170, 171*) thus demonstrating that the protein binding sites of those RNAs are conserved during evolution; (b)

TABLE II
RECONSTITUTION OF 50 S RIBOSOMAL SUBUNITS WITH DIFFERENT 5 S RNAs[a]

Source of 5 S RNA	5 S RNA incorporated into reconstituted particles of	
	B. stearothermophilus	E. coli
Escherichia coli	+	+
E. coli denatured B form	NT	−
Proteus vulgaris	+	NT
Bacillus stearothermophilus	+	+
Bacillus subtilis	+	NT
Micrococcus lysodeikticus	+	NT
Staphylococcus aureus	+	NT
Pseudomonas fluorescens	+	NT
Azotobacter vinelandii	+	NT
Halobacterium cutirubrum	+	NT
Chlorella chloroplast	NT	−
Chlorella cytoplasmic	NT	−
Yeast	−	−
Bean	−	NT
Wheat germ	−	NT
Artemia salina	−	NT
Xenopus laevis oocyte	NT	−
Xenopus laevis renal	NT	−
HeLa cell	NT	−
Rat liver	−	NT
Horse liver	−	NT

[a] A plus sign (+) indicates that 5 S RNA is incorporated and a minus sign (−) that it is not; NT indicates that this 5 S RNA was not tested. Experiments involving reconstituted B. stearothermophilus 50 S ribosomes were taken from References 6, 9, and 21 and those for E. coli from References 74, 87, 150, 170, and 171.

since the 5 S RNAs incorporated into B. stearothermophilus 50 S subunits yield active ribosomal particles (21), possible functional parts of these RNAs must also be conserved; (c) prokaryotic and eukaryotic 5 S RNAs must be classified into two separate groups, since none of the eukaryotic 5 S RNAs (21, 171) can be incorporated into bacterial ribosomes.

C. Functional Activities of 50 S Ribosomal Subunits with Modified 5 S RNAs

E. coli precursor 5 S RNA may contain up to five additional bases at its 5′ end (53, 56; see also Section II, D). The final maturation of the 5 S

RNA can take place on the 50 S ribosome, and it is therefore possible to isolate subunits containing the precursor of 5 S RNA. Since the biological activity of these particles is not decreased, it may be concluded that the exact sequence at the 5' end of the mature 5 S RNA is not crucial for the biological function.

An interesting hypothesis proposed that 5 S RNA participates in protein synthesis by formation of a peptidyl-5 S RNA intermediate (*175, 176*), but subsequent attempts to isolate such intermediates have failed (*177, 178*). Another hypothesis required a uridine with free hydroxyl groups at the 3' terminus of 5 S RNA (*176*). To test this hypothesis, 5 S RNA from *E. coli* (*78*) and *B. stearothermophilus* (*78, 179*) from which the 3-terminal nucleosides had been removed, leaving terminal 3'-phosphates were treated. The altered 5 S RNAs were equally active in reconstituted ribosomes (*78, 179*), and, since regeneration of the 3' end during protein synthesis did not take place (*78*), it can be concluded that hypotheses involving a biological function of the 3' terminus of 5 S RNA are incorrect.

IMPORTANCE OF INTERNAL NUCLEOTIDES OF 5 S RNA

Several experiments have been carried out to test whether or not nucleotides internally located in 5 S RNA are important for biological function. In one of these experiments, *E. coli* 5 S RNA was formylated under limiting conditions, so that only G_{41} was modified (*123*). Reconstitution experiments involving this modified *E. coli* 5 S RNA and *B. stearothermophilus* 50 S ribosomal components yielded functional particles (unpublished data cited in Reference *123*). These results receive support from other reconstitution experiments with *E. coli* 5 S RNA modified with kethoxal in positions G_{41}, and probably G_{13}, while still part of the 70 S ribosome (*128*). Therefore G_{41} and most probably G_{13} are not essential for protein binding and biological function.

In another series of experiments, *E. coli* 50 S ribosomal subunits were allowed to react with monoperphthalic acid, specific for unpaired adenines (*9, 120*). Analysis of the 5 S RNA revealed that two adenines are readily modified and that prolonged reaction time does not increase the number of 1-N-oxides of adenine per 5 S RNA. Such modified 5 S RNA could still be incorporated into *B. stearothermophilus* 50 S ribosomal subunits, but yielded particles that had lost nearly 60% of their activity in a poly(U)-directed polyphenylalanine synthesizing system (*9*). In addition, it should be noted that specific 5 S RNA · protein complexes bind T-Ψ-C-G, whereas complexes containing this modified 5 S RNA do not (see Section IV, D).

The experiments cited above imply that two adenines of 5 S RNA are on the surface of the ribosome and important for biological function, and that they may be involved in tRNA binding (see also Section V, D). The localization of the two modified adenines in 5 S RNA has been difficult, because N-oxidation does not permit introduction of a radioactive label. Since chemical modification shows that G_{41} of 5 S RNA is also on the surface of the 50 S (*129*) and 70 S (*128*) ribosome, it seems quite likely that the two adenines in positions 45 and 46 are the ones modified. This interpretation is supported by oligonucleotide binding to 5 S RNA · protein complexes of *E. coli* and *B. stearothermophilus* (Section IV, D).

D. Importance of 5 S RNA in Binding tRNA to Ribosomes

From early sequence studies of 5 S RNA, it was proposed that loop IV of tRNA, i.e., parts of the common G-T-Ψ-C-R sequence, can interact with complementary regions of 5 S RNA on the ribosome (*17, 23*). Subsequently, this hypothesis was supported by the observation that T-Ψ-C-G inhibits nonenzymic binding of tRNAs to ribosomes (*180, 181*). Oligonucleotide binding (Section IV, D) and chemical modification studies (Section V, C) involving 5 S RNA · protein complexes and reconstituted 50 S subunits further strengthened the case for the interaction of tRNA (sequence T-Ψ-C-G) with 5 S RNA on the ribosome (*9*).

It has been assumed that prokaryotic and eukaryotic 5 S RNAs have identical functions, and sequence information of pro- and eukaryotic initiator tRNAs, as well as of 5 S RNAs, has been used to support the hypothesis that the tRNA–5 S RNA interaction takes place at the ribosomal P-site (*39, 182, 183*). Recent experiments contradict partially the above hypothesis. The tRNA fragment T-Ψ-C-G inhibits the enzymic binding of aminoacyl-tRNA to the A-site of prokaryotic (*184, 185*) and eukaryotic (*186*) ribosomes. The P-site binding of initiator tRNAs is not affected. Also, T-Ψ-C-G can replace uncharged tRNA at the ribosomal A-site and trigger the stringent-factor-dependent synthesis of pppGpp and ppGpp (*187, 188*). In this context, we note that T-Ψ-C-G inhibits tRNA binding by associating only with the 50 S ribosomal subunit (*184*) and that a maximum of one mole of the tRNA fragment is bound per mole of 50 S ribosome (*189*).

In view of the results summarized here and of the elongation-factor-dependent GDP-affinity-label experiments described in Section V, F, it seems highly probable that tRNA interacts through its G-T-Ψ-C-R (loop IV) sequence with the C-G-A-A-C sequence of 5 S RNA in prokaryotic ribosomes.

Several observations argue against a similar function of eukaryotic 5 S RNAs. A survey of sequence analysis shows that eukaryotic 5 S

RNAs do not possess the sequence C-G-A-A-C at a specific site but instead the conserved sequence Y-G-A-U (see Section II, B), which is complementary to the sequence A-U-C-G of loop IV found only in eukaryotic initiator tRNAs (3). There are also differences in the primary structures of eukaryotic and prokaryotic 5 S RNAs, and a number of eukaryotic 5 S RNAs cannot be incorporated into bacterial ribosomes whereas all prokaryotic 5 S RNAs tested can (Section V, B). Therefore, it seems possible that eukaryotic 5 S RNA plays a role in the initiation phase of protein synthesis by interacting with initiator tRNA, as previously proposed (39, 182, 183, 190).

E. Enzymic Activities Associated with 5 S RNA·Protein Complexes

Because of the increasing evidence in favor of tRNA–5 S RNA interaction on the prokaryotic ribosome, studies were initiated to test the specific 5 S RNA · protein complexes for possible biological activities. These studies led to the discovery of GTPase and ATPase activities associated with B. stearothermophilus (191–195), E. coli (196) and rat liver (161, 197) 5 S RNA · protein complexes. Kinetic studies of B. stearothermophilus (191, 192) and rat liver (197) complexes suggest that the hydrolysis of ATP and GTP occurs at two different sites. The hydrolytic activities are independent of elongation factors.

Both GTPase and ATPase activities of the B. stearothermophilus (191, 195) complex are sensitive to fusidic acid and thiostrepton, whereas the E. coli complex is not (196). Erythromycin inhibits the GTPase but not the ATPase of B. stearothermophilus (195). The rat liver 5 S RNA · protein complex behaves like the B. stearothermophilus complex, in that it is also sensitive to fusidic acid (197). Since the antibiotics that inhibit the hydrolytic activities of the 5 S RNA · protein complexes also inhibit the ribosome- and elongation factor-dependent GTPase (198–200), these results suggest a possible involvement of 5 S RNA proteins in this function.

The implication that the E. coli 5 S RNA binding-proteins exhibit GTPase activity is further supported by 23 S RNA fragment studies. Such fragments contain, among other proteins, E-L18 and traces of E-L5 and show significant GTPase activity (201).

Experiments in which a photoaffinity label of GDP was bound, in the presence of fusidic acid and elongation factor G, to E. coli (202, 203) and B. stearothermophilus (167) ribosomes permitted the labeling of E-L5, E-L11, E-L18 and E-L30 and B-L10 and B-L22. Since it has been shown for all these proteins, except E-L11, that they can be part of 5 S RNA · protein complexes (see Section IV, A and B), these results support the earlier notion that the 5 S RNA binding proteins are involved in the elongation-factor-G-dependent GTPase. Because the affinity-labeling experiments were carried out under EF-G-dependent conditions, the results further support the hypothesis that 5 S RNA and its associated proteins are located at the ribosomal A-site (see Section V, D).

It also seems important to point out that a GDP affinity label experiment carried out with the *B. stearothermophilus* 5 S RNA · protein complex resulted in the labeling of B-L10 and B-L22, and that the comparative study mentioned above indicates structurally related GTPase sites in *E. coli* and *B. stearothermophilus* (*167*). Nevertheless, caution is required before drawing final conclusions about the GTPase center of the ribosome, as 50 S core particles that apparently lack E-L5, E-L18 and E-L25 but contain 5 S RNA have residual EF-G-dependent GTPase activity (*204*).

Another open question is the significance of ATPase activities present in prokaryotic (*191, 196*) and eukaryotic (*197*) 5 S RNA · protein complexes. Since the only ATPase activity so far identified with prokaryotic ribosomes involves the stringent-factor-dependent synthesis of pppGpp and ppGpp, which is associated with the ribosomal A-site, activity associated with the 5 S RNA · protein complex could possibly be involved in this function. On the other hand, the only ATPase so far detected with eukaryotic protein synthesis involves the initiation phase (*205, 206*). It is therefore interesting that eukaryotic 5 S RNA · protein complexes exhibit ATP hydrolytic activity and that eukaryotic 5 S RNA appears to be involved in chain initiation (see Section V, D).

F. Function of 5.8 S RNA

Unfortunately, there is no experimental evidence available as to the function of eukaryotic 5.8 S RNA. Nevertheless, there are several interesting aspects regarding the primary structure of 5.8 S RNA worth discussing, since they may point to its possible biological function.

Some experimental evidence on eukaryotic 5 S RNA points toward its interaction with initiator tRNA during the initiation phase of protein synthesis, not toward A-site tRNA interaction (Section V, D). Therefore the question arises with which other rRNA the G-T-Ψ-C sequence (loop IV) of amino-acyl-tRNA interacts at the ribosomal A-site (*186*). Since 5.8 S RNA contains an unusually high number of G-A-A-Y sequences, it would be a good candidate to have a function in eukaryotic ribosomes similar to that of the 5 S RNA in prokaryotic ribosomes. This hypothesis is partially supported by comparison of the primary sequence of yeast 5.8 S RNA with those of prokaryotic 5 S RNAs (Section II, C). In addition, prokaryotic 5 S RNA and eukaryotic 5.8 S RNA are secondary transcription products, while eukaryotic 5 S RNA is not.

VI. Outlook

The survey presented here of information available on the structures and functions of 5 S and 5.8 S RNAs indicates that prokaryotic 5 S RNAs have been studied in more detail and that they play an important role in ribosome structure and function. The information on eukaryotic 5 S and especially 5.8 S RNA is scarce and calls for more experimentation.

For the near future, more detailed investigations concerning the structure of 5 S RNA · protein complexes seem most promising, since they represent one of the best systems in which to study RNA–protein interactions. Certainly attempts will be made to crystallize these complexes to obtain the intimate information required for knowing the forces that govern nucleic acid–protein interaction.

ACKNOWLEDGMENTS

I am indebted to Drs. H. G. Wittmann and M. Achtman for critical discussions concerning this manuscript. I wish to thank Drs. R. Monier, H. Noller, E. G. Richards, P. Sigler, J. A. Steitz and S. M. Weissman, who have been kind enough to send me their preprints or other unpublished materials. In addition I am most grateful to my wife (H. Erdmann) for having so patiently drawn and redrawn the illustrations for this manuscript.

REFERENCES

1. R. Rosset, R. Monier and J. Julien, *Bull. Soc. Chim. Biol.* **46**, 87 (1964).
2. G. G. Brownlee, F. Sanger and B. G. Barrell, *Nature (London)* **215**, 735 (1967).
3. B. G. Barrell and B. F. C. Clark, "Handbook of Nucleic Acid Sequences." Joynson-Bruvvers Ltd., Oxford, England, 1974.
4. R. W. Holley, J. Apgar, G. A. Everett, J. T. Madison, M. Marquisee, S. H. Merrill, J. R. Penswick and A. Zamir, *Science* **147**, 1462 (1965).
5. H. G. Zachau, D. Dütting and H. Feldmann, *Angew. Chem.* **78**, 392 (1966); *Angew. Chem., Int. Engl. Ed.* **5**, 422 (1966).
6. M. Nomura and V. A. Erdmann, *Nature (London)* **228**, 744 (1970).
7. P. N. Gray and R. Monier, *FEBS Lett.* **18**, 145 (1971).
8. J. R. Horne and V. A. Erdmann, *Mol. Gen. Genet.* **119**, 337 (1972).
9. V. A. Erdmann, M. Sprinzl and O. Pongs, *BBRC* **54**, 942 (1973).
10. J. Marcot-Queiroz, J. Julien, R. Rosset and R. Monier, *Bull. Soc. Chim. Biol.* **47**, 183 (1965).
11. F. Galibert, C. J. Larsen, J. C. Lelong and M. Boiron, *Nature (London)* **207**, 1039 (1965).
12. D. G. Comb, N. Sarkar, J. DeVallet and C. J. Pinzino, *JMB* **12**, 509 (1965).
13. P. J. Payne and T. A. Dyer, *BJ* **124**, 83 (1971).
14. C. J. Leaver and M. A. Harmey, in "Ribosomes and RNA Metabolism" (J. Zelinka and J. Balan, eds.), p. 407. Publ. House Slovak Acad. Sci., Bratislava, CSSR, 1973.
15. P. Borst and L. A. Grivell, *FEBS Lett.* **13**, 73 (1971).

16. P. M. Lizardi and D. J. L. Luck, Nature (London) 229, 140 (1971).
17. B. G. Forget and S. M. Weissman, Nature (London) 213, 878 (1967).
18. J. J. Pene, E. Knight and J. E. Darnell, JMB 33, 609 (1968).
19. G. M. Rubin, JBC 248, 3860 (1973).
20. A. Viotti, E. Sala, R. Nucca and C. Soave, in "Ribosomes and RNA Metabolism" (J. Zelinka and J. Balan, eds.), p. 377. Publ. House Slovak Acad. Sci., Bratislava, CSSR, 1973.
21. P. Wrede and V. A. Erdmann, FEBS Lett. 33, 315 (1973).
22. G. G. Brownlee, F. Sanger and B. G. Barrell, JMB 23, 337 (1967).
23. G. G. Brownlee, F. Sanger and B. G. Barrell, JMB 34, 379 (1968).
24. B. du Buy and S. M. Weissman, JBC 246, 747 (1971).
25. C. A. Marotta, C. C. Levy, S. M. Weissman and F. Varricchio, Bchem 12, 2901 (1973).
25a. J. R. Stanley and J. D. Penswick, FEBS Abstr., 10th Meeting p. 421 (1975).
26. C. D. Pribula, G. E. Fox and C. R. Woese, FEBS Lett. 44, 322 (1974).
26a. C. R. Woese, C. D. Pribula, G. E. Fox and L. B. Zablen, J. Mol. Evol. 5, 35 (1975).
27. M. Y. Corry, P. I. Payne and T. A. Dyer, FEBS Lett. 46, 63 (1974).
28. M. Y. Corry, P. I. Payne and T. A. Dyer, FEBS Lett. 46, 67 (1974).
29. S. Y. Sogin, M. L. Sogin and C. R. Woese, J. Mol. Evol. 1, 173 (1972).
30. B. Yarry and R. Rosset, Mol. Gen. Genet. 113, 43 (1971).
31. B. Yarry and D. Rosset, Mol. Gen. Genet. 126, 29 (1973).
32. B. G. Forget and S. M. Weissman, Science 158, 1695 (1975).
33. G. G. Brownlee, E. Cartwright, T. McShane and R. Williamson, FEBS Lett. 25, 8 (1972).
34. M. Wegnez, R. Monier and H. Denis, FEBS Lett. 25, 13 (1972).
35. P. Y. Ford and E. M. Southern, Nature NB 241, 7 (1973).
36. R. Monier, in "Ribosomes" (M. Nomura, P. Lengyel and A. Tissières, eds.), p. 141. Cold Spring Harbor Press, New York, 1974.
37. J. Hindley and S. M. Page, FEBS Lett. 26, 157 (1972).
38. K. Nishikawa and S. Takemura, J. Biochem. Tokio 76, 925 (1974).
39. K. Nishikawa and S. Takemura, J. Biochem. Tokio 76, 935 (1974).
40. L. E. Hatlen, F. Amaldi and G. Attardi, Bchem 8, 4989 (1969).
41. M. J. Averner and N. R. Pace, JBC 247, 4491 (1972).
42. R. Williamson and G. G. Brownlee, FEBS Lett. 3, 306 (1969).
43. K. Takai, S. Hashimoto and M. Muramatsu, Bchem 14, 536 (1975).
44. F. Labrie, as cited in Williamson and Brownlee (42).
45. H. D. Robertson, E. Dickson, P. Model and W. Prensky, PNAS 70, 3260 (1973).
46. B. R. Jordan, G. Galling and R. Jourdan, FEBS Lett. 37, 333 (1973).
47. B. R. Jordan, G. Galling and R. Jourdan, JMB 87, 205 (1974).
48. B. E. H. Maden and J. S. Robertson, JMB 87, 227 (1974).
49. D. N. Nazar, T. O. Sitz and H. Busch, FEBS Lett. 45, 206 (1974).
50. G. M. Rubin, EJB 41, 197 (1974).
51. F. Galibert, J. C. Lelong and C. J. Larsen, C.R. Acad. Sci., Ser. D. 265, 279 (1967).
52. M. Adesnik and C. Levinthal, JMB 48, 187 (1970).
53. B. R. Jordan, B. G. Forget and R. Monier, JMB 55, 407 (1971).
54. H. A. Raué and M. Gruber, BBA 232, 314 (1971).
55. F. Galibert, P. Tiollais, F. Sanforche and M. Boiron, EJB 20, 381 (1971).
56. J. Feunteun, B. R. Jordan and D. Monier, JMB 70, 465 (1972).

57. B. E. Griffin and D. L. Bailie, *FEBS Lett.* **34**, 273 (1973).
58. N. R. Pace, M. L. Pato, J. McKibbin and C. W. Radcliffe, *JMB* **75**, 619 (1973).
59. T. J. Stoof, V. C. H. F. de Regt, H. A. Raué and R. J. Planta, *FEBS Lett.* **49**, 237 (1974).
60. M. L. Sogin and N. D. Pace, *Nature (London)* **252**, 598 (1974).
61. D. Ginsburg and J. A. Steitz, *JBC* **250**, 5647 (1975).
62. N. R. Pace, *Bacteriol. Rev.* **37**, 562 (1973).
63. S. Altman, *Cell* **4**, 21 (1975).
64. L. E. Hatlen, F. Amaldi and G. Attardi, *Bchem* **8**, 4989 (1969).
65. C. Soave, R. Nucca, E. Sala, A. Viotti and E. Galante, *EJB* **32**, 392 (1973).
66. A. Viotti, C. Soâve, E. Sala, R. Nucca and E. Galante, *BBA* **324**, 72 (1973).
67. D. D. Brown, P. C. Wensink and E. Jordan, *PNAS* **68**, 3175 (1971).
68. G. G. Brownlee, E. M. Cartwright and D. O. Brown, *JMB* **89**, 703 (1974).
69. R. D. Leibowitz, R. A. Weinberg and S. Penman, *JMB* **73**, 139 (1973).
70. J. Speiss and M. Birnstiel, *JMB* **87**, 237 (1974).
71. B. E. H. Maden and J. S. Robertson, *JMB* **87**, 227 (1974).
72. S. A. Udem, K. Kaufman and J. R. Warner, *J. Bact.* **105**, 101 (1971).
73. F. Galibert, J. C. Lelong, Ch. J. Larsen and M. Boiron, *BBA* **142**, 89 (1967).
74. M. Reynier, M. Aubert and R. Monier, *Bull. Soc. Chim. Biol.* **49**, 1205 (1967).
75. M. Aubert, J. F. Scott, M. Reynier and R. Monier, *PNAS* **61**, 292 (1968).
76. V. A. Erdmann, Ph.D. Thesis, Technische Universität Braunschweig, Germany (1968).
77. M. E. Geroch, E. G. Richards and G. A. Davies, *EJB* **6**, 325 (1968).
78. V. A. Erdmann, M. G. Doberer and M. Sprinzl, *Mol. Gen. Genet.* **114**, 89 (1971).
79. R. Rosset, R. Monier and J. Julien, *Bull. Soc. Chim. Biol.* **46**, 87 (1964).
80. P. Morell, I. Smith, O. Dubnan and J. Marmur, *J. Biochem.* **6**, 258 (1967).
81. S. Fahnestock, V. A. Erdmann and M. Nomura, in "Methods in Enzymology," Vol. 30: Nucleic Acids and Protein Synthesis, Part F (L. Grossman and K. Moldare, eds.), p. 554. Academic Press, New York, 1974.
82. E. G. Richards, R. Lecanidou and M. E. Geroch, *EJB* **34**, 262 (1973).
83. J. Hindley, *JMB* **30**, 125 (1967).
84. G. P. Philipps and J. L. Timko, *Anal. Biochem.* **45**, 319 (1972).
85. E. G. Richards and R. Lecanidou, in "Electrophoresis and Isoelectric Focusing in Polyacrylamide Gels" (R. C. Allen and H. R. Maurer, eds.), p. 82. de Gruyter, Berlin and New York, 1974.
86. D. I. Wolfrum, R. Rüchel, S. Mesecke and V. Neuhoff, *Z. Prakt. Chem.* **355**, 1415 (1975).
87. M. Aubert, G. Bellemare and R. Monier, *Biochemie* **55**, 135 (1973).
88. R. A. Weinberg and S. Penman, *JMB* **38**, 289 (1968).
89. G. Galling and B. R. Jordan, *Biochemie* **54**, 1257 (1972).
90. P. Wrede and V. A. Erdmann, unpublished.
91. D. G. Comb and T. Zehavi-Willner, *JMB* **23**, 441 (1967).
92. H. Boedtker and D. G. Kelling, *BBRC* **29**, 758 (1967).
93. L. P. Visentin, C. Chow, A. T. Matheson, M. Yaguchi and F. Rollin, *BJ* **130**, 103 (1972).
94. P. G. Connors and W. W. Beeman, *JMB* **71**, 31 (1972).
95. S. H. Kim, G. J. Quigley, F. L. Suddath, A. McPherson, D. Sneden, J. J. Kim, J. Weinzierl and A. Rich, *Science* **179**, 285 (1973).
96. S. H. Kim, F. L. Suddath, G. J. Quigley, A. McPherson, J. L. Sussman, A. M. J. Wang, N. C. Seeman and A. Rich, *Science* **185**, 435 (1974).

97. J. O. Robertus, J. E. Ladner, J. T. Finch, D. Rhodes, R. O. Brown, B. F. C. Clark and A. Klug, *Nature (London)* **250**, 546 (1974).
98. F. Cramer and V. A. Erdmann, *Nature (London)* **218**, 92 (1968).
99. J. F. Scott, R. Monier, M. Aubert and M. Reynier, *BBRC* **33**, 794 (1968).
100. G. Bellemare, R. J. Cedergren and G. H. Cousineau, *JMB* **68**, 445 (1972).
101. P. N. Gray and G. F. Saunders, *ABB* **156**, 104 (1973).
102. R. Lecanidou and E. G. Richards, *EJB* **57**, 127 (1975).
103. C. R. Cantor, *PNAS* **59**, 478 (1968).
104. C. R. Cantor, *Nature (London)* **216**, 513 (1967).
105. E. G. Richards, M. E. Geroch, M. Simpkins and R. Lecanidou, *Biopolymers* **11**, 1031 (1972).
106. P. N. Gray and G. F. Saunders, *BBA* **254**, 60 (1971).
107. M. J. Waring, *JMB* **13**, 269 (1975).
108. J. B. LePecq and C. Paoletti, *JMB* **27**, 87 (1967).
109. J. Feunteun, R. Monier, R. Garrett, M. LeBret and J. B. LePecq, *JMB* **93**, 535 (1975).
110. P. B. Sigler, *Annu. Rev. Biophys. Bioeng.* **4**, 477 (1975).
111. Y. P. Wong, D. R. Kearns, B. R. Reid and R. G. Shulman, *JMB* **72**, 741 (1972).
112. D. R. Kearns and Y. P. Wong, *JMB* **87**, 755 (1974).
113. B. R. Jordan, *J. Mol. Biol.* **55**, 423 (1971).
114. R. Vigne and B. R. Jordan, *Biochemie* **53**, 981 (1971).
115. G. Bellemare, B. R. Jordan and R. Monier, *JMB* **71**, 307 (1972).
116. A. D. Mirzabekov and B. E. Griffin, *JMB* **72**, 633 (1972).
117. D. Vigne, B. R. Jordan and R. Monier, *JMB* **76**, 303 (1973).
118. B. G. Forget and M. Reynier, *BBRC* **39**, 114 (1970).
119. J. Feunteun and R. Monier, *Biochemie* **53**, 657 (1971).
120. F. Cramer, V. A. Erdmann, F. von der Haar and E. Schlimme, *J. Cell. Physiol* **74** (Suppl.), 163 (1969).
121. F. van der Haar, E. Schlimme, V. A. Erdmann and F. Cramer, *Bioorg. Chem.* **7**, 282 (1971).
122. J. C. Lee and V. M. Ingram, *JMB* **41**, 431 (1969).
123. G. Bellemare, B. R. Jordan, J. Rocca-Serra and R. Monier, *Biochemie* **54**, 1453 (1972).
124. M. Litt, *Fed. Proc.* **32**, 586 (1973).
125. M. Aubert, G. Bellemare and R. Monier, *Biochemie* **55**, 135 (1973).
126. N. Delihas, G. A. Zorn and E. Strobel, *Biochemie* **55**, 1227 (1973).
127. G. Zorn, E. Strobel, R. Greenberg and N. Delihas, *J. Cell Biol.* **59**, 378a (1973).
128. N. Delihas, J. Dunn and V. A. Erdmann, *FEBS Lett.* **58**, 76 (1975).
129. M. F. Noller and W. Herr, *JMB* **90**, 181 (1974).
130. P. N. Gray, G. Bellemare, R. Monier, R. A. Garrett and G. Stöffler, *JMB* **77**, 133 (1973).
131. O. C. Uhlenbeck, J. Baller and P. Doty, *Nature (London)* **225**, 508 (1970).
132. J. B. Lewis and P. Doty, *Nature (London)* **225**, 510 (1970).
133. V. A. Erdmann, O. Pongs, P. Wrede and J. Zimmermann, in "Ribosomes and RNA Metabolism-2" (J. Zelinka and J. Balan, eds.), p. 425. Publ. House Slovak Acad. Sci., Bratislava, CSSR, 1976.
134. M. B. Soot, M. J. Saarma, R. L. E. Willems, A. J. Lind and S. K. Vasilenko, *Mol. Biol. USSR* **8**, 723 (1974).
134a. J. T. Madison, *ARB* **37**, 131 (1968).

134b. B. R. Jordan, *J. Theoret. Biol.* **34**, 363 (1971).
134c. I. D. Raacke, *BBDC* **31**, 528 (1968).
135. F. Cramer, This series **11**, 391 (1971).
136. V. A. Erdmann, S. Fahnestock, K. Higo and M. Nomura, *PNAS* **68**, 2932 (1971).
137. R. F. Gesteland, *JMB* **18**, 356 (1966).
138. M. Aubert, R. Monier, M. Reynier and J. F. Scott, *Proc. 4th FEBS Meeting* **3**, 151 (1967).
139. P. Morell and J. Marmur, *Bchem* **7**, 1141 (1968).
140. M. A. Siddiqui and K. Hosokawa, *BBRC* **36**, 711 (1969).
141. R. A. Garrett, C. Schulte, G. Stöffler, P. Gray and R. Monier, *FEBS Lett.* **49**, 1 (1974).
142. J. Marcot-Queirot and R. Monier, *Bull. Soc. Chim. Biol.* **49**, 477 (1967).
143. M. Reynier and R. Monier, *Bull. Soc. Chim. Biol.* **50**, 1583 (1968).
144. P. N. Gray, R. Garrett, G. Stöffler and R. Monier, *EJB* **28**, 412 (1972).
145. R. S. T. Yu, Hoppe-Seylers, *Z. Prakt. Chem.* **354**, 125 (1973).
146. Y. Yogo, H. Fujimoto and D. Mizuno, *BBA* **240**, 564 (1971).
147. J. R. Gormly, C. H. Yang and J. Horowitz, *BBA* **247**, 80 (1971).
148. K. Hosokawa, *JBC* **245**, 5880 (1970).
149. R. S. T. Yu and H. G. Wittmann, *BBA* **319**, 388 (1973).
150. D. S. T. Yu and H. G. Wittmann, *BBA* **324**, 375 (1973).
151. H. R. Burrell and J. Horowitz, *FEBS Lett.* **49**, 306 (1975).
152. E. Kaltschmidt and H. G. Wittmann, *PNAS* **67**, 1276 (1970).
153. D. G. Comb and N. Sarkar, *JMB* **25**, 317 (1967).
154. M. L. Petermann and A. Pavlovec, *Fed. Proc.* **28**, 725 (1969).
155. T. Zehavi-Willner, *BBRC* **39**, 161 (1970).
156. B. Lebleu, G. Marbaix, G. Huez, J. Temmerman, A. Burny and H. Chantrenne, *EJB* **19**, 264 (1971).
157. M. L. Petermann, M. G. Hamilton and A. Pavlovec, *FP* **30**, 1204 (1971).
158. G. Blobel, *PNAS* **68**, 1881 (1971).
159. M. L. Petermann, M. G. Hamilton and A. Pavlovec, *Bchem* **11**, 2323 (1972).
160. M. L. Petermann, A. Pavlovec and M. G. Hamilton, *Bchem* **11**, 3925 (1972).
161. F. Grummt and I. Grummt, *FEBS Lett.* **42**, 343 (1974).
162. M. Welfe, J. Stahl and H. Bielka, *FEBS Lett.* **26**, 228 (1971).
163. C. C. Sherton and I. G. Wool, *JBC* **247**, 446 (1972).
164. O. Pongs, K. H. Nierhaus, V. A. Erdmann and H. G. Wittmann, *FEBS Lett.* **40**, 528 (1974).
165. M. Nomura, A. Tissieres and P. Lengyel, eds., "Ribosomes." Cold Spring Harbor Laboratory, New York, 1974.
166. R. Brimacombe, K. H. Nierhaus, R. A. Garrett and H. G. Wittmann, This series, 1976.
167. J. A. Maassen and W. Möller, *BBRC* **64**, 1175 (1975).
168. P. N. Gray, G. Bellemare and R. Monier, *FEBS Lett.* **24**, 156 (1972).
169. J. Zimmermann, Ph. Thesis, Freie Universität Berlin, Berlin, 1975.
169a. J. Zimmermann and V. A. Erdmann, unpublished.
170. K. Hosokawa, *Int. Congr. Microbiol., 10th*, Ba-2 (1970).
171. G. Bellemare, R. Vigne and B. R. Jordan, *Biochemie* **55**, 29 (1973).
172. W. Herr and M. F. Noller, *FEBS Lett.* **53**, 248 (1975).
173. V. A. Erdmann, S. Fahnestock, K. Higo and M. Nomura, *PNAS* **68**, 2932 (1971).
174. H. Maruta, T. Tsuchiya and D. Mizuno, *JMB* **61**, 123 (1971).

174a. K. H. Nierhaus and F. Dohme, *PNAS* **71**, 4713 (1974).
175. Y. Kuriki and A. Kaji, *BBRC* **26**, 95 (1967).
176. I. D. Raacke, *PNAS* **68**, 2357 (1971).
177. Y. Kuriki and A. Kaji, *Bchem* **8**, 3029 (1969).
178. J. Jonák and I. Rychlik, *Collect. Czech. Chem. Commun.* **35**, 1613 (1970).
179. S. Fahnestock and M. Nomura, *PNAS* **69**, 363 (1972).
180. J. Ofengand and C. Henes, *JBC* **244**, 6241 (1969).
181. N. Shimuzu, H. Hayashi and K. Miura, *J. Biochem. (Tokyo)* **67**, 373 (1970).
182. S. Chladek, *BBRC* **45**, 695 (1971).
183. S. K. Dube, *FEBS Lett.* **36**, 39 (1973).
184. D. Richter, V. A. Erdmann and M. Sprinzl, *Nature NB* **246**, 132 (1973).
185. V. A. Erdmann, M. Sprinzl, D. Richter and S. Lorenz, *Acta Biol. Med. Germ.* **33**, 605 (1974).
186. F. Grummt, I. Grummt, H. J. Gross, M. Sprinzl, D. Richter and V. A. Erdmann, *FEBS Lett.* **42**, 15 (1974).
187. D. Richter, V. A. Erdmann and M. Sprinzl, *PNAS* **71**, 3226 (1974).
188. D. Richter, V. A. Erdmann and M. Sprinzl, *Acta Biol. Med. Germ.* **33**, 609 (1974).
189. V. A. Erdmann, S. Lorenz, M. Sprinzl and T. Wagner, *in* "Control of Ribosome Synthesis" (N. O. Kjeldgaard and O. Maaløe, eds.), p. 427. Munksgaard, Copenhagen, 1976.
190. N. G. Avadhani and D. E. Buetow, *BBRC* **50**, 443 (1973).
191. J. R. Horne and V. A. Erdmann, *PNAS* **70**, 2870 (1973).
192. J. R. Horne, Ph.D. Thesis, Freie Universität Berlin, Germany (1974).
193. V. A. Erdmann, J. R. Horne, O. Pongs, J. Zimmermann and M. Sprinzl, *in* "Ribosomes and RNA Metabolism" (J. Zelinka and J. Balan, eds.), p. 363. Publ. House Acad. Sci., Bratislava, CSSR, 1973.
194. V. A. Erdmann, J. R. Horne, R. Bald, T. Wagner and O. Pongs, *Z. Physiol. Chem.* **74**, 757 (1974).
195. J. R. Horne and V. A. Erdmann, *FEBS Lett.* **42**, 42 (1974).
196. M. Gaunt-Klöpfer and V. A. Erdmann, *BBA* **390**, 226 (1975).
197. F. Grummt, E. Grummt and V. A. Erdmann, *EJB* **43**, 343 (1974).
198. J. Lucas-Lenard and F. Lipmann, *ARB* **40**, 409 (1971).
199. S. Pestka, see this volume (1976).
200. D. Vazquez, *FEBS Lett.* **40**, S 63 (1974).
201. H. E. Roth and K. H. Nierhaus, *JMB* **94**, 111 (1975).
202. J. A. Maassen and W. Möller, *PNAS* **71**, 1277 (1974).
203. J. A. Maassen and W. Möller, *Acta Biol. Med. Germ.* **33**, 621 (1974).
204. G. Sander, R. C. Marsh, J. Voigt and A. Parmeggiani, *Bchem* **14**, 1805 (1975).
205. A. Marcus, *JBC* **245**, 955 (1970).
206. M. H. Schreier and T. Staehelin, *in* "Regulation of Transcription in Eukaryotes" (E. K. F. Bautz, P. Karlson and H. Kersten, eds.), p. 335. Springer-Verlag, Berlin and New York, 1973.

Note added in proof: Recent reconstitution experiments have shown that eukaryotic 5.8 S RNA interacts specifically with the *E. coli* 5 S RNA binding proteins E-L18 and E-L25, while eukaryotic 5 S RNA did not interact with *E. coli* 50 S ribosomal proteins. This observation suggests that eukaryotic 5.8 S RNA may have a similar function to that of prokaryotic 5 S RNA (P. Wrede and V. A. Erdmann, unpublished results).

High-Resolution Nuclear Magnetic Resonance Investigations of the Structure of tRNA in Solution

DAVID R. KEARNS

Department of Chemistry
Revelle College
University of California, San Diego
La Jolla, California

I. Introduction 91
II. NMR Studies of Mononucleotides 96
   A. Stacking Interactions and Ring-Current Effects 97
   B. Hydrogen-Bonding Effects 99
   C. Proton Exchange with Solvent 100
   D. Rules for Predicting Low-Field NMR Spectra 101
III. NMR Studies of Oligonucleotide Fragments 105
   A. Simple Helices 105
   B. Dimerization of Oligonucleotides 110
   C. Melting Behavior of Short Helices 112
IV. High-Resolution NMR or tRNA 113
   A. Quantitative Determination of the Number of Secondary- and Tertiary-Structure Base-Pairs 113
   B. Evidence for Common Tertiary-Structure Base-Pairs . . . 117
   C. Assignment and Interpretation of the tRNA Spectra . . . 123
V. Application of NMR to the Investigation of Problems of tRNA Structure 130
   A. Denatured Conformers of tRNA 130
   B. Dimerization of E. coli tRNA^{Tyr} 133
   C. Investigation of the Thermal Unfolding of tRNA 134
   D. Effect of Anticodon-Loop Modifications on tRNA Conformation 138
   E. Effect of Aminoacylation on tRNA Conformation 139
   F. Photocrosslinking of s^{4}U$_8$ and C$_{13}$ in E. coli tRNA . . . 140
   G. Interaction of tRNA with Drugs 141
   Summary 145
   References 146

I. Introduction

High-resolution NMR has been used for over 17 years in the investigation of protein structures (1), and it may therefore seem surprising that NMR studies of the base-pairing structure of polynucleotides are of more recent origin (2). Two reasons for this delay are that most interesting RNA and DNA molecules are of such high molecular weight that they cannot conveniently be studied by NMR (3, 4), and that polynucleotides contain only four different major components; consequently, their spectra are usually crowded with unresolved, overlapping lines that cannot be assigned and interpreted.

Two important developments served to open up the polynucleotide field for high-resolution NMR studies. First, a class of low-molecular-weight, well-characterized RNAs, the transfer RNAs, became available in the necessary quantities (4–5 mg) (5, 6). Second, high-field (50 and 72 kGauss) spectrometers, using superconducting magnets, were developed, which made it possible to resolve individual resonances in the spectrum and, perhaps even more important, to measure resonances from hydrogen-bonded protons of bases in aqueous solutions even though the signal from the protons of water is 10^5 times more intense (2).

It has been 18 years since tRNAs were first isolated (7) but only 10 years since the first one was sequenced (8). These molecules were interesting at first because of their central role in protein synthesis (9). More recently, it has been recognized that tRNAs also function in the regulation of other biochemical pathways (10, 11). Paralleling the developments in our understanding of the biochemical properties of tRNA, there has been substantial progress during the past five years in the determination of the structure of tRNA molecules. Two techniques, X-ray diffraction and high-resolution NMR, have provided very detailed information about the two-dimensional (secondary) and three-dimensional (tertiary) structure[1] of molecules in the crystalline and solution state, respectively. The X-ray diffraction studies on yeast tRNA^{Phe} have recently been reviewed by Sigler (12) and by Kim (13).[2] In this article, we describe the application of high-resolution NMR to an investigation of the structure of tRNA in solution.

When NMR studies of tRNA molecules were first initiated, there was a great body of circumstantial evidence supporting the validity of Holley's "cloverleaf" model for the initial folding of these molecules (8). In addition, various experiments indicated that tRNA molecules are further folded in solution and held in a three-dimensional structure by additional interactions (3, 5, 9). A wide variety of models have been proposed to describe the folding, and, depending upon the experimental data selected, the number of additional base-pairs proposed to be involved ranged from 8 (14) to 10 (15, 16).

To a first approximation, the structure (secondary and tertiary) of a tRNA molecule can be specified by indicating which bases in the primary sequence are hydrogen-bonded to each other (see Fig. 1) and how they

[1] See article by Cramer in Volume 11 of this series.
[2] See article by Kim in Volume 17 of this series.

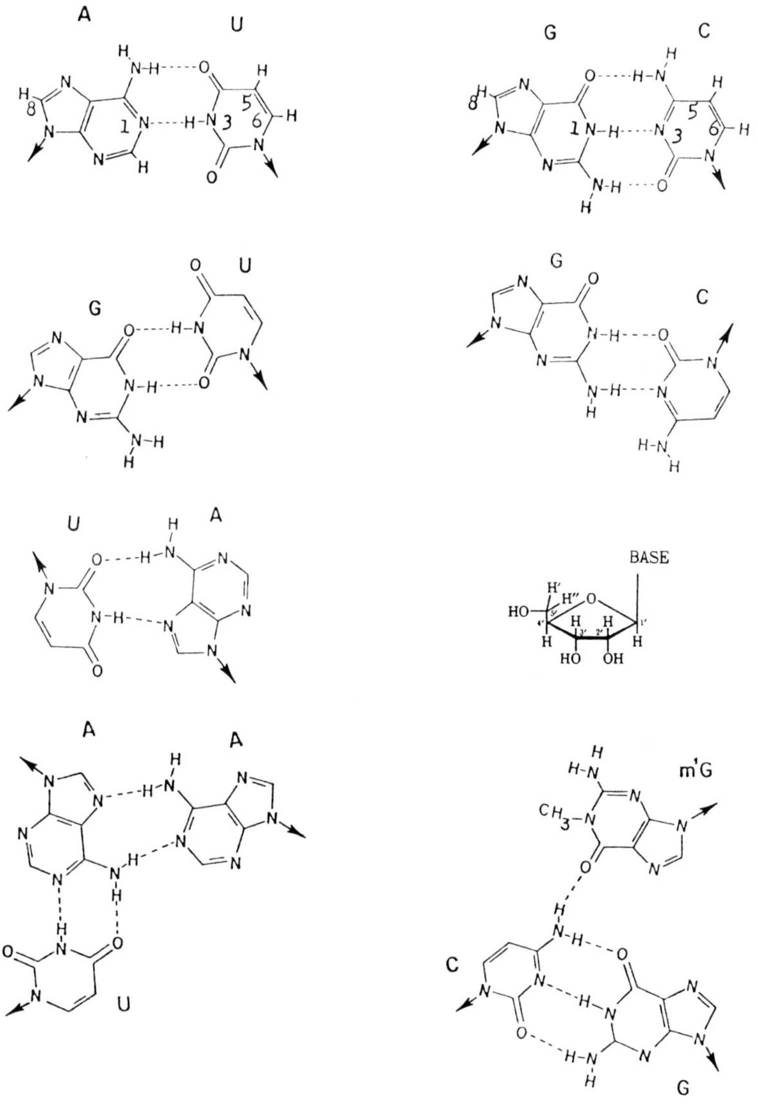

HIGH-RESOLUTION NMR OF tRNA   93

Fig. 1. A schematic diagram showing the standard Watson–Crick base-pairs and some of the tertiary-structure base-pairs and base-triples observed in yeast tRNA^{Phe}. The numbering system used to indicate the various protons of the bases and the sugar group are also indicated.

are stacked (placed one above another). Since the ring nitrogen protons and the NH_2 protons of the bases are the ones most affected by base-pairing, the initial NMR studies of tRNA have been concerned with resonances from these protons. Prior NMR studies of proteins in H_2O had shown that it is possible to observe resonances from hydrogen-bonded protons at low fields well resolved from other resonances of the molecule (17, 18). To find the analogous resonances from the hydrogen-bonded protons of nucleic acid base-pairs, we dissolved tRNA in H_2O and looked for resonances of hydrogen-bonded protons that would not be observed in D_2O. In our initial experiments with tRNA samples, we found the expected resonances between 11 and 15 ppm (referred to as the "low-field" region) (2). An example of the low-field 300 MHz NMR spectrum of a pure yeast tRNA^{Phe} is shown in Fig. 2. Besides the fact that it was even possible to observe resonances from protons in Watson–Crick base-pairs, the most exciting result of these studies is that the spectra are (partially) resolved, and it was immediately apparent that resonances from different A · U and G · C base-pairs are spread out over a rather large spectral range. From this spreading, we inferred there are local sequence effects on the chemical shifts of the resonances from the different base-pairs and that, once these sequence effects were understood, NMR could be used to provide very detailed information about the structure of tRNA molecules in solution.

In NMR spectroscopy, as in other forms of spectroscopy, there are three characteristic spectral features that are usually of interest (19), namely, the position, the intensity and the line width of a particular resonance.

1. *Position of a resonance.* The location of a resonance maximum is usually specified relative to the position of a resonance from some suitable standard. This may be expressed either in terms of magnetic field differences for a spectrometer operating at a fixed frequency or in terms of a frequency difference for spectrometers operating at a fixed magnetic field. In practice, it is customary to express shifts relative to the standard as fractional (parts per million, ppm) differences, referred to as "chemical shifts" (19). In most proton NMR experiments, 3-trimethylsilylpropane sulfonate or some water-soluble derivative is used as the reference; its position represents zero chemical shift. Using this scale, the resonance from water appears displaced by almost 5 ppm downfield, and the resonances from protons involved in Watson–Crick base-pairs are almost 15 ppm downfield from the reference. In the following discussion of proton NMR, *all* chemical shifts mentioned refer to shifts downfield relative to the reference.

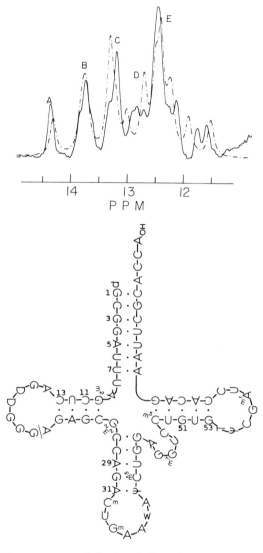

Fig. 2. (a) A comparison of the low-field nuclear magnetic resonance (NMR) spectrum of yeast tRNA^{Phe} at 300 MHz (solid curve) with the theoretical spectrum computed using the ring-current-shift theory described in the text. The NMR spectrum was obtained at 35°C in a solvent containing 0.1 M NaCl, 10 mM Mg^{2+} (free) at pH 7. (b) The "cloverleaf" model for yeast tRNA^{Phe}.

2. *Intensity.* Under most conditions, the intensity, which is the area under the spectral curve, is directly proportional to the number of nuclei in a particular chemical environment. Hence, NMR can be used to quantitate the number of bases involved in Watson–Crick base-pairs. No other solution technique can presently provide this information.

3. *Line width.* The width of a resonance line, $\Delta\nu$, is usually taken as the width at half the maximum peak height. For molecules that are tumbling rapidly, so that magnetic field inhomogeneities arising from neighboring protons are rapidly averaged out, the line width is determined by the lifetime of a proton in a particular chemical environment, τ, according to the expression $\tau = 1/(\pi\Delta\nu)$. (The reader will recognize this as a manifestation of the uncertainty principle, where $\Delta E = h\Delta\nu$ and $\Delta t = \tau$.)

Thus, from measurements of chemical shifts positions, intensities and line widths, NMR provides information about the number and types of bases in a particular environment, and the lifetime of a particular base-pair or structural feature in the molecule. In this way, NMR can be used to provide information about the static structure and the dynamics of the RNA structure.

In the following sections, we summarize the results of a number of NMR studies of a range of questions concerning the structural properties of tRNA molecules in solution. We begin (Section II) by first reviewing some of the basic properties of the mononucleotides required for the interpretation of the NMR spectra of the tRNA. In Section III are the results of studies of base-pairing of oligonucleotide fragments derived from tRNA molecules, since these systems serve as models for many of the features that are expected in the intact tRNA molecules and the interpretation of their spectral properties is usually more straightforward than that of the spectra of intact tRNA molecules. Section IV is a discussion of the use of NMR to study the structure of intact tRNA under normal (pH $\sim$7, $\sim$35°C, 0.1 M NaCl) conditions; these results are compared with recent X-ray diffraction studies on yeast tRNA^{Phe} (20–22). In Section V, we describe the application of NMR to the study of a number of changes in tRNA structure in response to different experimental conditions (salt, temperature, drugs, metals and chemical modifications).

II. NMR Studies of Mononucleotides

The structures of the common bases, here represented by A, U, G, C and T, are shown in the standard Watson–Crick base-pairs in Fig. 1. The numbering system used to indicate the different protons of the ribose group is also given in this diagram. The NMR spectra of 5'-AMP, 5'-CMP and 5'-GMP are shown in Fig. 3 and the assignments of the various resonances are indicated in Table I along with the assignments of resonances for the other bases. (Only the proton at C-1' of the ribose group are of interest here because the resonances from other ribose protons overlap markedly with each other or resonances from water.)

The chemical shifts of the base-proton resonances are sensitive to a host of environmental effects, including phosphate shielding, intermolecular associations, hydrogen bonding, exchange with solvent, temperature, and pH (23, 24). However, insofar as the tRNA spectra are concerned, only three major effects need to be considered at this point: stacking interactions, hydrogen bonding, and exchange with solvent.

A. Stacking Interactions and Ring-Current Effects

Experimentally, stacking interactions are manifested in the NMR spectra as upfield shifts of the resonances from base protons, resulting from base–base interactions in concentrated solutions or in single-stranded polynucleotides. Thus, relative to their position in dilute solution, the aromatic resonances from adenine can be shifted upfield by as much as 0.4 ppm as the result of stacking of the bases (19, 23–26). The shifts produced by the stacking of G residues are somewhat smaller, and

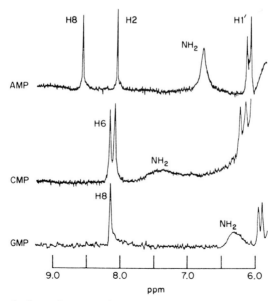

FIG. 3. The high-resolution nuclear magnetic resonance spectra of 5'-AMP and 5'-CMP and 5'-GMP in H$_2$O at 10°C, pH 7.

TABLE I
THE CHEMICAL SHIFTS OF RESONANCES FROM PROTONS OF THE FIVE COMMON RIBONUCLEOSIDES

Nucleoside	Assignment	Chemical shift (ppm)
Adenosine	C2-H	8.13
	C8-H	8.30
	C1'-H	6.1
Uridine	N3-H	11.5
	C6-H	7.9
	C5-H	5.9
	C1'-H	—
Ribothymidine	C6-H	7.65
	Methyl-H	1.9
	C1'-H	6.2
Guanosine	N1-H	11.2
	C8-H	8.0
	C1'-H	5.9
Cytidine	C6-H	7.9
	C5-H	6.1
	C1'-H	6.2

the shifts produced by C and U are even smaller (24). Studies of short oligonucleotides show also that there are important geometrical effects in that the magnitude of the shift induced by a base depends upon whether it is the 3' or 5' neighbor of the nucleotide of interest.

Most of the shifts induced by neighboring A and G residues can be attributed to the so-called "ring-current effect" (19, 23), which is well known from earlier studies of aromatic hydrocarbons (27). In the case of shifts produced by benzene, the effect is interpreted in terms of a magnetic dipole moment induced in the benzene π-electron system as a result of the applied magnetic field (19, 27). In a simplified model, the induced magnetic dipole moment is viewed as arising from a circulating electric current induced in the plane of the ring (hence the name, ring current). The magnetic lines of force arising from the induced "ring current" are depicted schematically in Fig. 4a. With complex aromatic hydrocarbons, the three-dimensional shape of the field is more complicated, but the simple theory indicates that the magnitude of the field should vary approximately as $1/R^3$, where R is the distance away from the center of the base. Thus the effect is short-range. The angular dependence of the induced field varies as $(1-3\cos^2\theta)$; therefore, a second important feature of the induced field is that the resonance from the proton located directly *above* the plane of the molecule experiences an *upfield* shift, whereas the resonance from a proton located in the *same* plane of the molecule experiences a *downfield* shift. Qualitatively, this explains why resonances from protons directly attached to an aromatic ring appear at such low fields (*19*).

B. Hydrogen-Bonding Effects

To a first approximation, hydrogen bonding has relatively little effect on the chemical shifts of resonances from nonexchangeable protons not directly involved in hydrogen bonding. However, there are substantial effects on resonances from the exchangeable ring nitrogen protons and the amino protons (*28*). The magnitudes of these effects are indicated by the results presented in Table II, which shows that in those solvents that form stronger hydrogen bonds, the resonances shift progressively to lower fields. The downfield shift of resonances due to hydrogen bonding has been well studied and, to a large extent, the effect can be attributed to removal of electron density from the protons involved in the hydrogen bond (*29–32*). Insofar as the tRNA studies are concerned, three points should be noted. (i) The downfield shifts of resonances from the ring nitrogen protons are much larger than are the shifts of the amino protons. (ii) The stronger the hydrogen bonding interaction, the larger is the downfield shift. (iii) The resonances from the ring-nitrogen protons start out at lower fields than the aromatic protons and move even further downfield, away from the position of the water resonance.

On the basis of these results, it is easy to understand the classic experiments of Katz and Penman (*28*), who used NMR to demonstrate the formation of A · U and G · C base-pairs in organic solvents. When A and U, or G and C, are mixed in organic solvents, large downfield shifts of the N3-H resonance of U and the N1-H resonance of G are observed. There are relatively smaller shifts of the resonance from the amino pro-

TABLE II
Effect of Hydrogen-Bonding Solvents on the Chemical Shift of Resonances from the Ring Nitrogen Protons of Uridine and Guanine

Solvent mixture	Chemical shift (ppm)	
	Uridine, N3-H	Guanosine, N1-H
100% $CDCl_3$	9.80	
50% $CDCl_3$–Me_2So	11.3	10.61
110% Me_2SO	11.42	10.7
74% H_2O–Me_2SO	11.42	11.08

tons. Resonances from the nonexchangeable protons are virtually unaffected by the base-pairing.

C. Proton Exchange with Solvent

Another important property of the ring nitrogen and amino protons of mononucleotides is that they exchange rapidly with water. At relatively low temperatures and at appropriate pH, the exchange is slow enough for resonances from the amino protons to be observed in aqueous solutions (33). We have carried out analogous measurements on the ring-nitrogen protons and find that the exchange rates of these protons are much faster than those of amino protons. Even at relatively low temperatures and optimum pH, the resonances from these protons are substantially broadened because the lifetime, τ, of the proton bound to the base is so short. The exchange rate may be slowed by adding some organic solvent (in our case Me_2SO); it is then possible to observe the resonance from the ring NH proton and follow its broadening and chemical shift, δ, as a function of the water concentration (Table III). From such data, we estimate that the lifetime of the G ring-nitrogen proton in water is about 2.6 msec with a resonance position of 11.2 ppm. The corresponding results for U are very similar, with $\tau = 2.5$ msec and $\delta = 11.4$ ppm (34).

Because of this rapid exchange with water, we expect to observe resonances in the low-field region of the spectrum from the ring-nitrogen

TABLE III
EFFECT OF WATER ON THE CHEMICAL SHIFTS AND LINE WIDTH AT 20°C OF THE AROMATIC AND EXCHANGEABLE PROTON RESONANCES OF GUANOSINE AND URIDINE IN Me_2SO

H_2O, fraction	Guanosine					Uridine		
	Chemical shifts (ppm)			Line widths (Hz)		Chemical shifts (ppm)		Line widths (Hz) N3-H
	NH_2	H8	N1-H	N1-H	NH_2	H6	N3-H	
0.00	6.50	8.00	10.70	14.5	8.0	8.02	11.35	7.2
0.15	6.50	8.00	10.77	15.0	9.0	8.02	11.38	9.0
0.43	6.50	8.00	10.90	25.0	10.0	8.02	11.44	32.3
0.59	—[a]	7.99	11.06	43.2	10.0	8.01	11.44	51.6
0.74	—[a]	7.98	11.08	58.5	10.0	8.00	11.44	64.5
0.77	—[a]	7.98	11.12	63.3	10.0	—[b]		
0.83	—[b]					7.98	11.44	77.5

[a] Signals adjacent to the strong H_2O proton resonance and were not detected.
[b] Sample was not prepared at this H_2O mole fraction.

protons of G and U (T) only if they are base-paired (or otherwise protected from exchange), and then only one resonance for each Watson–Crick base-pair should be observable. It is this simple numerology and the fact that the resonances occur in a spectral region virtually free of complications from other resonances that make it possible to use NMR to quantitate the number of base-pairs per tRNA and to provide information about the base-pairing structure of tRNAs and other polynucleotides in solution.

D. Rules for Predicting Low-Field NMR Spectra

We outline here the semiempirical procedure used to analyze the low-field NMR spectra of tRNA and oligonucleotide fragments derived from cleavages of whole tRNA, or prepared synthetically.

If we assume that the sequence-dependent chemical shifts of the A · U and G · C base-pairs originally seen in the low-field NMR spectra of tRNA (2) can be attributed entirely to ring-current effects exerted by neighboring bases, and that all RNA helices are in a standard RNA helix, then all that is required to predict the low-field spectra are: (i) the intrinsic position of the low-field resonance from a Watson–Crick G · C or A · U base-pair in the absence of neighboring base effects; (ii) the magnitudes of the ring-current shifts exerted by neighboring bases; and (iii) information about the stacking of bases in single-stranded regions adjacent to the terminus of a helix.

1. Determination of the Intrinsic Positions

A single A · U or G · C base-pair is not stable in aqueous solution, so it is not possible to determine directly the intrinsic positions for their resonances. However, from experimental data on tRNAs and synthetic samples it has been possible to establish reasonable values for the intrinsic positions [denoted $(A \cdot U)°$ and $(G \cdot C)°$, respectively].

From an examination of the low-field NMR spectra of 8–10 different tRNAs, we noticed that the number of resonances in the 14.5–13.5 ppm region is greatest with those tRNAs in which the $(A + U)$ content is highest (e.g., tRNA^{Phe}) whereas in very $(G + C)$-rich tRNA, such as E. coli tRNA^{fMet}, there is very little intensity in this region and the intensity in the 12–13 ppm region is relatively much stronger (35–38). From such data and the results of Katz and Penman (28), we concluded that the intrinsic position of the resonance from A·U is at lower fields than that from G · C. Since yeast tRNA almost never exhibits resonances below 14.3 ppm, we further concluded that the intrinsic position of A·U must be around 14.5–14.8 ppm. The spectrum of the $(G + C)$-rich tRNA^fMet indicated that the probable intrinsic position for a G · C base-pair is between 13.5 and 13.7 ppm. Values outside this range can be eliminated by the experimental data (39); we now believe that the best values are 14.5 and 13.6 ppm for $(A \cdot U)°$ and $(G \cdot C)°$, respectively.

2. Ring-Current Effects

To determine the magnitudes of the ring-current shifts exerted on the protons of a particular base-pair by neighboring bases, we have used the theoretical results of Giessner-Prettre and Pullman (40)[3]. If the RNA double-helix is assumed to have the standard A′-RNA structure (41), then the relative overlap of bases viewed down the helix axis will be as shown in Fig. 4c.

In this particular diagram, the hydrogen-bonded ring-nitrogen proton of the Pur_1-Pyr_1 base-pair is directly over a portion of the purine base, Pur_2, but it is considerably farther from the pyrimidine, Pyr_2. This same diagram also indicates that the hydrogen-bonded ring-nitrogen proton from the Pur_2-Pyr_2 base pair is directly under the Pur_1, but rather far removed from Pyr_1. From these and similar diagrams, the location of a specific hydrogen-bonded ring-nitrogen proton with respect to the bases of adjacent base-pairs can be determined. The theoretical maps of Giessner-Prettre and Pullman (40), shown in Fig. 4b can then be used to compute the magnitudes of the ring-current shifts exerted on each proton by neighboring bases. A tabulation of these results is presented in Table IV for ring-nitrogen protons for all possible combinations of nearest-neighbor bases.

Initially, we chose to view these calculated values as starting points for analyzing spectra, with the notion that they might have to be modified if difficulties were encountered (39). After examining the spectra of a number of different tRNAs, we are reasonably convinced that the values shown in Table IV give good results when used with the values 14.5 and 13.6 ppm for the intrinsic positions for A · U and G · U, respectively. When we used $(A \cdot U)° = 14.8$ and $(G \cdot U)° = 13.7$ ppm, we found that the ring-current parameters shown in Table III must be increased by about 20% in order to achieve reasonable fits of the experimental data (36, 37). The lower field value for $(A \cdot U)°$ was chosen specifically to account for some anomalously low-field resonances observed in the spectra of some *E. coli* tRNAs (36). However, we subsequently discovered that several of these resonances are due to $s^4U_8 \cdot A_{14}$ base-pairs (42, 43) or anomalously shifted A·U resonances, so that the principal argument for changing from $(A \cdot U)° = 14.5$ to $(A \cdot U)° =$

---

[3] See article by Pullman and Saran in this volume.

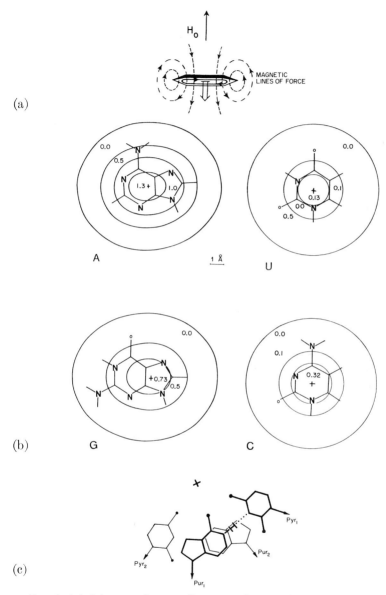

Fig. 4. (a) Schematic diagram illustrating the spatial distribution of the magnetic lines of force about a benzene molecule induced by the application of an external field H_o. (b) Intermolecular shielding values (in parts per million) due to the ring current in the four common bases in planes parallel to the molecular surface at a distance of 3.4 Å (40). (c) A schematic diagram depicting the overlap of bases in an RNA helix when viewed down the helix axis (41).

14.8 in the first place was removed. Therefore, in all of our more recent work, we have been using the set of parameters shown in Table IV.

3. Application to RNA Double Helices

The parameters given in Table IV provide all the information needed to compute the low-field NMR spectrum of any standard A'-RNA double helix of known sequence. As an example, consider the ring-current shifts exerted on an A · U base-pair sandwiched between the two G · C base-pairs in the structure shown below.

```
     5'   3'              Shift Contribution
     G · C                G = 0, C = 0.2
    |A · U|
     G · C                G = 0.6, C = 0
     3'   5'
```

From Table IV we see that the low-field A · U resonance would be shifted 0.2 ppm by the C above it and 0.6 ppm by the G below it, resulting in a net 0.8 ppm upfield displacement of the A · U resonance from its unshifted position. In an analogous manner, the low-field NMR spectrum for any RNA double-helix can be predicted, provided the sequence is known.

Since many assumptions go into the calculations, it seems desirable

TABLE IV

A Summary of Ring-Current-Shift Parameters Used in the Calculation of the Low-Field Proton NMR Spectra of tRNA Molecules[a]

5'	3'	5'	3'
U = 0	A = 1.1	U = 0	A = 1.1
C = 0	G = 0.6	C = 0	G = 0.7
G = 0	C = 0.1	G = 0	C = 0.2
A = 0.1	U = 0	A = 0	U = 0.1
U·A		**C·G**	
A·U (inverted)		**G·C** (inverted)	
U = 0.1	A = 0	U = 0.1	A = 0
C = 0.2	G = 0	C = 0.25	G = 0
G = 0.6	C = 0	G = 0.7	C = 0
A = 0.6	U = 0	A = 1.0	U = 0
3'	5'	3'	5'

[a] The notation 5' and 3' refers to the sugar positions at the ends of chains containing the neighboring bases indicated. The unshifted positions for the usual base-pairs are: $(A \cdot U)° = 14.5$ ppm, $(G \cdot C)° = 13.6$ ppm and $(A \cdot \Psi)° = 13.5$ ppm.

to list them so that when disagreements between experiments and theory are encountered, it may be possible to identify the source of the problem: (i) all RNA helices are assumed to have the same conformation (A′); (ii) the ring-current shift values of Table IV are assumed to be valid; (iii) all nearest-neighbor effects, whether of ring-current origin or not, are lumped into a single parameter; and (iv) second nearest-neighbor effects are not directly included in the theory, although they too may be absorbed in the parameters of Table IV. Disagreements between experiment and theory can therefore be expected to occur as a result of any of the following factors: (a) sequence-dependent conformation of the RNA double helix; (b) irregular stacking of adjacent bases in single-stranded regions or changes in the conformation of the double helix at the ends of the helix; (c) tertiary interactions, which affect the positions of resonances from secondary-structure base-pairs; (c) incorrect parameterization of the ring-current theory; (d) significant second-neighbor contributions to the ring-current shifts.

Detailed analysis of the spectra of oligonucleotide fragments and intact tRNA molecules provides evidence that some of these neglected effects may be important.

III. NMR Studies of Oligonucleotide Fragments

A. Simple Helices

Oligonucleotide fragments derived from tRNA are attractive systems for NMR study because they serve as models for certain features observed in intact tRNA and provide useful systems for testing assignments. Various physical chemical techniques had been used to demonstrate that oligonucleotide fragments, derived from enzymic or chemical cleavage of intact tRNA molecules, form small "hairpin" helices (44–51). Extensive studies of Bayev (49) and Römer (48), for example, indicated that tRNA^{Phe} fragments containing the nucleotides Nos. 1–35 and Nos. 36–76 are substantially base-paired at low temperatures, and Zachau, et al. (50) demonstrated that these recombined half-molecules (as well as other fragment combinations) can be aminoacylated enzymically. The fragment containing the nucleotides in the T-ψ-C stem of yeast tRNA^{Ser} has an optical melting temperature, t_m, greater than that of the intact molecule (51). A thorough review of the optical melting properties of various oligonucleotides has appeared (52).

While these earlier studies provided evidence that the tRNA fragments do base-pair, all are deficient in that it is necessary to make assumptions about the probable base-paired structures present. With NMR, it has been possible to determine these structures and to follow the details of the thermal disruption of simple double helices. NMR has also provided information about the stacking of bases and about intermolecular association of fragments. The results of some of the NMR studies are briefly summarized below.

Analysis of the NMR spectra was aided by constructing a matrix for each fragment in which the horizontal and vertical indices cover the complete sequence of the molecule and slashes (/) indicate the bases in particular rows and columns that are complementary (i.e., A and U or G and C); an X is used for G and U (53, 54). A double helix then appears as a run of contiguous slashes. After eliminating helices containing fewer than three good Watson–Crick base-pairs, and/or fewer than three bases in a loop, we obtain matrices such as the one shown in Fig. 5a for the anticodon arm of *E. coli* tRNA^{fMet}. This matrix indicates that there is only one stable helix and that it is identical to that in the anticodon stem of the intact tRNA^{fMet} molecule. All other possible base-

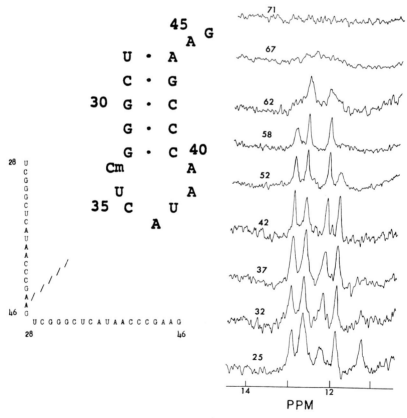

FIG. 5a.

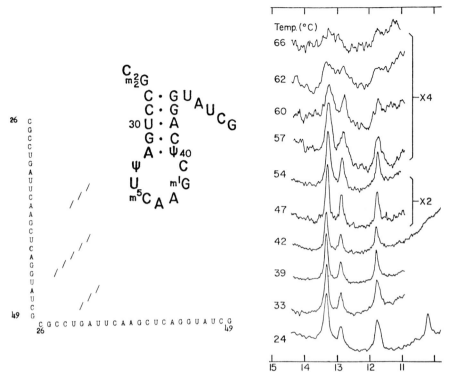

Fig. 5. (a) Base-pairing matrix and low-field nuclear magnetic resonance (NMR) spectrum of the anticodon arm of *Escherichia coli* tRNA^{fMet} in a solution containing 10 mM Mg^{2+}, 0.1 M NaCl and 10 mM cacodylate buffer at pH 7. (b) Base-pairing matrix and low-field NMR spectrum of the anticodon arm of yeast tRNA$^{Leu}_{UUG}$ in low-salt solution containing 10 mM cacodylate at pH 7.

pairing schemes involve either two base-pair helices (which would be marginally stable) or helices with interruptions (bulges). The NMR spectrum of this fragment at various temperatures is also shown in Fig. 5a. While the low-temperature spectrum contains five peaks, the total intensity corresponds to six, rather than just five, base-pairs (the peak at 12.8 has about twice the intensity of the other peaks in the spectrum). Because of its high field position and early-melting characteristics, and the fact that an "extra" resonance is also observed in the anticodon arms of other tRNA, we believe that the ~11.3 ppm resonance is the "extra" resonance and that it can be assigned to a protected U_{34} in the anticodon loop. The remainder of the resonances in the spectrum can be assigned base-pairs in the anticodon stem. By direct application of the semiempirical ring-current-shift theory described in the preceding section, the expected positions of the resonances from the five cloverleaf-structure

TABLE V

COMPARISON OF THE OBSERVED AND CALCULATED RESONANCE POSITIONS FOR THE LOW-FIELD PROTONS OF VARIOUS tRNA FRAGMENTS

tRNA Species (Cloverleaf helix)	Cloverleaf base-pair	Ring-current predicted resonance position (ppm)	Observed resonance position (ppm)	Error Calc-Obs (ppm)
E. coli tRNA^{fMet}	A · U$_{28}$^{a}	13.2	12.75	+0.45
	G · C$_{29}$	11.8	11.9	−0.1
(Anticodon stem)	G · C$_{30}$	12.2	12.3	−0.1
	G · C$_{31}$	12.6	12.7	−0.1
	G · C$_{32}$^{a}	13.1	12.9	+0.2
	G · C$_{10}$^{a}	11.8	12.0	−0.2
(hU stem)	A · U$_{11}$	13.7	13.45	+0.25
	G · C$_{12}$	13.2	13.25	−0.05
	G · C$_{13}$^{a}	12.5	12.5	0.0
Yeast tRNA^{Asp}	G · C$_{48}$^{a}	12.9	13.2	−0.3
	G · C$_{49}$	12.2	12.2	0
(T-Ψ-C stem)	G · C$_{50}$	12.6	∼12.6	0
	G · C$_{51}$	12.6	12.6	0
	G · C$_{52}$^{a}	13.1	12.7	+.4
Yeast tRNA^{Phe}	G · C$_{10}$^{a}	12.7	12.8	−.1
(hU stem)	G · C$_{11}$	13.3	13.4	−.1
	A · U$_{12}$	13.7	14.1	−.4
	G · C$_{13}$^{a}	11.5	11.8	−.3
	G · C$_{27}$^{a}	12.3	12.5	−.2
	G · C$_{28}$	11.9	12.0	−.1
(Anticodon stem)	A · U$_{29}$	13.3	13.3	0
	G · C$_{30}$	12.4	12.5	−.1
	A · Ψ$_{31}$^{a}	13.2	13.35	−.15
-3'-¾	G · C$_{49}$^{a}	12.4	12.6	−.2
	A · U$_{50}$	13.3	13.3	0
(T-Ψ-C stem)	G · C$_{51}$	12.5	12.4	+.1
	A · U$_{52}$	13.8	13.8	0
	G · C$_{53}$^{a}	12.5	12.5	0
Yeast tRNA$^{Leu}_{UUG}$	A · U$_{30}$	13.3	13.4	−.1
	G · C$_{28}$^{a}	13.25	13.4	−.15
(Anticodon stem)	A · Ψ$_{32}$^{a}	13.3	13.4	−.1
	G · C$_{29}$	12.8	13.0	−.2
	G · C$_{31}$	11.5	11.8	−.3

^{a} Terminal base pairs.

base-pairs were computed on the assumption that bases in single-stranded regions adjacent to terminal base-pairs are stacked in the helix. These results are presented in Table V along with the experimental results. At elevated temperatures, intensity at 12.8 ppm is lost, and on this basis one resonance in this peak is assigned to the $A \cdot U_{28}$ base-pair at the free terminus of the stem. The resonance from the adjacent $G \cdot C_{29}$ base-pair melts next, and resonances from the interior $G \cdot C_{30}$ and $G \cdot C_{31}$ base-pairs are the last to melt.

The base-pairing matrix for the anticodon "hairpin" derived from yeast $tRNA_3^{Leu}$ is shown in Fig. 5b, and again the anticodon helix with five base-pairs is expected to be the only stable helix. The observed NMR spectrum shown in Fig. 5b, contain peaks at ~13.3, 12.8 and ~11.8 with relative intensities 3, 1 and 2, respectively. A comparison between the predicted and observed spectra is given in Table V (55). It is interesting to note that this anticodon fragment also contains an extra resonance at about 11.6 ppm that melts quite early (~40°C) and cannot be assigned to any of the Watson–Crick base-pairs. The following considerations indicate the "extra" 11.6 ppm resonance should be assigned to U_{34}. The second base in the anticodon loop of all tRNAs (except mammalian $tRNA^{fMet}$) is U. All of the four anticodon arms we have studied contain an extra resonance at about 11.5 ± 0.2 ppm that cannot be accounted for in terms of expected Watson–Crick base-pairs. The NMR spectra of intact tRNAs invariably contain a common resonance at ~11.5 ppm that melts at very low temperatures in the absence of Mg^{2+} (56). Because of the high-field position of the resonance, U_{34} (U_{33} in yeast $tRNA^{Phe}$) cannot be involved in an $A \cdot U$ base-pair. Rather, we infer from model system studies (34) that the N3-H of U is hydrogen-bonded to some oxygen group, and recent crystal structure data on yeast $tRNA^{Phe}$ suggest it is hydrogen-bonded to the phosphate group joining A_{36} and A_{35} (20–22).

A comparison of the predicted and observed positions for several other tRNA fragments are also shown in Table V. In general, we find that resonances from internal base-pairs are rather well predicted, but occasionally there are relatively large errors associated with resonances from terminal base-pairs. This suggests that bases in single-stranded regions adjacent to a terminal base-pair may not be stacked as though they were part of the helix, and we may expect similar problems with intact tRNA. Taking the data collectively, we find that the standard deviation between the predicted and observed positions of the resonances is 0.16 ppm if terminal base pairs are included, but 0.14 ppm if they are excluded. Since the deviations between experiment and theory are both positive and negative, it is clear that simply adjusting the intrinsic positions, $(A \cdot U)°$ and $(G \cdot C)°$, will not improve the fit. Similarly, we do not find evidence for systematic errors, which would suggest that the ring-current parameters should be altered. Rather, we believe that the discrepancies are probably due to end effects, irregularities in the small RNA helices and limitations of the ring-current-shift theory (neglect of second-neighbor effects, for example). It would obviously be desirable to have a much larger catalog of fragment spectra to further refine the ring-current theory.

While the fragments discussed above have been useful in testing the validity of the ring-current shift calculations, other fragments have presented new problems. The half-molecule fragment containing the T-Ψ-C arm of tRNA[Phe] is one case in point. In contrast to other fragments studied, its spectrum is extremely diffuse and poorly resolved (37). In our initial analysis of this fragment, we attempted to interpret this spectrum in terms of the five-base-pair T-Ψ-C stem (37), but an examination of the base-pair matrix revealed a number of possible pairing schemes, all with rather similar stabilities. It now seems likely that the poorly resolved spectrum is due to a mixture of base-paired structures. This may represent an example where the base-paired structure is quite different from that assumed in earlier optical and physical chemical studies of this fragment (52).

B. Dimerization of Oligonucleotides

Fragments containing long single-stranded ends can form additional base pairs by dimerization. Experimentally, this is manifested in the

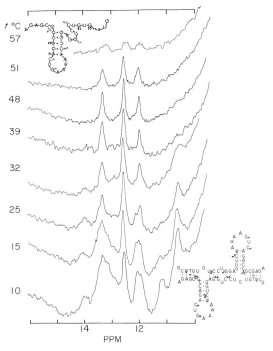

FIG. 6a.

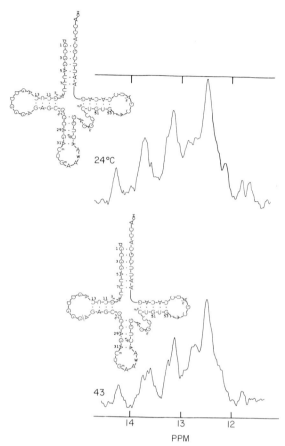

FIG. 6. (a) Temperature dependence of the low-field nuclear magnetic resonance (NMR) spectrum of the anticodon half-fragment of yeast tRNA^{Phe} (upper left), illustrating the effect of dimerization on the spectrum at low temperatures. A model for the dimer is shown at the lower right. The solution contained 0.1 M NaCl, 10 mM MgCl$_2$ and 10 mM cacodylate at pH 7. (b) A comparison of the low-field NMR spectra of intact tRNA^{Phe} (see Fig. 2) and a 1:1 mixture of the 5' quarter and 3' three-quarter fragments of tRNA^{Phe} (lower) illustrating the recombination of the fragments to regenerate the base-pairs of the intact molecule.

NMR spectra by the appearance of extra resonances in the low-field spectrum at low temperatures that disappear when the temperature is raised. Several examples of dimerization of yeast tRNA^{Phe} fragments are shown in Fig. 6. Again, a base-pairing matrix was used to identify possible dimer structures and ring-current calculations were then used to compute their NMR spectra. From the number and positions of the new resonances added, it was possible to develop a plausible structure for the dimer (Fig. 6) (57). While other structures cannot be rigorously eliminated, this gives the best fit of the NMR data. Similar structures were first proposed by Römer (48), who presented optical evidence for the dimerization, but he was unable to determine either the number or type of new base-pairs in the dimer.

The reassociation of the two fragments derived from breaking the hU loop of yeast tRNA^{Phe} is of special interest in light of the fact that these two fragments can recombine to give a molecule that can be aminoacylated (50). Optical studies of this recombination have been published (52), and the NMR results (57) are shown in Fig. 6b. A comparison of these results with the spectrum of the intact tRNA^{Phe} molecules demonstrates that all or most of the base-pairs originally present in the intact molecule are also present in the fragment spectrum. It will now be interesting to examine the spectra of fragments derived from other tRNAs that do not recombine to produce biologically active molecules. We suspect that, in these cases, the base-pairing structures will differ from those in the intact molecule.

C. Melting Behavior of Short Helices

The temperature dependence of the spectrum of the tRNA^{fMet} anticodon arm (shown in Fig. 5) illustrates the use of NMR to investigate the thermal denaturation of an RNA double helix. Between 17° and 42°C, the resonance assigned to $A \cdot U_{28}$ (see Table V) broadens and then disappears. Resonances from the four $G \cdot C$ base pairs are unaffected up to ca. 50°C where broadening of the resonance from $G \cdot C_{29}$ (11.9 ppm) and $G \cdot C_{32}$ (12.9 ppm) becomes evident. At 58°C, these resonances are substantially broadened, but the resonance from the two interior base-pairs ($G \cdot C_{31}$ and $G \cdot C_{30}$) are still unaffected. These latter two resonances are finally lost at 71°C, which is within 7°C of the optical t_m.

The fact that the resonances disappear by line broadening is attributed to the uncertainty-principle broadening that occurs when the lifetime of a proton in a base-pair hydrogen bond is reduced to the millisecond range. If the lifetime of a proton in a Watson–Crick base-pair is τ, then the uncertainty-principle contribution to the line widths is given by the expression $\Delta \nu = 1/\pi \tau$ (58, 59). Therefore, a line width of 150 Hz (0.5 ppm at 300 MHz) corresponds to a lifetime in the base-pair of approximately 2 msec. Using this line-broadening criterion, we can assign an NMR t_m to each individual base-pair in the "hairpin." We find that the terminal $A \cdot U_{28}$ base-pair has a melting point over 25°C below those of some of the interior base-pairs, and there is about a 10°C difference between the melting behavior of other $G \cdot C$ base pairs and $G \cdot C_{30,31}$. Similar results are also obtained for the melting behavior of anticodon hairpins from other tRNAs. Thus, according to the NMR results, the melting of individual base-pairs in the segment is sequential rather than cooperative, and proceeds with the terminal base-pairs opening first and interior base-pairs melting last. This has important implications with regard to the use of NMR to investigate the thermal unfolding of intact tRNA molecules, which is discussed later (Section V).

IV. High-Resolution NMR of tRNA

A. Quantitative Determination of the Number of Secondary- and Tertiary-Structure Base-Pairs

In earlier studies of the structure of tRNA in solution using other techniques, it was virtually impossible to determine the total number of base-pairs per molecule. This can now be accomplished by high-resolution NMR. Since each Watson–Crick base-pair contributes one, and only one, resonance to the low-field spectrum, integration of the low-field spectra provides a direct determination of the number of base-pairs. In practice, it has turned out to be difficult to integrate spectra accurately, and during the past four years, we have used several different methods, including: (i) comparison with an external standard (metcyanomyoglobin) (2); (ii) comparison with certain peaks in the low-field spectra, which are *assumed* to correspond to an integral number of resonances per molecule (37, 38, 60, 61); (iii) comparison with the intensity of methyl resonances (e.g., from thymine) observed in the high-field region (0–3 ppm) of the spectrum (61, 62); (iv) most recently, comparison with the intensity in the aromatic (8.5–6.5 ppm) region of the spectrum (63, 64). We have recently discussed these different methods and indicated why the use of the aromatic proton resonances to integrate the low-field spectra is superior (65).

1. THE METHOD

For a tRNA of known sequence, the number of resonances that should occur in the aromatic region (9–6.5 ppm) of the spectrum can be computed. Resonances in this region are due to "nonexchangeable" protons and therefore can serve as internal standards for integrating the low-field spectra. Because of contributions from amino protons in the aromatic region, it is necessary to measure the low-field region in H_2O and the aromatic region (or ribose region) separately in D_2O under identical experimental conditions (same instrument conditions, sample, and sample tube). Repeated measurements made on the same sample are very reproducible ($\pm 5\%$), but this does introduce a possible source of error. Even this potential error can be eliminated simply by using the methyl resonance to monitor the sample before and after conversion from H_2O to D_2O. In this way, it is possible to demonstrate that there is no change in the sample produced when the solvent is changed from H_2O to D_2O.

As an example of this method of integration, consider the spectra of *E. coli* tRNA$_1^{Val}$ shown in Fig. 7 (*64*). It is clear that, for the aromatic spectra measured in D_2O, there is no ambiguity as to where to draw the baseline in the aromatic region. The only problem is to set the spectral range to be included in the integration. From our studies of aromatic resonances in synthetic DNA samples and ring-current calculations (T. Early, unpublished results), we estimate that the resonance shifted most upfield (from the C-2 proton of the adenine sandwiched between two other adenines) would be located around 6.5 ppm. We therefore take 6.5 ppm as the high field limit for the aromatic resonances. The low-field limit is chosen as 9 ppm, since none of the common bases have aromatic resonances outside this range (see Table I).

In the low-field region of the spectrum shown in Fig. 7, the choice of the base line is again fairly obvious, although a second, less desirable base line has been included to illustrate the way it would affect the integration. Comparison of the integrated intensity in the aromatic region (assumed to correspond to 89 protons per molecule) with the intensity

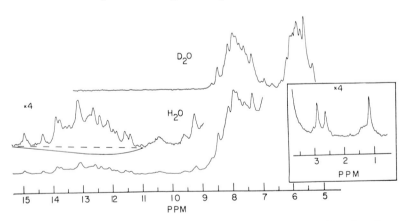

FIG. 7. The 300 MHz nuclear magnetic resonance spectrum of *Escherichia coli* tRNA$_1^{Val}$, illustrating the use of resonances in the aromatic region (9–6.5 ppm) to integrate the low-field spectra. The spectrum in the low-field region was obtained in H_2O. To obtain spectra in the aromatic and ribose regions free of contributions from exchangeable amino-protons, measurements were also carried out in D_2O under conditions identical with those used with the H_2O sample. The spectrum in the methyl region (1–3 ppm) is shown as an inset.

in the low-field region of the spectrum yields the number of resonances in the low-field spectrum; these results are shown in Table VI for *E. coli* tRNA^{Val} and other tRNAs.

Use of the aromatic resonances offers a number of advantages over the other integration methods mentioned above. First, the intensity in the aromatic region corresponds typically to 85–100 protons per molecule, so that the signal-to-noise ratio is high. Second, the presence of contaminating material is much less serious in the aromatic region owing to its larger intensity, and the presence of rare bases will have only a small effect on the total intensity of the aromatic region. A very important advantage in using the aromatic region in calibrating the low-field region is that the two regions are integrated over similar spectral widths, 2 ppm versus 3.5 ppm, and both regions contain overlapping resonance. In this way, the line-shape problems encountered using methyl resonances or single resonances in the low-field region are eliminated (*65*).

However, the method is not without possible sources of error. Special

TABLE VI
A Summary of the Results of Integrations of the Low-Field NMR Spectra of tRNA

tRNA Species (number of cloverleaf Watson–Crick base pairs)	Magnesium ion concentration[a]	t (°C)	Integration method (ref.)	Number of low-field resonances
yeast tRNA^{Phe} (*20*)	~2 mM	24°	External (*2*)	21 ± 3
	10 mM dialyzed	36°	Internal	19.6
	10 mM total	37°	Internal (*37*)	18.5
	10 mM dialyzed	40°	Aromatic (*65*)	22.2 ± 1
E. coli tRNA^{fMet} (*19*)	3 mM	40°	External (*35*)	23 ± 2
	5 mM	24°	Internal (*36*)	19
	5 mM dialyzed	37°	Internal (*60*)	27
	10 mM dialyzed	37°	Aromatic (*65*)	23 ± 1
E. coli tRNA$^{Glu}_{1,2}$ (*20, 19*)	10 mM	40°	Internal (*36*)	20
	10 mM dialyzed	40°	Aromatic (*65*)	21.5 ± 1
E. coli tRNA$^{Val}_1$ (*19*)	15 mM	37°	Internal (*62*)	26 ± 3
	10 mM dialyzed	35°	Aromatic (*65*)	23 ± 1
	15 mM	37°	Internal, methyl (*61*)	26 ± 1
yeast tRNA^{Asp} (*17*)	15 mM	37°	Internal (*78*)	20
E. coli tRNA^{Asp} (*18*)	10 mM dialyzed	40°	Aromatic (*65*)	20.2 ± 1
E. coli tRNA$^{Trp}_{su+}$ (*20*)	40 mM	40°	Aromatic (*65*)	22 ± 1
	3 mM	40°	Aromatic (*65*)	19 ± 1
E. coli tRNA$^{Tyr}_{1,2}$ (*23*)	10 mM dialyzed	37°	Aromatic (*65*)	26 ± 2

[a] Ten mM dialyzed means dialyzed against 10 mM Mg^{2+}.

folding of the tRNA molecules could induce the anomalous downfield shifting of some of the ribose protons from their expected position in the 6.4 ppm into the aromatic region, and, conversely, some resonances from aromatic protons could be shifted upfield into the ribose region. This possibility was eliminated by comparing the relative intensities of the aromatic and ribose regions and showing that the intensities observed are in agreement with the expected values. Differential saturation of aromatic or low-field resonances could lead to errors, but the power levels used in our experiments were below those needed for saturation. If the aromatic resonances were partially saturated, we would see an apparent intensity in the low-field that is too large.

We conclude that the use of the aromatic protons to calibrate the intensity of the low-field region avoids most of the difficulties associated with methods previously used and should lead to highly accurate values for the number of resonances in low-field region of the spectrum. It is difficult to determine the accuracy of the method, but various considerations indicate the error is less than ± 2 base-pairs per molecule and probably closer to ± 1 base-pair per molecule.

2. Discussion of Integration Results

For samples dialyzed against Mg^{2+}, the number of resonances in the low-field region varies from 20 (yeast $tRNA^{Asp}$) to 26 ($E.\ coli\ tRNA^{Tyr}$) (see Table VI). With a couple of exceptions, most tRNA contain ca. 3 more resonances (base-pairs) than required by the cloverleaf model. We believe these extra resonances are due to tertiary structure base-pairs; the evidence for this is discussed in the next section. However, before leaving the integration results, it is necessary to comment on some of the apparent and real discrepancies that still exist.

Magnesium-deficient samples are more susceptible to thermal denaturation than are samples that have adequate levels of magnesium, and resonances for tertiary base-pairs are especially sensitive to the presence of magnesium. Thus, some of the differences between the values obtained recently (64, 65) and those reported previously (2, 36–39) can be attributed to differences in Mg^{2+} concentration. For example, in yeast $tRNA^{Phe}$ (Fig. 2) samples containing adequate levels of Mg^{2+}, there is more intensity at 14.4 ppm than in the spectra of the Mg^{2+}-deficient samples reported earlier (37). The differences between the values originally reported for $E.\ coli\ tRNA^{fMet}$ (36) and those obtained using the improved integration method (63, 64) can also be attributed to differences in the magnesium concentrations, and to the use of samples in which the s^4U_8 residue was not photocrosslinked or photooxidized. The results on $E.\ coli\ tRNA^{Trp}_{su^+}$ are particularly interesting since they show that there are 2–3 more base-pairs than predicted by the cloverleaf model in the presence of Mg^{2+}, but when the Mg^{2+} is removed, these extra base-pairs are lost.

Although this effect accounts for most of the differences between the earlier values and those obtained using the improved integration methods, some notable discrepancies still exist. Recently, there have been several reports indicating that two class-I *E. coli* tRNAs have about 26–27 resonances in the low-field region (*60*, *61*). These studies were based either on the use of methyl resonances as internal standards or the use of resolved low-field resonances as internal standards. The use of low-field peaks as internal standards is questionable and there are many problems with the use of methyl resonances as internal standards, most of which will lead to intensities for the low-field region that are too high (*65*).

In summary, the integration results indicate that there are two to four more base-pairs per molecule than required by the cloverleaf model, provided the tRNAs contain adequate levels of Mg^{2+}. With Mg^{2+}-deficient samples (2–5 Mg^{2+} per tRNA) the number of base-pairs at about 40° is, within experimental error, the same as that required by the cloverleaf model; this was the basis for our earlier suggestion that we had no evidence for extra base-pairs (*39*). Under the experimental conditions used in those experiments (low Mg^{2+} and relative high temperatures), most resonances from tertiary-structure base-pairs have partially or completely melted out. The results support our original analysis of these spectra in terms of cloverleaf base-pairs (*36, 37, 39*).

B. Evidence for Common Tertiary-Structure Base-Pairs

While the most accurate integration results provided evidence for two to four extra base-pairs, there was other independent NMR evidence that the most class-I tRNAs contain a common set of such base pairs.

1. The $A_{14} \cdot s^4U_8$ Base-Pair

Several years ago, Levitt (*14*) analyzed the sequence homologies in tRNA and on the basis of these and other data proposed a number of tertiary-structure base-pairs which he expected would be common to most class I tRNAs. The one between the common A at position 14 and U (or s^4U) in position 8 was of particular interest to us, since we often noticed anomalously low-field resonances (14.8 ± 1 ppm) in spectra of *E. coli* tRNA, but not yeast tRNA (compare Fig. 2a and Fig. 7). Yeast tRNAs have no thiouridine at position 8, whereas *E. coli* tRNAs usually do, and this suggested that s^4U_8 might be responsible for the 14.8 ppm resonance. To test this possibility, the spectrum of unfractionated *E. coli* tRNA was measured and, as expected, a broad, rather weak resonance, is located around 14.8 ppm (see Fig. 8) was observed. To prove that the s^4U_s residue was involved, three different chemical procedures were used to convert s^4U to U (*42, 43*); in each case the 14.8 ppm resonance was lost and a new resonance appeared at around 14.3 ppm (see Fig. 8). [The fact that the new intensity appears at 14.3 ppm coincident with the loss of intensity at 14.8 ppm is consistent with studies on s^4U and U mononucleotides in Me_2SO, which suggested that the resonance from an $s^4U \cdot A$ base-pair should appear further downfield than the corresponding resonance from an $A \cdot U$ base pair (*34*).] These observations indicate that most (ca. 50%) *E. coli* tRNAs have a tertiary-structure base-pair involving s^4U_s and an A located elsewhere in the molecule, and that conversion of s^4U_s to U_s results in formation of a new $A \cdot U$ base-pair. Since various studies have demonstrated that s^4U_s can be crosslinked photochemically with C_{13} in several *E. coli* species without significantly affecting biological activity (*66–69*) we assume that it is A_{14} that is base-paired to s^4U_s. Bergstrom and Leonard's analysis of the photoproduct obtained from photocrosslinking of C and s^4U (*70, 71*) and the electron density map found in the crystallographic studies (*72, 73*) suggest a reverse Hoogsteen pairing between s^4U_s and A_{14}. Furthermore, sequence data show that U_s (or s^4U_s) and A_{14} are present in all tRNAs and a reverse Hoogsteen $A \cdot U$ base pair cannot isomorphously be replaced by G and C. Since studies of the unfractionated *E. coli* tRNA demonstrated that the $A \cdot s^4U$ base-pair is common to most *E. coli* tRNAs, there was little doubt that this base-pair would also be found in pure tRNA species; subsequent studies of *E. coli* tRNA^{Arg} and tRNA^{Val} have proved this to be so (*43, 62*).

In the light of these recent findings we should mention that resonances

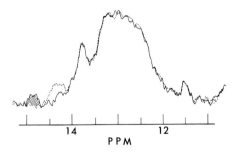

FIG. 8. A comparison of the low-field spectrum of unfractionated *Esherichia coli* tRNA before (———) and after (- - - -) conversion of the s^4U_s to U_s. The only changes in the spectrum correspond to a loss of intensity at 14.8 ppm (shaded) and a gain of intensity at about 14.3 ppm.

at about 14.8 ppm in the spectra of *E. coli* tRNA^{Arg} and tRNA^{fMet} were incorrectly assigned to secondary-structure base-pairs (*36*). The reassignment of the tRNA^{fMet} spectrum has been discussed elsewhere (*74*).

The NMR observations, which confirm the existence of the $A_{14} \cdot s^4U_8$ base-pair, are interesting in connection with optical studies of tRNA conformation. Several groups have previously proposed that the hyperconformation at 340 nm of s^4U can be used to monitor the tetiary folding of the molecule (*75, 76*), and our results support this.

2. OTHER COMMON TERTIARY-STRUCTURE BASE-PAIRS

In earlier studies, using Mg^{2+}-deficient samples and less accurate integration methods, we had little or no evidence for the presence of any resonances from teritary-structure base-pairs. In fact, we had begun to suspect that these resonances might never be seen because of rapid exchange with the solvent. However, once we found experimental conditions where the resonance from the $s^4U \cdot A$ base-pair could be observed, this gave hope that others might be found. Equally important, the studies of the resonance from $s^4U \cdot A$ indicated that resonances from other *common* tertiary-structure base-pairs would have *common* chemical shifts (*42*).

With these ideas in mind, unfractionated *E. coli* and yeast tRNAs were examined at several different temperatures with and without Mg^{2+} (*56*). These experiments were based on the following notions. First, if the proposed tertiary-structure base-pairs are identical, or nearly so in all class I tRNAs, they should give rise to a set of relatively sharp resonances that would stand out above the broad background of unresolved overlapping resonances from secondary-structure base-pairs. Second, tertiary-structure base-pairs are usually weaker than secondary-structure base-pairs, hence resonances from them should be more temperature sensitive than the secondary ones. The NMR spectra of unfractionated *E. coli* and yeast tRNA are shown in Fig. 9. As expected, they do exhibit several sharp resonances in the low-field region, at 50°C, but only in high concentrations of Mg^{2+}. The effect of Mg^{2+} on the spectrum is particularly clear in the difference spectrum, which shows pronounced peaks at 13.8, 13.1 and 11.5 ppm. The behavior of unfractionated yeast tRNA is similar to *E. coli* tRNA, although the peaks in the difference spectra are not as sharp as in *E. coli* (*56*). (This can probably be attributed to the greater variability in the homologous base-pairs.)

The peaks in the difference spectrum share three important characteristics that indicate they should be assigned to tertiary-structure base-pairs: (i) common resonance positions in the low-field spectrum, indicating that they correspond to base-pairs common to most tRNAs; (ii) early

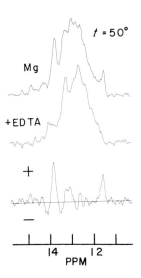

FIG. 9. Nuclear magnetic resonance evidence for common tertiary structure base-pairs in unfractionated *Escherichia coli* tRNA. The spectra with (upper) and without (middle) Mg^{2+} and the difference spectrum (lower) are shown. The principal losses in intensity that occur upon removal of Mg^{2+} are located at 13.8, 13.1 and 11.5 ppm; these are assigned to tertiary-structure base-pairs.

melting, indicating that they are among the weakest base-pairs in the molecule; (iii) intensities corresponding to about 0.9 proton per peak. On the basis of these observations and the integration results discussed earlier (Section IV, A) we conclude that the three main features observed in the *E. coli* and yeast difference spectra ($\pm Mg^{2+}$) correspond to tertiary-structure base-pairs (56). These results, taken in conjunction with the earlier studies of the base-pairing involving s^4U_8 (42, 43) indicate that most tRNAs contain about 3–4 common tertiary-structure base-pairs that contribute resonances to the 11.5 to 15 ppm region.

Some tentative assignments of these resonances to tertiary-structure base-pairs observed (72, 73) in X-ray diffraction studies of yeast tRNA[Phe] (see Fig. 10) can be expected. These assignments, first suggested by Bolton and Kearns (56), have been used in a more recent work by Reid and Robillard (61) to interpret the spectrum of *E. coli* tRNA[Val].

14.8 or 14.3 ppm Resonance. The results discussed above in this section provide strong evidence that the 14.8-ppm resonance in *E. coli* tRNA should be assigned to the $A_{14} \cdot s^4U$ base-pair. Correspondingly, the 14.3-ppm resonance observed in yeast tRNA is assigned to an $A_{14} \cdot U_8$ base-pair.

13.8 ppm Resonance. The peak at 13.8 ppm in the *E. coli* difference

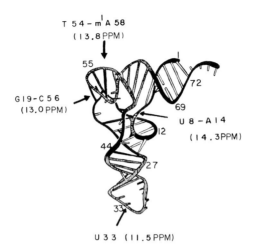

FIG. 10. The three-dimensional structure of yeast tRNA^{Phe}. Taken from data published in ref. 21.

spectrum is tentatively assigned to the invariant A · T base-pair in the T-Ψ-C loop on the following grounds. Of the various base-pairs proposed to occur in all tRNAs, the A · T base-pair is the only one that can reasonably be expected to yield a resonance as low as 13.8 ppm. Furthermore, in unfractionated yeast tRNA, this resonance is split with intensities distributed between 13.8 and 13.6 ppm (56). In yeast tRNA m^{1}A and A appear in about equal frequency and m^{1}A · T is predicted to resonate at higher field than A · T.

13.1 ppm Resonance. The relatively sharp and intense resonance observed in both *E. coli* and yeast tRNAs is assigned to the common invariant G_{19} · C_{56} base-pair. Hilbers and Shulman first observed early-melting resonances in the spectrum of *E. coli* tRNA^{Glu} (at 13.1 and 12.9 ppm) and assigned one of them (12.9 ppm) to a G_{15} · C_{48} (77). [This particular assignment cannot be correct, as they later noted (62), since X-ray diffraction data indicate that this particular pair forms a reverse Watson–Crick base-pair in which the ring-nitrogen proton of G_{15} is hydrogen-bonded to the exocyclic oxygen of C_{48} (20–22).] In the light of our results, we suggest that the early-melting 13.1-ppm resonance of tRNA^{Glu} is the one that should be attributed to a tertiary-structure base-pair.

11.5 ppm Resonance. Initially (56) we assigned the 11.5 ppm resonance to the relatively common G_{22} · C_{13} secondary-structure base-pair, but new data argue against this assignment. G_{22} · C_{13} does not occur frequently enough to account for the intensity of the common resonance at 11.5 ppm. Also, the spectra of many pure tRNAs show a second resonance in the 11.5–11.8 ppm region in addition to the resonance from $G_{22} \cdot C_{13}$ predicted to be in this same region. The possible origin of this resonance is suggested by studies of the anticodon fragments of *E. coli* tRNA^fMet (54) yeast tRNA$_1^{Val}$ and tRNA$_3^{Leu}$ (55). All these fragments exhibit an extra resonance at 11.8–11.5 ppm, which, as we discussed in Section III, can be assigned to a protected U_{33} in the anticodon loop.

Resonances in the 11.5–9.5 ppm Region. The spectra of yeast and *E. coli* mixed tRNAs, in the region between 11.5 and 9.5 ppm, show two additional common resonances at 10.5 and ca. 9.5 ppm at 35°C (65). Yeast tRNA also shows intensity in this region, although it is distributed over a wider region than is that of *E. coli*, reflecting its greater variability. Model system studies suggest that resonances from $G \cdot U$ pairs might be located in this region and Robillard *et al.* (78) have assigned resonances in the 10–11 ppm region to $G \cdot U$ "wobble" base-pairs, in which the ring-nitrogen protons of both G and U contribute one resonance. In support of this assignment, they noted that the number of resonances correlates with the number of $G \cdot U$ pairs in the four tRNAs that they examined (62). However, the correlation is not good for other tRNAs. For example, *E. coli* tRNA^Asp is predicted to have six resonances in this region from 3 $G \cdot U$ pairs, whereas it has at most two or three. Furthermore, the spectrum of yeast tRNA tRNA^Asp reported (78) is anomalous in that there should only be 4 resonances from $G \cdot U_5$ and $G \cdot U_{30}$ whereas almost 12 resonances are observed in the 10–11 ppm region. Four of these could have been attributed to $G \cdot U_{10}$ and $G \cdot U_{13}$ in the hU stem, but this would have been inconsistent with their conclusion that the resonances from the two $A \cdot U$ base-pairs in the hU stem are absent. It is possible that some of the extra resonances at 10–11 ppm are due to aggregation, since yeast tRNA^Asp aggregates under NMR conditions and aggregation is known to give rise to resonances in the 10–11 ppm region (presumably due to the ring-nitrogen protons of G and U residues protected from exchange by aggregation). In addition, we have obtained spectra for yeast tRNA^Asp containing high Mg^{2+} levels that are essentially the same, in the 11–15 ppm region of the spectrum, as those published (78), but with reduced intensity in the 10–11 ppm region with high Mg^{2+} levels. The spectra of *E. coli* tRNA$_{su\pm}^{Trp}$ also raise questions about the $G \cdot U$ assignment. tRNA$_{su-}^{Trp}$ and tRNA$_{su+}^{Trp}$ have respectively $A \cdot U_{11}$ and $G \cdot U_{11}$ in their hU stems, and both species have an additional $G \cdot U_{50}$ base-pair. The spectrum of tRNA$_{su-}^{Trp}$ ($G \cdot U_{11}$) has about four resonances between 10 and 11 ppm as would be predicted

(78), but the spectra of tRNA$^{Trp}_{su+}$ (A · U$_{11}$) and tRNA$^{Trp}_{su-}$ are identical in the 10–11 ppm region (C. Jones, unpublished results). Furthermore, the resonances from *E. coli* tRNA$^{Val}_1$, tRNA^{Arg}, and tRNA^{phe}, which are assigned to G · U pairs (62), all occur at ca. 10.5 ppm; this is surprising, since nearest-neighbor ring-current shifts on the G · U pairs in the three tRNAs should be quite different (probably varying from 0.3 to 1.6 ppm). Thus, while we expect to find resonance from G · U pairs in the region about 10–11 ppm, their assignment to observed resonances is not confirmed. Consequently, at this time we can only make plausible suggestions for the assignment of the two common resonances that occur at ca. 10.0 and 9.5 ppm. One likely candidate is the resonance from the ring-nitrogen proton of G$_{15}$ hydrogen-bonded to the exocyclic carbonyl of C$_{48}$. Other possibilities would be amino protons hydrogen-bonded to ring nitrogens, or Ψ_{55}.

In addition to the tertiary-structure base-pairs discussed above, the X-ray structure of yeast tRNA^{Phe} (see Fig. 10) indicate triples involving m^{7}G$_{46}$ · G$_{22}$ · C$_{13}$ and A$_9$ · A$_{23}$ · U$_{12}$ (72, 73). We have recently examined the effect of depurination and removal of m^{7}G on the spectrum of mixed *E. coli* tRNA and have found no evidence for loss of a resonance that could be assigned to the proposed m^{7}G · G$_{22}$ · C$_{13}$ base-triple (Bolton and Kearns, unpublished results).

The fact that most tRNAs exhibit a set of common set of resonances in the range region between 9.5 to 15 ppm that can be assigned to tertiary-structure base-pairs provides very strong experimental evidence that most class-I tRNAs have a common three-dimensional structure. If the proposed assignments are correct, most would correspond to the tertiary-structure base-pairs proposed from X-ray diffraction studies of yeast tRNA^{Phe} (72, 73). Kim *et al.* (79)[2] and Klug *et al.* (80) have argued, on the basis of base homologies, that the structure they deduced for yeast tRNA^{Phe} applies quite generally and NMR measurements provide the experimental evidence that this is in fact the case. We have not located resonances from all the tertiary interactions seen in the crystal, but we suspect that these are located in other regions of the spectra, where they would be difficult to detect.

C. Assignment and Interpretation of the tRNA Spectra

1. APPLICATION OF THE RING-CURRENT-SHIFT THEORY

Once the total number of base-pairs per molecule has been determined and resonances from tertiary-structure base-pairs are located, the remaining resonances can be assigned to secondary-structure base-pairs. In our analysis of the tRNA spectra, intrinsic positions are assigned to

A · U (A · Ψ) and G · C base-pairs, and the cloverleaf model is then used in conjunction with a set of semiempirical nearest-neighbor ring-current-shift parameters (Table IV) to compute the observed spectra. As a first approximation, all bases adjacent to the terminus of a helix are assumed to be completely stacked on the terminal base-pair, and the T-Ψ-C stem is assumed to be stacked on the amino-acid-acceptor stem.

As an example of the application of this method, the computer and observed spectra of yeast tRNA^{Phe} are compared in Fig. 2 and in Table VII. The fit of the experimental spectrum is excellent (standard deviation = 0.09 ppm), but several points require further comment. First, it should be noted that a resonance from a tertiary-structure base-pair has been included at 13.75 ppm to account for the observed intensity in peak B. Second, the assignments given in Table VII are identical with those given originally (37–39) with the following exceptions. In an earlier calculation (37), where the intrinsic position $(A \cdot U)° = 14.8$

TABLE VII
The Assignments of the Low-Field Resonances in Yeast tRNA^{Phe}

Resonance (Fig. 2)	Intensity	Obs. (ppm)	Calc. (ppm)	Assignment	Error (Calc-Obs)
A	1.5	14.4	14.3	A · U$_6$,T^a	−0.1
B	3.0	13.8	13.8	A · U$_{52}$	0
		13.75	13.8	T	0
		13.7	13.7	A · U$_{12}$	0
C	3.8	13.4	13.3	A · U$_{50}$	−0.1
		13.3	13.3	A · U$_{29}$	0
		13.3	13.3	G · C$_{11}$	0
		13.2	13.2	A · U$_7$	0
E	~2	12.9	13.0	A · Ψ$_{31}$	0.1
		12.8	12.7	G · C$_{10}$	−0.1
	~1	12.75	12.7	G · C$_2$	0
		12.6	12.5	G · C$_{51}$	−0.1
		12.55	12.5	G · C$_{53}$	0
	10.5 ~5	12.5	12.4	G · C$_1$	−0.1
		12.5	12.4	G · C$_{30}$	−0.1
		12.5	12.4	G · C$_{49}$	−0.1
		12.45	12.2	G · C$_3$	−0.2
	~2	12.3	12.3	G · C$_{27}$	0
		12.15	11.9	G · C$_{28}$	−0.2
	~1.1	11.8	11.5	G · C$_{13}$	0
		11.6		U$_{33}$	

^{a} T, tertiary.

ppm was used, the resonance from $A \cdot U_7$ was predicted to lie at 13.5 ppm, which is in the minimum between peaks B and C in the spectrum of Fig. 2. We therefore had the option of assigning $A \cdot U_7$ to a resonance in either B or C, and, in order to account for intensity observed in the 13.7 ppm region, we assigned $A \cdot U_7$ to a resonance in peak B. With our present set of parameters, the resonance from $A \cdot U_7$ is predicted to be at 13.2 ppm (assuming that the acceptor stem and the T-ψ-C stem are stacked), so it would be natural to assign $A \cdot U_7$ to one of the resonances in peak C. [Although several studies assigned the resonance from $A \cdot U_7$ to peak B rather than to peak C in the tRNA^{Phe} spectrum, a reexamination of the consequences of our reassignment shows that none of the conclusions drawn earlier needs to be altered.]

Since integration of the spectrum shown in Fig. 2 indicates that three resonances are present in peak B (65), this assignment leaves one resonance at 13.8 unaccounted for. Integration of the complete low-field spectrum of tRNA indicates a total of about 21 to 22 resonances, so that a couple of tertiary-structure base-pair resonances must be present, and 13.8 ppm is where the resonance from a common one occurs. This assignment is also supported by early-melting resonances at 13.7 and 13.2 ppm (81). The intensity at 14.4 ppm corresponds to about 1.5 protons per molecule. One of these resonances can be attributed to $A \cdot U_6$, and the other partial resonance is attributed to a tertiary-structure base-pair between A_{14} and U_8 for reasons already discussed.

The assignments shown in Table VII also reverse the ordering of the resonance from $G \cdot C_{11}$ and $A \cdot \Psi_{31}$, since the resonance from $G \cdot C_{11}$ is predicted to be about 0.1 ppm lower field and the resonance from $A \cdot \Psi_{31}$ is predicted to be about 0.3 ppm higher field, primarily as a result of the assumed change in the intrinsic position for $A \cdot U$ base-pairs. No other changes in assignment are required since the majority of the resonances are predicted to have virtually the same positions, using the present parameterization or one described earlier (36).

The results with tRNA^{Phe} seem to support the semiempirical theory used to compute the spectrum. To further test the theory, we have used the same set of assumptions and computed the spectra of other tRNAs and compared these with the observed spectra. The results of these comparisons are summarized in Table VIII in terms of the standard deviation between the computed and observed spectra. In most cases, the standard deviation is about 0.11 to 0.14 ppm, but it must be realized that this method of assignment simply minimizes the disagreements between the computed and observed errors. The true error could be somewhat larger. For example, in the case of *E. coli* tRNA^{Trp}, the assignment of the two lowest-field resonances given simply by maximizing the fit between the

TABLE VIII

A Summary of the Standard Deviation between the Observed and Computed Low-Field NMR Spectra of Various tRNAs

tRNA species	Standard deviation between computed and observed resonances (ppm)
Yeast: tRNA^{Phe}	0.090
tRNA$^{Leu}_{UUG}$	0.11
tRNA$^{Leu}_{CUA}$	0.14
tRNA$^{Val}_1$	0.14
E. coli: tRNA^{Glu}	0.12[a]
tRNA$^{Val}_1$	0.12
tRNA^{Phe}	0.12
tRNA^{Trp}	0.085[a] (0.17)[b]
tRNA^{Tyr}	0.135

[a] Omits one anomalous resonance.
[b] Value obtained if the assignments of the low-field resonances are reversed.

computed and observed results is reversed on the basis of other experimental data (C. Jones, unpublished results). In view of these and other possible anomalies that may be present, it would clearly be desirable to have additional assignment criteria.

One approach is to compare the spectra of two different tRNAs that have a number of homologous base-pairs, but some differences in their sequences. Two tRNA^{Leu} species from yeast have been used for this purpose. Based on their sequences, ten of the base-pairs common to both species (termed "equivalent" base-pairs) have identical nearest neighbors and therefore should give rise to common resonances in the two different spectra. However, the resonances from ca. ten of the base-pairs in each molecule are predicted to be different. When the two spectra were compared (Fig. 11), we find that, with one exception, resonances from all the equivalent base-pairs match up quite well in the two spectra (82). The one resonance assigned to an equivalent base-pair that appears to be in different positions in the two spectra is assigned to the terminal base-pair of the T-Ψ-C stem, and this is predicted to be very sensitive to the alignment of the T-Ψ-C and amino-acid-acceptor stem.

Another way to test the accuracy of the ring current calculations is to compute the spectrum of a tRNA using an incorrect sequence. We have done this for a number of possible combinations, and the results are summarized in Table IX. Depending upon the theoretical and ex-

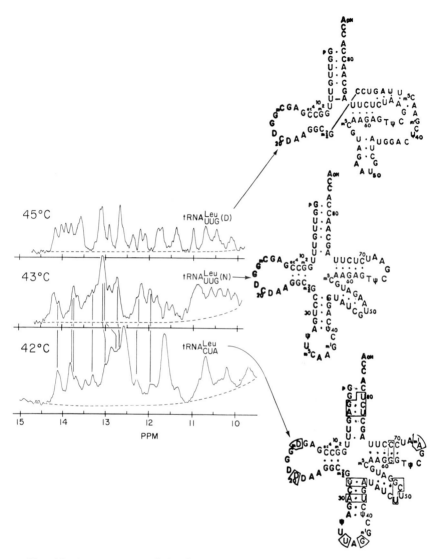

FIG. 11. A comparison of the low-field nuclear magnetic resonance spectra of the native and denatured conformers of yeast tRNA$^{Leu}_{UUG}$ and yeast tRNA$^{Leu}_{CUA}$ showing the correspondence of equivalent base-pairs in the two different species. A model for the secondary structure of the denatured conformer of yeast tRNA$^{Leu}_{CUA}$ is also shown (top).

TABLE IX
A Comparison of the Low-Field NMR Spectra Observed for One tRNA with the Theoretical Spectrum Calculated for a Second tRNA[a]

tRNA species		Standard deviation between obs. and calc. spectrum (ppm)
Obs. spectrum	Calc. spectrum	
(y) Asp	(y) Val	0.22
(y) Phe	(E · c) Trp	0.12
(E · c) Glu	(E · c) Trp	0.17
(E · c) Trp	(y) Leu$_{CUA}$	0.185
(y) Leu$_{CUA}$	(E · c) Trp	0.12
(y) Leu$_{CUA}$	(y) Leu$_{UUG}$	0.17
(y) Phe	(E · c) Glu	>0.18

[a] The results are shown in terms of the standard deviation between the computed and observed spectrum for each pair. y = yeast; E · c = *Escherichia coli*.

perimental spectra compared, we find the standard deviations run from 0.12 to over 0.20 ppm. It is interesting to note that, for tRNA pairs where the standard deviation is small, the spectrum calculated using the sequence of one tRNA with the spectrum of the other tRNA agree quite well (e.g., compare *E. coli* tRNA^{Trp} and yeast tRNA^{Phe}). However when the spectra are computed to be rather different (standard deviations of the order of 0.18 or larger), similarly large disagreements are obtained when the spectrum computed using one sequence is compared with the spectrum measured for the other tRNA. From these comparisons, we would expect the standard deviation between the properly computed and observed spectrum of any given species to be not larger than about 0.12 ppm. This probably represents the present limitation of the theory, and we do not expect to be able to improve it significantly until more is known about some of the other factors contributing to shifts of the low-field resonances in addition to the ring-current effects.

2. Anomalies

Before discussing some of the implications of the results with regard to the structure of tRNA molecules, we need to point out some of the anomalies that still exist and indicate some aspects of the low-field spectra that remain to be understood.

Many of the resonances in the tRNA are predicted to occur between 13.5 and 12.5 ppm, where they overlap one another, so it is difficult to test the assignment of the resonances in this region. However, at the high- and low-field extremes, we have a better chance of testing the theory. In most cases, there is reassuringly good agreement between the predicted and observed positions in these regions; when disagreements are observed, they often can be attributed to end effects. However, there are at least three tRNAs that definitely have anomalies in their low-field region of the spectrum. These include the tRNAs for Glu, Arg and Gly, each of which exhibits resonances at lower fields than predicted by ring-current-shift theory. For example, *E. coli* tRNA^{Glu} has resonances at 14.7 and 14.3 ppm whereas the only resonances in this region (from $A \cdot U_{11}$ and $A \cdot U_2$) are both predicted to be at 14.2 ppm. Similar anomalies are observed with tRNA^{Arg} and tRNA^{Gly}, and it is interesting to note that in each case there is an $A \cdot U$ base pair in position 11. This suggests that some sort of structural perturbation (e.g., a base-triple, or ion binding) may be responsible for this anomaly.

3. SUMMARY OF STRUCTURAL FEATURES IDENTIFIED BY NMR IN RELATION TO THE CRYSTAL STRUCTURE OF YEAST tRNA^{Phe}

a. The Secondary Structure Corresponds to the Cloverleaf Model. Evidence for the cloverleaf secondary structure has been so strong that it would have been very surprising if the NMR results had indicated otherwise (5, 9, 12). Fortunately, this was not the case, and in fact the NMR results provide the strongest confirmation of the details of the cloverleaf model. Within the accuracy of the measurements (±1 base-pair), the NMR results show that almost all cloverleaf secondary-structure base-pairs are present in class I tRNAs. The possible exceptions may be those cases where an $A \cdot U$ base pair is located next to a $G \cdot U$ pair (37). Even if the ring-current-shift theory is less accurate than we have suggested it to be, there is no doubt that the total number of secondary-structure base-pairs is very nearly exactly that required by the cloverleaf model, and, moreover, the relative number of $A \cdot U$ and $G \cdot C$ base-pairs is the same as that predicted by the cloverleaf model. (This number can be obtained by taking 13.2 ppm as the approximate dividing line between $A \cdot U$ and $G \cdot C$ base-pairs.)

b. Alignment of the Amino-Acid-Acceptor Stem and the T-Ψ-C Stem. From the analyses, it is evident that the positions of resonances associated with terminal base-pairs are sensitive to the stacking of adjacent bases in single- (or double-) stranded regions of the molecule. In particular, resonances from two interior terminal base-pairs of the amino-acid stem and the T-Ψ-C stem are sensitive to the alignment of these two stems. In most tRNAs, the two resonances of interest overlap with other resonances, so it has not been possible to check this feature carefully in a number of tRNAs. However, the resonances of interest in *E. coli* tRNA^{Phe} and tRNA^{Glu} identified, and they clearly indicate that the T-Ψ-C stem is stacked on the acceptor stem (83). Furthermore, it should be noted that a good account of the spectra of other tRNAs can be given if we assume that these two helices are stacked. In the earlier analyses of some individual tRNA spectra, it was sometimes assumed that there was incomplete stacking of bases on terminal base-pairs. However, using the ring-current-shift parameters given in Table IV, we can give a consistent interpretation of the spectra of these same tRNAs if we assume maximum stacking of all bases.

c. *Tertiary-Structure Base-Pairs.* In the presence of excess Mg^{2+}, the integrations of the low-field spectra between 11 and 15 ppm indicate that most tRNAs contain between two and four additional resonances, which can be attributed to tertiary-structure base-pairs. Most *E. coli* tRNAs, which have an s^4U residue at position 8, also have a characteristic resonance at 14.8 ppm that can be attributed to the $A_{14} \cdot s^4U_8$ base-pair. In yeast tRNA the corresponding $A_{14} \cdot U_8$ base-pair gives rise to a resonance at 14.3 ppm.

In addition to the $A \cdot U_8$ base-pair, there is evidence for resonances from three additional common tertiary-structure base-pairs, these being located at 13.8, 13.0 and 11.5 ppm. Some tentative assignments of these were made in terms of the pairs that Levitt (14) predicted would be common to most class I tRNAs, and that now have been found in crystalline yeast tRNA^{Phe} (72, 73). In the region between 11 and 9.5 ppm, there are resonances at 10.5 and 9.5 ppm that are also common to most tRNAs and that apparently arise from some common tertiary interactions. Hence, the NMR measurements provide strong experimental evidence that most class I tRNAs have the same set of tertiary-structure base-pairs, and therefore have the same three-dimensional structure in solution as in the crystalline state. In subsequent sections, we discuss other NMR measurements that support the conclusion that virtually all tRNAs have the same three-dimensional structure.

V. Application of NMR to the Investigation of Problems of tRNA Structure

A. Denatured Conformers of tRNA

A number of tRNAs can exist in a metastable, denatured conformer that cannot be aminoacylated (84–86). These "denatured" tRNAs are of special interest in connection with studies of the recognition of tRNA by the cognate synthetases and of the conformational mobility of the tRNA, and it would be especially desirable to identify the structural changes responsible for loss of activity. Progress toward this goal has been hampered because so little was known about the secondary structures of the denatured conformers and virtually nothing was known about their tertiary structures.

High-resolution NMR has now been used to investigate the denatured conformers of two tRNAs, yeast tRNA$_3^{Leu}$ (87–89) and *E. coli* tRNA^{Trp} (90). The spectra of the native and denatured conformers of the latter are shown in Figs. 11 and 12. In the native conformer of yeast tRNA$_3^{Leu}$ there are 21 ± 2 base-pairs (44°C, low Mg^{2+}), and analysis of the spectra is consistent with the cloverleaf model. The denatured conformer of tRNA$_3^{Leu}$ contains only three fewer base-pairs than the native conformer, but its spectrum is quite different (87). To find a suitable model for it, all possible base-pairing schemes (generated by construction of a base-pairing matrix) consistent with the number of base-pairs found by NMR and with the oligonucleotide binding data of Uhlenbeck *et al.* (91) were examined to find ones with predicted NMR spectra consistent with the observed spectrum. This process led to a model for the denatured conformer of yeast tRNA^{Leu} (Fig. 11) in which the hU stem and the anticodon stem are opened up and bases from the original anticodon stem and loop (bases 31–35) are paired with bases of the T-Ψ-C loops (bases 66–70) (89). In addition to accounting for the room-temperature spectrum, the model also provides a logical account of the sequential melting observed by NMR (89).

Two subsequent studies provide additional support for the proposed model. The pattern of kethoxylation (only "exposed" G residues react) of the denatured conformer agrees with our model (92). A closely (sequence) related tRNA^{Leu} has recently been found that cannot be converted to a denatured conformer, and the changes in sequence are such that our model would predict that it is incapable of forming the denatured tRNA$_3^{Leu}$ conformer (93).

The low-field spectrum of *E. coli* tRNA$_{su-}^{Trp}$ indicates that there are 22 ± 1 base-pairs present in the native form, and the observed resonances can be accounted for in terms of 18 cloverleaf-structure base-pairs, one protected G or U, and two tertiary-structure base-pairs, including $A_{14} \cdot s^4U_8$. The denatured conformer contains approximately two fewer base-pairs than the native conformer, and its spectrum can be accounted for in terms of a model in which the acceptor stem, the T-Ψ-C stem, and the anticodon stem of the cloverleaf model are retained, and some bases of the hU stem are paired with bases from the minor loop (see Fig. 12) (90). This model is consistent with the ease of interconversion of native and denatured conformers, the pattern of kethoxalation of the native and denatured conformers (94) and the retention of the tertiary structure $A_{14} \cdot s^4U_8$ base-pair in the denatured conformer (90).

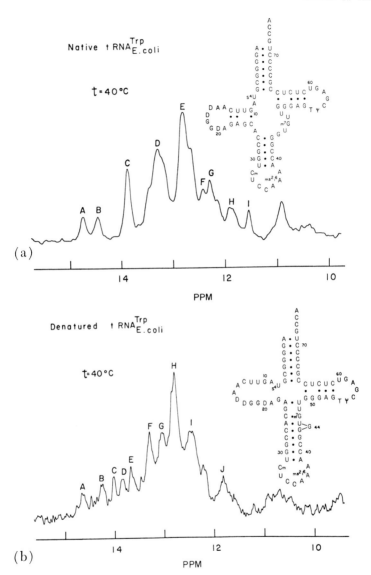

FIG. 12. A comparison of the low-field nuclear magnetic resonance spectra of the native (a) and the denatured (b) conformers of *Escherichia coli* tRNA^Trp. The models for the secondary structure of the native and denatured conformers are also shown. Both samples were dialyzed against 0.18 M NaCl and 10 mM cacodylate at pH 7. In addition, the native sample contained 10 mM Mg^{2+} (free).

Alternative models for the denatured conformer were considered, but none adequately accounted for the observed spectrum.

It thus appears that, when applied in conjunction with other measurements, NMR can be used to work out the base-pairing structure of small RNA molecules even though they do not contain all the simplifying structural features present in the cloverleaf model. If limited digestion were used to reduce the molecular weight of large RNAs, NMR may also be useful determining the structures of RNAs of much higher molecular weight.

B. Dimerization of *E. coli* tRNA^{Tyr}

Studies of the intermolecular association of RNA molecules are of interest in a number of different contexts, including tRNA binding to ribosomes (95), intermolecular association of different and identical tRNA in solution (96, 97), the binding of small oligonucleotides to tRNA (98, 99) and the formation of stable or metastable denatured tRNA conformers. The dimerization of *E. coli* tRNA^{Tyr} has been extensively studied by a number of different methods (thermodynamic, kinetic and spectroscopic), and a model for the secondary structure of the dimer has been proposed (100). NMR has now been applied to this problem (101) and some of these results are shown in Fig. 13. Integration of the monomer spectrum (11.5–15 ppm) indicates 26 ± 2 base-pairs (65) of which the cloverleaf secondary structure accounts for 23. The ca. three extra resonances (located at 13.80, 13.75 and 13.0 ppm, respectively) are assigned to tertiary-structure base-pairs, and additional resonances observed in the spectrum between 11 to 11.5 ppm are assigned to ring NH protons from protected G or U residues. Surprisingly enough, a resonance from $A_{14} \cdot s^4U_8$, commonly observed at 14.8 ppm in spectra of other class I *E. coli* tRNAs, is *absent* in tRNA^{Tyr}. This tRNA differs from class I tRNA in that the hU stem contains only three base-pairs, a G_{13} residue is interposed between A_{14} and the terminal base-pair of the hU stem, and the hU loop contains 11 nucleotides. It is possible, therefore, that class III tRNAs such as *E. coli* tRNA^{Tyr} have a tertiary structure quite different from that of class I tRNA.

Comparisons of monomer and dimer low-field spectra, as well as internal comparison of the low-field spectra with the intensity in the aromatic region indicate a *loss* of six base-pairs per molecule upon dimer formation (101). Despite this reduction in base-pairs, the temperature dependence of the dimer low-field NMR spectrum (19 resonances at 55°C) points to a surprisingly stable dimer structure. This suggests that dimer formation may involve melting of weaker helices (hU stem and minor stem) and some tertiary-structure base-pairs, and the formation

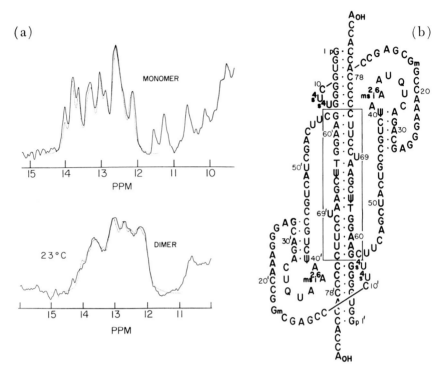

FIG. 13. (a) A comparison of the low-field nuclear magnetic resonance spectra of the monomer (upper) and dimers (lower) of *E. coli* tRNA^{Tyr} (———) with computer-stimulated spectra (··········), which contain 25 resonances for the monomer and 19 resonances for the dimer. (b) A model for the secondary structure of the dimer.

of six new intermolecular base-pairs per tRNA dimer. On the basis of the NMR measurements, optical data, and a consideration of the base-pairing matrix, a structure for the dimer was proposed in which the T-Ψ-C stem and the T-Ψ-C loop regions are involved in intermolecular base-pairing (Fig. 13). Alternative models proposed are inconsistent with the NMR observations. The dimer model is consistent with the thermodynamic, kinetic, optical and NMR data, is topologically feasible, and accounts for formation of higher aggregates containing an even number of monomer units (*101*).

C. Investigation of the Thermal Unfolding of tRNA

In the preceding section, the NMR spectra of the native tRNA obtained at one specific temperature were analyzed in terms of secondary- and tertiary-structure base-pairs of the cloverleaf model. The temperature dependence of these spectra offers an independent method for testing assignments and provides information about the conformational changes during the thermal unfolding of tRNA molecules.

The thermal denaturation of a tRNA molecules is usually manifested in low-field spectra by a broadening of resonances, which occurs when the lifetime of a proton in a base-pair is reduced to 2 msec (corresponding to a line width of ~200 Hz). [Other experimental situations that might be realized experimentally with regard to helix dissociation rates and proton exchange with solvent have been discussed in detail by Crothers et al. (58, 59).] Optical measurements and temperature-jump studies indicate that the thermal unfolding of the tRNA molecules can occur in a sequential fashion (59, 102, 103, 104). Since resonances in the NMR spectrum can be assigned to individual base-pairs in the molecule, this would then permit NMR to be used to identify various structural features as they melt out.

The temperature dependence of the denatured conformer of tRNA^{Leu} spectrum is one of the better examples of sequential melting of a tRNA (Fig. 14) (89). Between 25° and 45°C there is a slight loss of intensity at 13.5 ppm, but by 52°C substantial losses in intensity are observed. Comparison of the 35° and 55°C spectra indicates a loss of approximately 9 resonances by 52°C, which can be accounted for in terms of the loss of the minor stem and the noncloverleaf helix unique to the denatured conformer. Between 52° and 62°C there is a further reduction in intensity corresponding to a loss of about four more resonances, whose positions agree well with resonances assigned to the T-Ψ-C stem (which contains only four base-pairs in the model). The resonances that remain at 62°C are well accounted for in terms of base-pairs in the amino-acid-acceptor stem (89).

In contrast with yeast tRNA^{Leu}, the melting behavior of tRNA^{Phe} is more cooperative as the results shown in Fig. 15 indicate. A detailed discussion of the NMR melting behavior of tRNA^{Phe} was given by Hilbers et al. (81), who reported results quite similar to those shown here, and earlier by Wong et al. (105). At 49°C there is some selective loss of intensity at 13.2 ppm (assigned to loss of A · Ψ_{32}) and 12.5 (tentatively assigned to the resonance from G · C_{31} which is adjacent to A · Ψ_{32}) and some selective broadening of the resonance at 14.4 which is assigned to A · U_6. At higher temperatures there is a general loss of intensity throughout the spectrum, although the 54°C spectrum gives evidence that the final stages of the melting may become sequential.

The melting behavior of E. coli tRNA^{Glu} (77) and tRNA^{fMet} (59) and yeast tRNA^{Asp} (78) have been all analyzed in terms of sequential melting of the arms of the cloverleaf. For tRNA^{Glu}, the order of stability

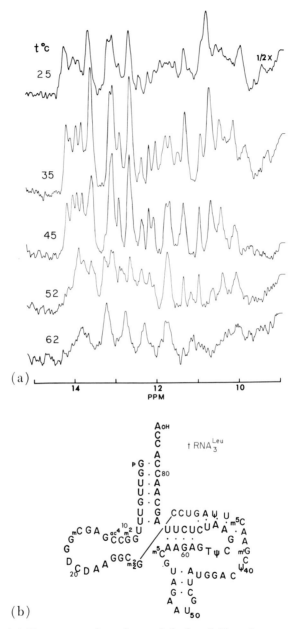

Fig. 14. (a) Temperature dependence of the low-field nuclear magnetic resonance spectrum of the denatured conformer of yeast tRNA$_3^{Leu}_{UUG}$ illustrating the sequential disruption of the secondary structure as a function of temperature. (b) Model for the secondary structure of the denatured conformer.

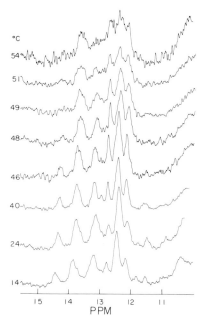

Fig. 15. Temperature dependence of the low-field nuclear magnetic resonance spectrum of yeast tRNA^{Phe} in a solution containing 0.1 M NaCl and 10mM cacodylate at pH 6.

of the cloverleaf stems is hU < T-Ψ-C < acceptor < anticodon (77). The interpretation of the melting behavior of tRNA^{fMet} given by Crothers et al. (59) was cast in doubt when the assignments of some resonances were revised [Wong et al. (74)], but despite these reassignments the sequence of events in the melting can still be interpreted in a manner consistent with that given originally (59). Robillard et al. (78) recently studied the melting of yeast tRNA^{Asp}, which has an unusual hU stem containing only two good Watson–Crick base-pairs (both A · U pairs) terminated at both ends by G · U or G · Ψ base-pairs. They found no evidence for resonances from any of these base-pairs and infer the following order of melting: resonances attributed to three tertiary-structure base-pairs are lost first, followed by loss of resonances from the acceptor stem and the antidocon stem; finally, at the higher temperatures, only resonances from the T-Ψ-C stem remain.

While it is possible to give a plausible account of the temperature dependence of the NMR spectrum of tRNA^{Leu}, tRNA^{fMet}, and tRNA^{Glu} in terms of sequential loss of tertiary structure base-pairs and stems of the cloverleaf structure, it should be noted that certain alternative interpretations cannot be eliminated. In particular, the possibility that resonances from terminal base-pairs of a stable helix may melt along with resonances from interior base-pairs of a second, weaker helix cannot be ruled out. In fact, studies of tRNA fragments (Section III) demonstrate that melting of a single helix can extend over a range of temperatures with resonances from terminal base-pairs melting 10–20°C before interior base-pairs melt.

Crothers and Shulman attempted to overcome this sort of uncertainty by combining optical temperature-jump data for tRNA^{fMet} with the NMR data (59). In the temperature-jump experiments, it was possible to identify at least five kinetically distinguishable processes and to measure the temperature dependence of the rate constant for each process. These data were extrapolated for each stem to determine the temperature at which the corresponding process would be observed in the NMR measurements as a broadening of resonances (i.e., rate constant ca. 500 sec^{-1} for disruption of base-pairs). In this way, they were able to associate various kinetic processes seen optically with the melting of various arms of the cloverleaf.

While there may be some ambiguity in interpreting the temperature dependence of low-field NMR spectra, the results of such studies in combination with measurements of the aromatic and methyl resonances may make it possible to use NMR to provide information about conformational changes in tRNAs that cannot be obtained by other solution-state techniques.

D. Effect of Anticodon-Loop Modifications on tRNA Conformation (105)

It has been speculated that the alteration of the conformation of a loop can affect the tRNA conformation in regions distant from that loop (106). The removal of the so-called Y-base[4] from the anticodon loop of yeast tRNA^{Phe} is one modification that is easily effected (mild acid hydrolysis) and that has interesting effects on the biochemical properties of the molecule. The *rate* of charging is slightly slower for tRNA^{Phe} minus its wye than for the unaltered tRNA^{Phe}, but the *extent* of charging is the same for both molecules (107). Ribosomal binding is completely abolished by removal of wye, and, rather surprisingly, yeast tRNA^{Phe} minus wye cannot be charged by *E. coli* synthetase although yeast tRNA^{Phe} can be charged by this heterologous enzyme. These observations led Thiebe and Zachau (107) to suggest that removal of wye might induce con-

[4] The name wye, with symbols Wyo and W for the nucleoside (wyosine) have been proposed ("Handbook of Biochemistry," (G. Fasman, ed.), 3rd ed., Vol. I, Chemical Rubber Co., Cleveland, Ohio, 1975) [Ed.].

formation changes both in the anticodon loop region, and elsewhere in the molecule. While there was other evidence for a change in the conformation of the anticodon loop (*108, 109*), there was no agreement as to whether removal of the base causes other changes in the conformation of tRNA^{Phe} (*110*).

The NMR of tRNA^{Phe} has been examined before and after removal of wye, and a surprisingly large number of changes in the low-field NMR spectrum appeared (*105*). Analysis of those changes indicates that all but one of the resonances affected by the alteration can be attributed to base-pairs in the anticodon stem (*105*). The other resonance affected, at 11.5 ppm, can now be attributed to a perturbation of U_{33}. It was concluded that a change in the anticodon loop affects the helical conformation of the entire anticodon stem (probably a change in pitch and tilt of the bases). An important change in the tRNA structure is also indicated by optical melting experiments where tRNA^{Phe} minus wye exhibited a substantial early melting not shown by tRNA^{Phe} (*105*). [More recent chemical modification studies also support the NMR results: while 7-methylguanine at position 46 is not easily attacked in native tRNA, in tRNA^{Phe} minus wye it is susceptible to irreversible degradation by sodium borohydride under mild conditions (*111*).]

Since removal of wye presumably alters the loop conformation by changing the stacking arrangements, it follows that incorporation of some suitable group in place of wye should largely reverse the effects. As expected, when proflavine is covalently introduced into the anticodon loop in place of wye, the changes in the NMR spectrum are largely reversed (*112*) and most of the original biochemical functions are restored (*113*).

Perhaps the most important conclusion to be drawn from these studies is that the conformation of a loop can have significant effects on the conformation of an entire helical stem and other more distant parts of the molecule. In the particular system studied, a change in the conformation of the anticodon loop propagates a structural change through about five base-pairs in the anticodon stem to the interior core of the molecule. This provides a mechanism whereby interactions with loops of tRNA molecules may induce conformational changes in double-helical stems that are propagated into other parts of the molecule. This has interesting biological implications that should be explored.

E. Effect of Aminoacylation on tRNA Conformation

There have been various proposals that aminoacylation induces changes in the secondary and/or tertiary structure of tRNA, but the experimental evidence for such changes has been conflicting (*114–117*).

NMR seems to be an ideal tool for testing this proposal, since it can be used to monitor secondary and tertiary structural features in tRNA molecules. Preliminary studies have already been carried out with yeast tRNA^{Phe}, but no changes in the low-field spectrum were observed (118). From this it was inferred that aminoacylation does not alter the secondary or tertiary structure of this particular tRNA. This work should be repeated using other tRNAs, since the original studies were carried out on samples containing relatively low levels of Mg^{2+} (six per tRNA) and solutions were prepared by adding tRNA to the buffer solutions rather than by dialysis, which gives more well-resolved spectra. Furthermore, other tRNAs (notably *E. coli* tRNA) have spectra that lend themselves better to monitoring changes in tertiary structure.

F. Photocrosslinking of s^{4}U$_8$ and C$_{13}$ in *E. coli* tRNA

Several years ago, Favre *et al.* (66) discovered that the s^{4}U$_8$ and C$_{13}$ residues in many *E. coli* tRNAs can be photochemically crosslinked; this reaction provided the first solid experimental evidence that these two residues, located some distance apart in the primary sequence, are actually in close proximity in the native molecule. The biological properties of native and photocrosslinked molecules have been compared and, in general, the photocrosslinking does not substantially affect activity (66–69). Daniel and Cohn (60) studied the effect of photocrosslinking on a portion (15.3 to 13.3 ppm) of the lowfield spectrum of *E. coli* tRNA^{fMet} and found that the resonance we assigned to the A$_{14}$ · s^{4}U$_8$ base-pair is lost. A second resonance at 14.6 ppm, which they tentatively assigned to the G$_9$ · G$_{12}$ bond of the G$_9$ · G$_{12}$ · C$_{24}$ triple, is shifted slightly upfield. We have now carried out studies of the effect of photocrosslinking on the low-field NMR spectra of several other pure *E. coli* tRNAs and a mixture of tRNAs. Mixed tRNAs were studied to determine whether or not resonances from the common tertiary-structure base-pairs were affected. The results are shown in Fig. 16, where it can be seen that photocrosslinking abolishes the 14.9 ppm resonance from the sU$_8$ · A$_{14}$ base-pair and affects a common resonance at 13.0 ppm, tentatively assigned to a common G$_{19}$-C$_{56}$ tertiary structure base-pair. None of the other resonances from common tertiary-structure base-pairs (e.g., peaks at 13.8, 11.5 or 10.5 ppm) seem to be affected. The implication that there is little effect of photocrosslinking on the tertiary structure of the tRNA correlates well with studies showing that the photocrosslinked molecules retain their biochemical activity (67–69). Studies of the pure *E. coli* tRNA^{Phe} and tRNA$_1^{\text{Val}}$ have disclosed additional spectral changes unique to each tRNA and tentatively interpreted in terms of changes in the stacking of the amino-acid-acceptor stem on the T-Ψ-C stem (P. Bolton,

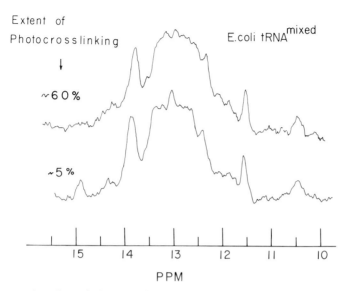

FIG. 16. The effect of photocrosslinking s^4U_8 and C_{13} on the low-field spectrum of unfractionated *Escherichia coli* tRNA^{mixed}. These results show that there are no changes in tertiary structure base-pairs giving rise to resonances at 13.8, 13.1, 11.5 and 10.5 ppm.

unpublished results). Studies of other tRNAs should serve to better identify the common structural alterations produced by the photocrosslinking reaction.

G. Interaction of tRNA with Drugs

Many potent antibiotics presumably function by binding to proteins (*119*), but a large class of drugs and dyes interact more strongly with DNAs and RNAs, including tRNAs (*120*). Interest in the binding of drugs to RNA in general, and to tRNA specifically, arises from several considerations. First, these drugs may selectively interfere with certain biochemical processes involving tRNA (e.g., aminoacylation, ribosome binding). If more were known about the nature of binding, the location of binding sites, and selectivity, they would be more useful technically, and perhaps drugs with better properties could be designed. Second, many dyes can be used as fluorescent probes of tRNA structure and in studies of the interaction of tRNA with other RNA and proteins.

Of the many intercalating drugs and dyes that have been examined, a large number involve ethidium bromide (Etd Br). This drug strongly interacts with polynucleotides (*121, 122*), and the chemical (*123*) and biological consequences of this interaction have been the subject of numerous investigations (124, 125). Its antiviral and antibacterial properties have been observed (126) and it inhibits nucleic acid synthesis (127), as well as DNA and RNA polymerase activity (128, 129). The observations of *in vivo* biological activity have raised interesting questions about the way in which it affects the properties of DNA and RNA. EtdBr is known to bind preferentially to double-stranded DNA and RNA (130, 131) by intercalating between adjacent base pairs in the double-helix (129, 132), but little is known about local site requirements.

In previous studies of the binding of EtdBr to tRNA^{Phe} where the concentration of Na$^+$ was low (typically less than 30 mM) and no Mg^{2+} was present (125, 133–135), six strong binding sites were observed (134). However, when the metal ion concentration is increased the number of binding sites (134) and the magnitude of the binding constants decreases (133). With the high concentration of metal ions used in our studies, only a single binding site is observed for tRNA^{Phe} (134). The decrease in binding with increasing metal ion concentration is taken as an indication that EtdBr binding destabilizes the native tertiary structure of tRNA, which is favored by increased concentrations of metal ion.

Although the binding of EtdBr and other drugs to tRNA has been studied by a number of different techniques, none of these have permitted the binding site(s) to be identified. This can now be accomplished by NMR. Measurements on the binding of Etd Br to a number of different tRNAs have been carried out (Jones and Kearns, unpublished results); the results for yeast tRNA^{Phe} (136) are shown in Fig. 17. Comparing spectra taken before and after addition the drug, it can be seen that there is a pronounced effect on resonances at 14.3 and 13.5 ppm and smaller changes at 13.9 and 13.3. The difference spectrum at the top of Fig. 17 shows that the intensity lost at 14.3 ppm is gained in the region from 13.3 to 13.7 ppm. It was expected that intercalation of aromatic molecules such as EtdBr into tRNA would give rise to large current shifts (22, 27). The observation of upfield shifts in the NMR spectrum is consistent with and further supports the conclusion that EtdBr binds to the tRNA by intercalation (129–132).

Referring to the assignments of the low-field NMR spectrum of tRNA^{Phe}, (Table VII) we see that all resonances in the 14.5 to 13.4 ppm region are due to A · U base-pairs. The most strongly perturbed peak at 14.4 ppm is assigned to A · U$_6$ with some contribution from an A$_{14}$ · U$_8$ base-pair. Since other evidence (discussed below) indicates that tertiary-structure base-pairs are not affected by EtdBr binding, we conclude that the site of binding is at the base of the acceptor stem adjacent to the A · U$_6$ base-pair.

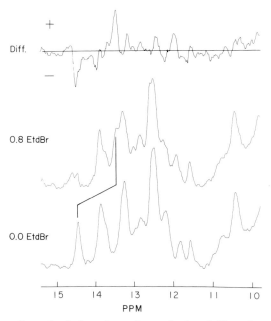

FIG. 17. The effect of ethidium bromide on the low-field nuclear magnetic resonance spectrum of yeast tRNA^{Phe} at 24°C. The top spectrum presents the difference between the two lower spectra. Both samples contained 0.38 M NaCl, 8 mM Mg^{2+} in Tris buffer at pH 7.5.

Having determined that the strongest site for EtdBr binding is adjacent to A · U$_6$, it is of interest to ask what aspect of this site is special. Conceivably the binding site in tRNA^{Phe} is unique because it is the only place in the secondary structure of this tRNA with two adjacent A · U base-pairs. Alternatively, the uniqueness of the binding site may arise from special constraints imposed by the tertiary folding of the molecule, which prevents binding at other potential sites. A study of eight other tRNA molecules confirms that, with the possible exception of tRNA^{Glu}, the results are consistent with a binding site adjacent to the sixth base-pair of the amino-acid-acceptor stem (Jones and Kearns, unpublished results).

The fact that there is only a single strong EtdBr binding site adjacent to the sixth base-pair can be interpreted in the following way. In the presence of Mg^{2+}, the tRNA molecules are folded in such a way that 3-4 additional tertiary structure base-pairs stabilize the native structure of the tRNA, provided certain precise geometrical constraints can be satisfied. Since the binding of EtdBr, and consequent unwinding (~26° in the case of DNA; 137) of the helix induced at the binding, could change the relative orientations of different portions of the molecule, the only binding sites allowed are those that permit the tertiary interactions to be maintained. If the binding of EtdBr disrupts these interactions, the binding strength should be substantially reduced. In terms of the crystal structure of tRNA^{Phe} (72, 73) (Fig. 10), it is evident that intercalation of EtdBr in either the T-Ψ-C stem or the hU stem would disrupt a number of the tertiary interactions. It is less evident why intercalation in the anticodon stem is not permitted, as none of the bases in this stem are involved in any tertiary interactions. We note, however, that the interior terminus of this stem is attached to the core of the molecule and the outer terminus is closed with a loop. Apparently this provides enough constraint to prevent binding in the anticodon loop. Binding to the amino-acid-acceptor stem is reasonable, as this is the only helical stem in the molecule not involved in tertiary structure and it has a free end. Why binding adjacent to base-pair No. 6 is favored is not clear.

The above interpretation of the factors that lead to binding of EtdBr to a unique site on the tRNA molecules implies that *all* drugs that bind by intercalation may exhibit quite similar properties. To test this hypothesis, the binding of the antimalarial drug chloroquine was compared to that of EtdBr. Both have the same effect on the spectrum of *E. coli* tRNA^{Glu}. One resonance in the peak around 13.2 ppm is further shifted upfield by about 0.6 ppm along with some other more subtle changes in the spectrum. Thus, there is strong evidence suggesting that other drugs may bind to the different tRNAs at the same site.

The binding of EtdBr to tRNA is also of interest since it permits the structure of the tRNA molecule to be perturbed in a known manner and then an investigation of the effect of this on its biochemical properties. A study of the effect of EtdBr on the rate of aminoacylation of different purified tRNAs reveals some rather interesting differences (*138*) (Thompson and Kearns, unpublished results). The binding of EtdBr to yeast tRNA^{Phe} almost completely inactivates this molecule, whereas in the case of *E. coli* tRNA^{Val}, the rate of aminoacylation is depressed by only a factor of about two. The effect on *E. coli* tRNA^{fMet} is intermediate in that at the rate of aminoacylation is reduced by a factor of about eight. Thus, depending on the particular tRNA, the binding of a single EtdBr molecule may have rather different effects on the rate of aminoacylation and hence ultimately on protein synthesis. It will be interesting to see whether or not other intercalating drugs will affect the aminoacylation reactions in the same way as this drug.

These initial studies suggest that it may be possible to correlate information about specific sites of binding with changes in the biological activity of tRNA induced by drug binding. The binding of EtdBr to tRNA also provides a very simple method for introducing a fluorescent label at a specific site on the molecule, and this should prove useful in future fluorescence studies of the interaction of tRNA with other molecules. In particular, it would be interesting to study the interaction of labeled synthetases with tRNAs containing fluorescent labels.

Summary

High resolution NMR has been used to provide information about the secondary and tertiary structure of tRNA molecules in solution. We have focused almost exclusively on low-field resonances from the ring nitrogen protons involved in base pairing because these have provided most of the information to date.

NMR complements and corroborates the recent X-ray diffraction studies[2] on one particular tRNA, yeast tRNA^{Phe} and, most important, permits the crystal results on this one tRNA to be generalized to an entire class of tRNA. In this way, the NMR results have served as a bridge to justify the extension of the crystallographic results to tRNA^{Phe} and other tRNAs in solution.

NMR can be used to treat tRNA structure and function problems that are currently unapproachable by X-ray diffraction analysis. NMR has been used to work out plausible structures for the denatured conformers of two tRNAs and for the dimer of another. The alteration of tRNA conformation as a result of several different chemical modifications removal or replacement of bases in the anticodon loop of tRNA^{Phe}, photocrosslinking of the s^4U_8 and C_{13} in $E.$ $coli$ tRNA) has also been analyzed. NMR experiments have provided interesting insight into the interaction of tRNA molecules with intercalating drugs and dyes, which may find application in other biochemical and physical chemical studies of tRNA–protein interactions. These, represent just a few of the potential problems that NMR may help solve. In essence, the NMR measurements provide a method for monitoring 20–25 different sites on a tRNA molecule. While not all of the resonances in any given tRNA molecule can be unambiguously assigned, the spectra can be used so see whether or not there is any evidence for a change in tRNA conformation in response to some perturbation or interaction, even though it may not be possible to exactly locate the site of interaction. In favorable cases, such as in the binding of drugs to tRNA, it is possible to actually locate the site of binding. The alteration of tertiary structure can also be easily monitored by NMR, as well as the conformation of the anticodon loop.

As we noted earlier in this review, there are other regions of the NMR spectra that can be used to provide information about the conformation of tRNA molecules. These include the resonances from aromatic protons (located between 8.5 and 6.5 ppm) and the methyl groups of minor bases (such as T, m⁷G, m¹A), which give rise to resonances in the region between 4.0 and 1 ppm. Work in these areas is just beginning to appear along with preliminary studies of ¹³C and ³¹P NMR studies. With these new developments and the expected progress in proton NMR, the stage is set for the application of NMR to a very wide range of tRNA structure–function problems. Moreover, we may anticipate the application of NMR to a study of the structure of other RNAs, such as ribosomal RNA, messenger RNA, and special fragments derived from very high-molecular-weight RNA, such as viral RNA.

ACKNOWLEDGMENTS

Our first NMR studies were initiated in the laboratory of and in collaboration with Dr. Robert Shulman of Bell Telephone Laboratories, and I especially want to acknowledge the excitement of our joint collaboration. I also want to express my appreciation to the following individuals with whom I have collaborated: S. Chang, F. Cramer, K. Muench, B. Reid, R. Wells and H. Zachau. I am particularly indebted to the students, P. Bolton, H. Chao, T. Early, C. Jones, C. Larkey, D. Lerner, B. Rordorf, F. Thompson, L. Wong and Y. Wong, who have contributed so much to the NMR studies. The work in my laboratory was supported by grants from the U.S. Public Health Service (GM-19313, GM-21431 and GM-22969).

REFERENCES

1. O. Jardetzky and N. G. Wade-Jardetzky, ARB **40**, 605 (1971).
2. D. R. Kearns, D. J. Patel and R. G. Shulman, Nature (London) **229**, 338 (1971).
3. M. J. Waring, Annu. Rep. Chem. Soc. (London) B **65**, 551 (1960).
4. C. C. McDonald, W. D. Phillips and J. Lazar, JACS **89**, 4166 (1967).
5. D. H. Gauss, F. von der Haar, A. Maelicke and F. Cramer, ARB **40**, 1045 (1971).
6. H. O. Weeren, A. D. Ryon, D. E. Heatherly and A. D. Kelmers, Biotech Bioeng. **12**, 889 (1970).
7. M. B. Hoagland, P. C. Zamecnik and M. L. Stephenson, BBA **24**, 215 (1957).
8. D. W. Holley, J. Apgar, G. A. Everett, J. T. Madison, N. Marquisee, S. H. Merrill, J. R. Penswick and A. Aamir, Science **147**, 1462 (1966).
9. H. G. Zachau, Angew. Chem., Int. Ed. Engl. **8**, 71 1(1969).
10. N. Sueoka and T. Kano-Sueoka, this series **10**, 23 (1971).
11. J. A. Lewis and B. N. Ames, J. Mol. Biol. **66**, 131 (1972).
12. P. B. Sigler, Annu. Rev. Biophys. Bioeng. **4**, 477 (1975).
13. S. H. Kim, this series **17**, 182 (1975).
14. M. Levitt, Nature (London), **224**, 759 (1969).
15. F. Cramer, H. Doepner, F. von der Haar, E. Schlimme and H. Seidel, PNAS **61**, 1384 (1968).
16. C. R. Cantor, S. R. Jaskunas and I. Tinoco, JMB **20**, 39 (1966).
17. J. D. Glickson, C. C. MacDonald and W. D. Phillips, BBRC **35**, 492 (1969).
18. B. Sheard, T. Yamane and R. G. Shulman, JMB **53**, 35 (1970).
19. J. A. Pople, W. G. Schneider and H. J. Bernstein, "High-Resolution Nuclear Magnetic Resonance." McGraw-Hill, New York, 1959.

20. J. E. Ladner, A. Jack, J. D. Robertus, R. S. Brown, D. Rhodes, B. F. C. Clark and A. Klug, NARes. **2**, 1629 (1975).
21. G. J. Quigley, N. C. Seeman, A. H.-J. Wang, F. L. Suddath and A. Rich, NARes. **2**, 2329 (1975).
22. J. L. Sussman and S. H. Kim, BBRC **68**, 89 (1976).
23. P. O. P. Ts'o, M. P. Schweizer and D. P. Hollis, Ann. N.Y. Acad. Sci. **158**, 256 (1969).
24. P. O. P. Ts'o, "Basic Principles in Nucleic Acid Chemistry," Vol. 1, p. 453. Academic Press, New York, 1974.
25. S. I. Chan and J. H. Nelson, JACS **91**, 168 (1969).
26. P. O. P. Ts'o, N. S. Kondo, M. P. Schwizer and D. P. Hollis, Bchem **8**, 997 (1969).
27. C. E. Johnson, Jr. and F. A. Bovey, J. Chem. Phys. **29**, 1012 (1958).
28. L. Katz and S. Penman, JMB **15**, 220 (1966).
29. R. S. Drago, N. O'Bryan and G. C. Vogel, JACS **92**, 3924 (1970).
30. J. G. C. M. van Duijneveldt-van de Rijdt and F. B. van Duijneveldt, JACS **93**, 5644 (1971).
31. G. C. Pimentel and A. L. McClellan, Annu. Rev. Phys. Chem. **21**, 347 (1971).
32. M. Jaszunski and A. J. Sadlej, Theoret. Chim. Acta **30**, 257 (1973).
33. B. McConnell and P. C. Seawell, Bchem **11**, 4382 (1972).
34. Y. P. Wong, Ph.D. Thesis, Univ. of Calif. Riverside, California, 1972.
35. D. R. Kearns, D. Patel, R. G. Shulman and T. Yamame, JMB **61**, 265 (1971).
36. R. G. Shulman, C. W. Hilbers, D. R. Kearns, B. R. Reid and Y. P. Wong, JMB **78**, 57 (1973).
37. D. R. Lightfoot, K. L. Wong, D. R. Kearns, B. R. Reid and R. G. Shulman, JMB **78**, 71 (1973).
38. D. R. Kearns, D. R. Lightfoot, K. L. Wong, Y. P. Wong, B. R. Reid, L. Cary and R. G. Shulman, Ann. N.Y. Acad. Sci. **222**, 324 (1973).
39. D. R Kearns and R. G. Shulman, Acc. Chem. Res. **7**, 33 (1974).
40. C. Giessner-Prettre and B. Pullman, J. Theor. Biol. **27**, 87 (1970).
41. S. Arnott, Progr. Biophys. Mol. Biol. **21**, 267 (1970).
42. K. L. Wong, P. H. Bolton and D. R. Kearns, BBA **383**, 446 (1975).
43. K. L. Wong and D. R. Kearns, Nature (London) **252**, 738 (1974).
44. D. Riesner and G. Maass, EJB **36**, 76 (1973).
45. K. Morikawa, M. Tsuboi and Y. Tyogoku, Nature (London) **223**, 537 (1969).
46. A. A. Schreier and P. R. Schimmel, JMB **86**, 601 (1974).
47. R. Thiebe and H. G. Zachau, BBRC **36**, 1024 (1969).
48. D. Römer, D. Riesner and G. Maass, FEBS Lett. **10**, 352 (1970).
49. A. D. Mirzabekov, D. Lastity and A. A. Bayev, FEBS Lett. **4**, 284 (1969).
50. R. Thiebe, K. Harbers and H. G. Zachau, EJB **26**, 144 (1972).
51. S. M. Coutts, BBA **232**, 94 (1971).
52. D. Riesner and R. Römer, "Physico-Chemical Properties of Nucleic Acids," Vol. 2, pp. 237–319. Academic Press, New York, 1973.
53. I. Tinoco, O. C. Uhlenbeck and M. D. Levine, Nature (London) **230**, 362 (1971).
54. K. L. Wong and D. R. Kearns, Biopolymers **13**, 371 (1974).
55. B. F. Rordorf, D. R. Kearns, E. Hawkins and S. H. Chang, Biopolymers **15**, 325 (1976).
56. P. H. Bolton and D. D. Kearns, Nature (London) **255**, 347 (1975).
57. B. F. Rordorf and D. R. Kearns, Bchem Submitted for publication.
58. D. M. Crothers, C. W. Hilbers and R. G. Shulman, PNAS **70**, 2899 (1973).

59. D. M. Crothers, P. E. Cole, C. W. Hilbers and R. G. Shulman, *JMB* **87**, 63 (1974).
60. W. E. Daniel, Jr. and M. Cohn, *PNAS* **72**, 2582 (1975).
61. B. R. Reid and G. T. Robillard, *Nature (London)* **257**, 287 (1975).
62. B. R. Reid, N. S. Ribeiro, G. Gould, G. Robillard, C. W. Hilbers and R. G. Shulman, *PNAS* **72**, 2049 (1975).
63. D. R. Kearns, C. R. Jones, K. L. Wong, P. H. Bolton and M. Wolfson, "Conformation of tRNA in Solution," presented at EMBO Workshop on tRNA Structure and Function. Nof Ginossar, Israel, 1975.
64. P. H. Bolton and D. R. Kearns, *Nature (London)* in press.
65. P. H. Bolton, C. R. Jones, D. Lerner, L. Wong, D. R. Kearns, *Bchem* Submitted for publication.
66. J. Ninio, A. Favre and M. Yaniv, *Nature (London)* **223**, 1331 (1969).
67. L. Chaffin, D. R. Omilianowki and R. M. Bock, *Science* **172**, 584 (1971).
68. F. Berthelot, F. Gros and A. Favre, *EJB* **29**, 343 (1972).
69. D. S. Carre, G. Thomas and A. Favre, *Biochimie* **56**, 1089 (1974).
70. D. E. Bergstrom and N. J. Leonard, *JACS* **94**, 6178 (1972).
71. D. E. Bergstrom and N. J. Leonard, *Bchem* **11**, 1 (1972).
72. S. H. Kim, F. L. Suddath, G. J. Quigley, A. McPherson, J. L. Sussman, A. H. J. Wang, N. C. Seeman and A. Rich, *Science* **185**, 435 (1974).
73. J. D. Robertus, J. E. Ladner, J. T. Finch, D. Rhodes, R. S. Brown, B. F. C. Clark and A. Klug, *Nature (London)* **250**, 546 (1974).
74. K. L. Wong, Y. P. Wong and D. R. Kearns, *Biopolymers* **14**, 749 (1975).
75. T. Seno, M. Kobayashi and S. Nishimura, *BBA* **174**, 71 (1969).
76. M. Saneyoshi, T. Anami, S. Nishimura and T. Samejima, *ABB* **152**, 677 (1972).
77. C. W. Hilbers and R. G. Shulman, *PNAS* **71**, 3239 (1974).
78. G. T. Robillard, C. W. Hilbers, B. R. Reid, J. Gangloff, G. Dirheimer and R. G. Shulman, *Bchem* **15**, 1883 (1976).
79. S. H. Kim, J. L. Sussman, F. L. Suddath, G. J. Quigley, A. McPherson, A. H. J. Wang, N. C. Seeman and A. Rich, *PNAS* **71**, 4970 (1974).
80. A. Klug, J. Ladner and J. D. Robertus, *JMB* **89**, 511 (1974).
81. C. W. Hilbers, R. G. Shulman and S. H. Kim, *BBRC* **55**, 953 (1973).
82. B. F. Rordorf, D. R. Kearns, E. Hawkins and S. H. Chang, *Biopolymers* In press.
83. R. G. Shulman, C. W. Hilbers, Y. P. Wong, K. L. Wong, D. R. Lightfoot, B. R. Reid and D. R. Kearns, *PNAS* **70**, 2042 (1973).
84. T. Lindahl and A. Adams, *Science* **152**, 512 (1966).
85. A. Adams, T. Lindahl and J. D. Fresco, *PNAS* **57**, 1684 (1967).
86. R. E. Streeck and H. G. Zachau, *EJB* **30**, 382 (1972).
87. Y. P. Wong, D. R. Kearns, R. G. Shulman, T. Yamane, S. Chang, J. G. Chirikjian and J. R. Fresco, *JMB* **74**, 403 (1973).
88. D. R. Kearns, Y. P. Wong, E. Hawkins and S. H. Chang, *Nature (London)* **247**, 541 (1974).
89. D. R. Kearns, Y. P. Wong, S. H. Chang and E. Hawkins, *Bchem* **13**, 4736 (1974).
90. C. R. Jones, D. R. Kearns and K. H. Muench, *Bchem* In press.
91. O. Uhlenbeck, J. G. Chirikjian and J. R. Fresco, *FP* **31**, 420 (1972).
92. E. Hawkins and S. H. Chang, *NARes*. **1**, 1531 (1974).
93. K. Randerath, L. S. Y. Chia, R. C. Gupta, E. Randerath, E. R. Hawkins, C. K. Brum and S. H. Chang, *BBRC* **65**, 157 (1975).

94. C. M. Greenspan and M. Litt, *FEBS Lett.* **41**, 297 (1974).
95. D. Richter, V. A. Erdmann and M. Sprinzl, *PNAS* **71**, 3226 (1974).
96. J. S. Loehr and E. B. Keller, *PNAS* **61**, 1115 (1968).
97. H. G. Zachau, *EJB* **5**, 559 (1968).
98. O. C. Uhlenbeck, *JMB* **65**, 25 (1972).
99. G. Hogenauer, *EJB* **12**, 527 (1970).
100. S. K. Yang, D. G. Soll and D. M. Crothers, *Bchem* **11**, 2311 (1972).
101. B. F. Rordorf and D. R. Kearns, *Bchem* In press.
102. D. Riesner, R. Romer and G. Maass, *BBRC* **35**, 369 (1969).
103. D. Riesner, R. Romer and G. Maass, *EJB* **15**, 85 (1970).
104. P. E. Cole and D. M. Crothers, *Bchem* **11**, 4368 (1972).
105. D. R. Kearns, K. L. Wong and Y. P. Wong, *PNAS* **70**, 3843 (1973).
106. R. Thiebe and H. G. Zachau, *BBRC* **33**, 260 (1968).
107. R. Thiebe and H. G. Zachau, *EJB* **5**, 546 (1968).
108. V. Cameron and O. Uhlenbeck, *BBRC* **50**, 635 (1973).
109. O. Pongs, *FEBS Lett.* **28**, 284 (1972).
110. M. P. Schweizer, S. I. Chang and J. E. Crawford, "Physico-Chemical Properties of Nucleic Acids," Vol. 2, p. 201. Academic Press, New York, 1973.
111. W. Wintermeyer, Presented at EMBO Workshop on tRNA Structure and Function. Nof Ginossar, Israel, 1975.
112. K. L. Wong, D. R. Kearns, W. Wintermeyer and H. G. Zachau, *BBA* **395**, 1 (1975).
113. W. Wintermeyer and H. G. Zachau, *FEBS Lett.* **18**, 214 (1971).
114. G. J. Thomas, Jr., M. C. Chen, R. C. Lord, P. S. Kotsiopoulos, T. R. Tritton and S. C. Mohr, *BBRC* **54**, 570 (1973).
115. A. Danchin and M. Grunberg-Managò, *FEBS Lett.* **9**, 327 (1970).
116. R. R. Gantt, S. W. Englander and M. V. Simpson, *Bchem* **8**, 475 (1969).
117. A. J. Adler and G. D. Fasman, *BBA* **204**, 183 (1970).
118. Y. P. Wong, B. R. Reid and D. R. Kearns, *PNAS* **70**, 2193 (1973).
119. H. Kersten, *FEBS Lett.* **15**, 261 (1971).
120. S. C. Jain and H. M. Sobell, *JMB* **68**, 1 (1972).
121. J. Paoletti and J-B LePecq, *JMB* **59**, 43 (1971).
122. R. Barzilai, *JMB* **74**, 739 (1973).
123. K. Harbers, R. Thiebe and H. G. Zachau, *EJB* **26**, 132 (1972).
124. R. de Nobrega Bastos and H. R. Mahler, *JBC* **249**, 6617 (1974).
125. L. Wheelis, M. K. Trembath and R. S. Criddle, *BBRC* **63**, 838 (1975).
126. L. Dickinson, D. H. Chantril, G. W. Inkley and M. J. Thompson, *Br. J. Pharmacol.* **8**, 562 (1953).
127. R. Tomchick and H. G. Mandel, *J. Gen. Microbiol.* **36**, 225 (1964).
128. W. H. Elliot, *BJ* **86**, 562 (1963).
129. M. J. Waring, *BBA* **87**, 358 (1964).
130. J. B. Le Pecq and C. Paoletti, *JMB* **27**, 87 (1967).
131. S. Aktipis and W. W. Martz, *Bchem* **13**, 112 (1974).
132. G. P. Kreishman, S. I. Chan and W. Bauer, *JMB* **61**, 45 (1971).
133. R. Bittman, *JMB* **46**, 251 (1969).
134. C. Urbanke, R. Romer and G. Maass, *EJB* **33**, 511 (1973).
135. T. Tao, J. H. Nelson and C. R. Cantor, *Bchem* **9**, 3514 (1970).
136. C. R. Jones and D. R. Kearns, *Bchem* **14**, 2660 (1975).
137. J. C. Wang, *JMB* **89**, 783 (1974).
138. P. Lurquin and J. Buchet-Mahieu, *FEBS Lett.* **12**, 244 (1971).

Premelting Changes in DNA Conformation

E. Paleček

Institute of Biophysics
Czechoslovak Academy of Sciences
612 65 Brno
Czechoslovakia

I. Introduction	151
II. Evidence for Premelting Changes in DNA Conformation	154
A. Early Studies	154
B. Optical Methods	155
C. Methods of Electrochemical Analysis	162
D. Formaldehyde Reaction	169
E. Enzymic Methods	171
F. Other Techniques	174
III. DNA Conformation in Solution	177
A. Classical Forms of the Double Helix	177
B. Dependence on the Nucleotide Sequence	178
C. Anomalies in the Primary Structure of Double-Helical DNA	181
D. Superhelical Turns	186
E. Nature of DNA Premelting	189
IV. Relation between DNA Conformation and Function	203
V. Concluding Remarks	204
References	206
Addendum	212

I. Introduction

Our knowledge of DNA function *in vivo* is limited in part by insufficient information concerning details of conformation of the DNA molecule (or of its segments) *in situ*, e.g., during DNA interaction with a specific protein, with a cell wall or other membranes, etc. Studies of DNA denaturation *in vitro* have yielded information that is very important for the elucidation of such basic molecular-biological processes as DNA replication, transcription of genetic information, etc., in spite of the fact that denaturation takes place under conditions that usually do not occur in living cells (e.g., high temperature and extreme pH values). It is highly probable that studies of changes in DNA conformation occurring under conditions much closer to those existing *in vivo* will produce information that could help us to understand better how DNA functions *in vivo*.

a. Melting of DNA. DNA denaturation is discussed in many reviews (e.g., *1–4*); therefore only a summary of basic facts will be given here. Two complementary polynucleotide strands form the double helix, whose stability is secured by noncovalent bonds consisting mainly of (a) hydrogen bonds between A and T or G and C, and (b) stacking forces between the pyrimidine and purine rings. The free energy of hydrogen bonding is about -3 kcal (G·C) or -2 kcal (A·T) per mole of base-pair while the stacking free energy corresponds to about -7 kcal per mole of base-pair. In solution under physiological conditions, the double helix is stable. Large increases in temperature or changes of pH, exchange of aqueous solutions for an organic solvent, etc., may destabilize the double helix and cause DNA denaturation. According to Szybalski (*4*), denaturation consists principally of three overlapping steps: (a) *collapse* of the hydrogen bonded double-helical structure; (b) *collapse* of base stacking; (c) *dissociation* or complete separation of the complementary strands. The denaturation is accompanied by marked changes in the physical, chemical and biological properties of the DNA, which may be followed by means of various techniques. The most frequently used technique in DNA denaturation studies has been absorption spectrophotometry, which has contributed considerably to our basic knowledge of DNA denaturation. If a solution of native DNA is heated in a cuvette and the optical density at 260 nm is measured as a function of temperature, a curve such as that shown in Fig. 1 is obtained. The increase of optical density in the region B-C is connected with the disturbance of interaction of π-electrons between bases, owing to the collapse of

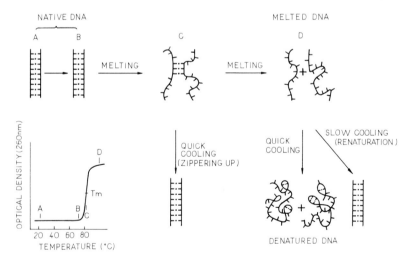

Fig. 1. Schematic presentation of DNA thermal helix-coil transition. Terminology according to Kohn, Spears and Doty (*5*).

the double-helical structure. Conformational changes occurring in the region B-C are quickly reversed if the temperature is decreased. Changes in the region C-D are basically irreversible; if, however, the DNA solution contains a relatively homogeneous population of molecules (viral or bacterial DNAs), these changes, under special conditions, can be slowly reversed to a great extent (renaturation). The fact that no changes in optical density are observed in the region A-B contributed significantly to the conclusion (considered in the literature up to recently) that no changes in the DNA secondary structure occur in this region. The sharp transition between the native and denatured states of DNA (in the B-C region) observed by means of spectrophotometric and other methods has been compared to the melting of a crystal, and the temperature at which this process took place was denoted as "temperature of melting" (t_m).

Premelting Changes. Further studies have shown that the dependence of the DNA structure on temperature is more complex than this. Various basically different methods show that changes in temperature in the premelting region (Fig. 2) result in changes in DNA properties. The evidence of premelting changes yielded by CD[1] and polarography is

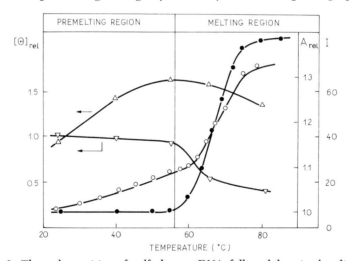

FIG. 2. Thermal transition of calf thymus DNA followed by circular dichroism (CD), derivative pulse polarography and spectrophotometry. CD: △——△, positive band (275 nm); ▽——▽, negative band (245 nm); ○——○, polarography; ●——●, absorbance at 269 nm. θ_{rel}, the ratio between the ellipticity at the given temperature and at 25°C. A_{rel}, the ratio between the absorbance (260 nm) at the given temperature and the absorbance at 25°C. I, the height of the pulse-polarographic peak in divisions. Adapted from Paleček and Frič (21).

---

[1] Nonstandard abbreviations: CD, circular dichroism; ORD, optical rotatory dispersion; ds, double-stranded; ss, single-stranded.

among the most important. In Fig. 2, the dependence on temperature of the height of the pulse-polarographic peak, of the magnitude of the positive and negative CD bands, and of the optical density at 260 nm in the premelting region are clearly shown; there is an increase in the height of the pulse-polarographic peak and in the positive CD band with increasing temperature in the premelting region, while the optical density remains unchanged. The measurements were performed in 4 M $NaClO_4$; basically the same results were obtained in other media and at lower ionic strengths (Section II, B, 1 and II, C). The changes observed by both CD and polarographic methods in the premelting region differ qualitatively from those caused by melting.

The existence of premelting changes in DNA conformation is supported also by the kinetics of the formaldehyde reaction (Section II, D), by observations of DNA cleavage with "single-strand-specific" or "region-specific" nucleases (Section II, E), and by the results of other techniques, of which the quite recent studies of Raman spectra (Section II, B, 2) and high-resolution proton nuclear magnetic resonance (Section II, F) appear to be of special importance.

The existence of conformational changes in DNA at premelting temperatures has long been overlooked in the literature. It is therefore useful to summarize the experimental data related to premelting; such a summary is contained in Section II. The tendency to interpret the data or to correlate the results of various methods in this section have been limited to a minimum. In Section III, the influence of the primary structure on DNA conformation is discussed and an attempt is made to explain qualitatively the nature of DNA premelting. It is shown that local structural changes in the vicinity of anomalies of DNA primary structure (ss-breaks,[1] chain termini, base modification) have certain features in common with premelting changes and that they differ from the changes caused by denaturation. In Section IV, certain examples of the relation between the function and structure of dsDNA[1] are shown.

II. Evidence for Premelting Changes in DNA Conformation[2]

A. Early Studies

The first notes on changes in DNA properties due to increase of temperature in the premelting region come from the end of the 1950s

[2] In this review, we confine ourselves mainly to structural changes preceding *thermal* denaturation of DNA. Structural changes in DNA due to changes in pH, ionic strength and solvent (in the premelting range) are mentioned only in connection with the main topic.

and the beginning of the 1960s. Doty et al. (6) showed the dependence of the optical rotation at the sodium D line of calf thymus DNA on temperature in the premelting zone. T'so and Helmkamp (7) confirmed their results. However, neither group (6, 7) interpreted these measurements in the sense of structural changes. Fresco (8) later suggested that there are subtle changes in the geometry of the molecule as the temperature is raised, perhaps in the pentose constituent to which ultraviolet absorption is insensitive. Independently of these observations (6–8), Paleček (9) demonstrated changes in polarographic behavior of DNAs (isolated from various sources) with temperature increase in the premelting region and explained his finding by changes in the distances between bases in the double helix. The dependence of formation of UV-induced cross-links on temperature (10) may be considered also as evidence for premelting change, although the authors did not interpret their measurements in this way. In 1963, Freund and Bernardi (11) found changes in the viscosity of a DNA solution with temperature in the premelting zone and ascribed them to changes in the DNA secondary structure.

Since the beginning of the 1960s, a large body of evidence has accumulated that speaks unambiguously for premelting changes in the DNA secondary structure. This evidence has been obtained by numerous basically different methods. A common feature of these methods was their applicability over a wide range of temperatures. The premelting changes in DNA conformation were shown to be reversible (11–14) and thus cannot be studied at a constant temperature after heating and cooling the sample. [Sarocchi and Guschlbauer (24) have recently shown that a certain portion of premelting may be irreversible (Section II, B, 1, c).]

B. Optical Methods

1. OPTICAL ROTARY DISPERSION AND CIRCULAR DICHROISM

Polarimetry is one of the methods that yielded the first evidence for changes of DNA properties in the premelting zone (Section II, A). Surprisingly, neither this method nor its modern variant, ORD, contributed to further development of premelting studies. [According to Samejima and Yang (15), the ORD profile of DNA remains almost unchanged at temperatures below 80°C.] On the other hand, the closely related method—CD (16–18)—yielded very important data.

a. *CD of DNA.* In 1964, Brahms and Mommaerts (19) reported CD spectra of DNA with one positive band at about 273 nm and one negative band at about 243 nm. Loss of the double-stranded structure (due to thermal denaturation) decreased the amplitude of both bands.

On the contrary, elevation of temperature in the premelting zone increased the positive CD band. Unfortunately, only one CD spectrum (at 45°C) was presented. The increase of the rotational strength of the positive band was explained by the formation of an intermediate form of DNA. From 1964 to 1970, CD seems not to have been used for premelting studies. A number of papers then appeared (14, 20–25), showing that the premelting behavior is qualitatively the same for all DNAs examined (phage, bacterial, salmon sperm and calf thymus), and that, if the temperature is increased, the CD spectra become more positive (Fig. 3). These CD changes are completely reversible (14, 22). In contrast to dsDNA, denatured DNA showed a linear decrease of the positive CD band (14, 20) with increase of temperature in the premelting zone. Large changes were observed in poly(dA — dT) (14, 24) and poly(dA · dT) (24), while in poly(dG · dC) (24, 26) and poly(dG — dC) (14) either no changes or substantially smaller ones were found (Fig. 4). The behavior of DNA samples with various (G + C)-content [(14), E. Paleček, J. Šponar and I. Frič, unpublished] was at first sight consistent with that of biosynthetic polynucleotides (Fig. 5a); however, with DNAs of

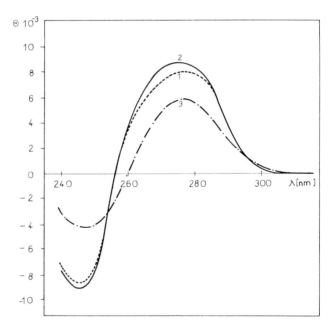

Fig. 3. Circular dichroism spectra of DNA at various temperatures. Calf thymus DNA in 0.015 M NaCl with 0.0015 M sodium citrate (pH 7). Curve 1, 22°C; 2, 46°C (premelting); 3, 87°C (melting) Roussel–Jouan Dichrograph, Model 185. θ, molar ellipticity.

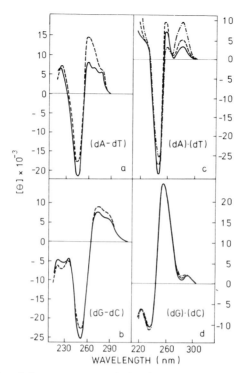

Fig. 4. Circular dichroism spectra of double-stranded polydeoxyribonucleotides at various temperatures. (a) Poly(dA-dT) in 0.001 M cacodylate buffer, pH 7, 0.001 M EDTA, 0.005 M NaCl, ———, 5°C; ———, 39°C. (b) Poly(dG-dC) in 0.005 M phosphate buffer, pH 7.3, 1×10^{-5} M EDTA, 0.01 M NaCl; ———, 1°C; ---, 48°C. (c) Poly(dA · dT) in cacodylate buffer, pH 7, 0.1 M Na⁺. (d) Poly(dG · dC) in cacodylate buffer, pH 7, 0.1 M Na⁺. (c) and (d) ———, 5°C; ---, 35°C; -·-·-, 55°C. (a) (b) From Gennis and Cantor (14); (c), (d) from Sarocchi and Guschlbauer (24).

nearly the same (G + C)-content (about 42–43%), differences in premelting behavior were observed (Fig. 5b). A great difference in the premelting behavior of DNA *Bacillus subtilis* and *Bacillus brevis* had been found earlier by polarographic measurements (Section II, C). These DNAs do not form hybrid molecules on denaturation and reannealing in solution (27), and therefore their nucleotide sequence is expected to be very different. Further experiments (24) showed poor correlation between the (G + C)-content and the ellipticity at 272.5 and 282.5 nm at various temperatures; however, a good correlation was obtained with the reduced all-pyrimidine nearest-neighbor frequencies.

The premelting behavior of poly(dA — dT) and a series of oligomers

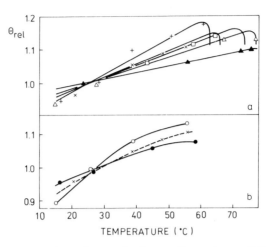

Fig. 5. Temperature dependence of relative ellipticities at 273 nm of various DNAs. (a) DNA samples with various G + C content: +——+, *Staphylococcus aureus* (31% GC); ×——×, calf thymus (42% G + C); □——□, *Escherichia coli* (50% GC); △——△, calf thymus satellite (57% G + C); ▲——▲, *Micrococcus luteus* (72% GC). (b) DNA samples with the same (42–43%) G + C content. ○——○, *Bacillus subtilis*; ●——●, *Bacillus brevis*; ×——×, calf thymus. For θ_{rel} see Fig. 2. Measurements were performed in 0.0075 M sodium phosphate with 0.01 M EDTA, pH 6.8 on a Roussel–Jouan Dichrograph Model 185. (E. Paleček, J. Šponar and I. Frič, unpublished observations.)

of $(dA - dT)_n$ with n ranging from 10 to 21 was compared (14). The curves of the molar ellipticity at 258 nm as a function of temperature were essentially parallel, and it was concluded that the change in enthalpy (ΔH) is not a function of chain length above $(dA - dT)_{10}$.

Changes in molar ellipticity of the positive band as a function of temperature were observed with calf thymus DNA at both high and low salt concentrations (14, 21, 22). No marked dependence on salt concentration was observed in the range 0.05–0.5 M NaCl (14). This observation applies also to poly(dA − dT) and some of the oligo(dA − dT) complexes. Increasing the salt concentration (of aqueous LiCl, NaCl, KCl, NH₄F) above 1.0 M or decreasing the temperature (22) causes a reduction of the positive CD band of DNA (Fig. 6a), with a shift in its maximum from 276 nm to 282 nm. For all salt concentrations and temperatures employed, the CD spectra display the major negative band around 245 nm. Even at the highest salt concentrations, no light scattering was observed; the possibility of large-aggregate formation could thus be excluded. DNA in concentrated LiCl shows the largest changes in CD with changing temperature; the extent of the variation decreases in the

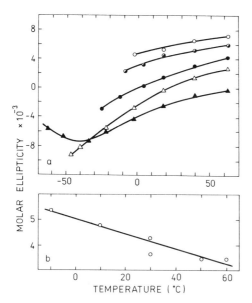

Fig. 6. The temperature dependence of the molar ellipticity of calf thymus DNA. (a) At various LiCl concentrations (at 276 nm): ○——○, 0.58 M; ◐——◐ 1.50 M; ●——●, 2.96 M; △——△, 4.52 M; ▲——▲, 7.26 M. (b) In 4.06 M guanidinium chloride. Adapted from Studdert, Patroni and Davis (22).

order Li, Na, K, NH_4. The dependence of CD spectra on temperature may be correlated with the heat of hydration of the cation (Li^+ has the highest heat of hydration, while NH_4^+ has the smallest). The dependence of the CD spectra of poly(dA — dT) on solvent composition and temperature resembles in principle that of DNA. DNA in guanidinium chloride solutions shows an opposite trend, with the positive CD band increasing slightly as temperature decreases (Fig. 6b).

Quite recently, it has been found (M. Daune, personal communication) that Me_4NCl (even at 10^{-2} M or 10^{-1} M) causes a disappearance of the premelting changes; i.e., the amplitude of the positive CD band becomes insensitive to temperature changes up to melting temperatures. It was further established that premelting is independent of ionic strength even in the range of 1–50 mM NaCl, and that it is uninfluenced by the DNA molecular weight in the range of about 5×10^5 to 10^7.

b. *Nucleoproteins.* In native deoxyribonucleoproteins, premelting changes in CD spectra, typical for DNA, have not been observed (23, 28, 29). Removal of nonhistone proteins and of the slightly lysine-rich histones H2A and H2B from the nucleoprotein results in the reappearance of the premelting changes (29); lysine-rich fraction H1 alone does not suppress the DNA premelting. In a poly(L-lysine,L-valine) · DNA complex, the premelting is only partially suppressed (30). The behavior of nucleoproteins (23, 28–30) clearly shows that the DNA premelting cannot be caused by a protein contamination of the DNA sample, as has been suggested (25).

c. *Double-Stranded RNA.* The molar ellipticity at 260 nm of rice-dwarf-virus dsRNA shows no changes in dependence on temperature in the premelting region (31). Recently, a decrease in the positive CD band of phage f2 dsRNA (replicative form) with increasing temperature has been observed (M. Vorlíčková and E. Paleček, unpublished). This decrease was more marked at lower ionic strengths. A similar decrease in the positive CD band (around 265 nm) in the range of −3.5 to +20°C occurs with the double-stranded oligonucleotide $(A)_6(U)_6$ (32).

d. *DNA in Ethanol.* The addition of ethanol to an aqueous solution of DNA gives an increase of the positive CD band (19, 33); at a concentration close to 80%, this reaches its maximum, while the negative band almost disappears. It follows from many observations that DNA may exist in an aggregated state in ethanolic solutions (34–38). However, the A-form-like CD of natural (39) and synthetic (40) DNAs can appear in such solutions even in the absence of aggregation. This supports the presumption of Ivanov et al. (33, 39) that a B-to-A transition occurs. This transition seems independent of temperature (39, 41). DNAs with less than 30% G + C are not transformed into the A form (41).

2. INFRARED AND RAMAN SPECTROSCOPY

The infrared spectra of DNA in the range of 950 to 1350 cm^{-1}, where the main bands represent the sugar–phosphate backbone vibrations (42, 43), show a gradual decrease and broadening of the absorption band at 1089 cm^{-1} as the temperature increases above room temperature (44). This fact was ascribed to small premelting structural changes in the —C—O—P—O— region of the DNA backbone. Absorption at 1056 cm^{-1} and 1019 cm^{-1} remained unchanged in the premelting region.

Raman scattering appears to be a very useful technique for studying the effect of base stacking, hydrogen bonding and backbone conformation in nucleic acids (45–47). Rimai et al. (48) and Erfurth and Peticolas (49) demonstrated premelting changes in the Raman spectra of calf-thymus DNA in D_2O. The temperature profiles of the DNA bands could be divided into four categories. (i) Certain bands show a gradual, reversible increase in intensity prior to melting; such changes were observed in the bands of all four bases (Fig. 7). The band at 1673 cm^{-1} (Fig. 7a) is of particular interest since it is related to the vibration of the thymine C=O involved in hydrogen bonding. (ii) The sugar–phos-

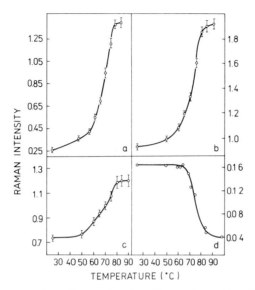

FIG. 7. Temperature dependence of various Raman intensities. Calf-thymus DNA in D$_2$O, 0.01 M sodium cacodylate, 0.001 M EDTA, pD 7.0. Raman intensity of (a) the 1658 cm^{-1} thymine band (relative to the 1091 cm^{-1} band); (b) the 1575 cm^{-1} guanine band; (c) the 765 cm^{-1} cytosine band; (d) the 1047 cm^{-1} backbone C=O band. Adapted from Erfurth and Peticolas (49).

phate backbone bands show no temperature dependence at least up to melting (Fig. 7d). (iii) The base bands show little or no dependence on temperature (e.g., 1260 cm^{-1}, cytosine and adenine; 1340 cm^{-1}, adenine). (iv) There are slow frequency changes due to deuteration of the C-8 hydrogen of guanine and adenine.

The results contained in category (i) were considered to indicate localized structural changes, probably associated with hydrogen-bond breaking and base unstacking. However, it was not possible to distinguish between these two events in the carbonyl region of thymine. The independence of sugar–phosphate backbone vibrations of temperature in the premelting region indicate that the backbone conformation of the helix is not substantially disordered by premelting. The sugar–phosphate backbone vibrations (828–1061 cm^{-1}) reappeared at almost full sensitivity and appropriate frequency (49) when the DNA sample was thermally denatured and quickly cooled. This fact suggests, in our opinion, that the method cannot be very sensitive to local changes in the backbone conformation. It is, however, sufficiently sensitive to exclude an overall transformation of the DNA molecule into the A form (45–47, 50, 51) during premelting.

3. Absorption Spectrophotometry

The independence of DNA absorption spectra from temperature in the premelting region was long considered to indicate intactness of DNA secondary structure in this region. Only recently did Sarocchi and Guschlbauer (24) demonstrate the dependence of the UV difference spectra of various DNAs on temperature. Three important features were observed: (a) hypo- and hyperchromic changes at various wavelengths amounting to less than 1% of the total absorbance at 260 nm and to relatively greater values at higher wavelengths; (b) the presence of isosbestic points; (c) the irreversibility of the spectrum at 22°C once the sample is cooled to 3°C. Qualitatively similar results have been obtained with commercial calf-thymus DNA (48); however, the presence of traces of ssDNA, usually present in commercial DNA samples, was not excluded. Changes in the degree of aggregation of ssDNA material could significantly influence the difference spectra. Irreversible changes in difference spectra resulted also from a short exposure of T7 phage DNA to acidic pH values (3.5 to 5.0) (24, 52). In agreement with the CD measurements, a correlation with reduced all-pyrimidine frequencies was found. However, the irreversibility of the observed optical density changes suggests that they might be caused by events other than those responsible for CD changes in which the reversibility was demonstrated (14, 52).

C. Methods of Electrochemical Analysis

1. DNA

Denatured DNAs as well as other single-stranded polynucleotides containing adenine and/or cytosine are polarographically reducible (53–56). In native DNA, the double bonds of adenine and cytosine, representing the primary sites of the electroreduction (53, 54, 57, 58), are involved in the Watson–Crick type of hydrogen bonding and are not accessible for electroreduction (Fig. 8); native DNA (at moderate ionic strengths and neutral pH) shows no polarographic reduction step at room temperature. If the temperature is gradually raised, a small reduction step appears around 50°C (59, 60) and increases with temperature. Essentially similar results have been obtained by oscillographic polarography at controlled current (9, 61, 62) and by alternating current polarography (59, 63). The most sensitive variant of polarography is derivative (differential) pulse polarography (54), with which the polarographic signal (peak II) of dsDNA can be observed even below

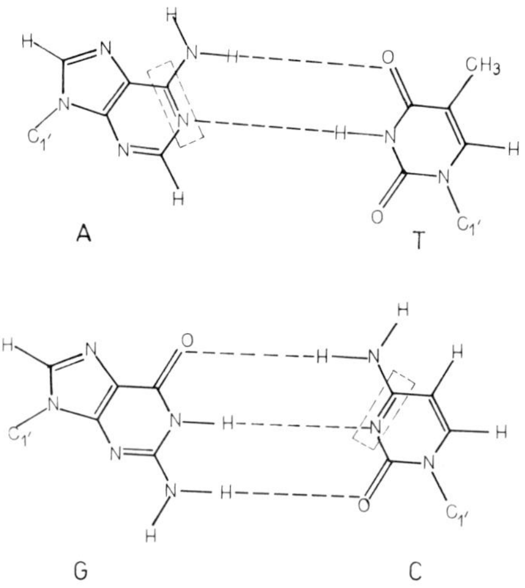

Fig. 8. Schematic presentation of hydrogen bonding between complementary bases in the DNA double helix and primary sites of the polarographic reduction. The reduction sites of adenine and cytosine are indicated by rectangles (interrupted lines).

room temperature. The peak II seen with native calf-thymus DNA at room temperature is about a hundred times smaller and appears at more positive potentials than the peak III of denatured DNA (Fig. 9). The height of peak II increases with temperature (Fig. 10) in the premelting zone (21, 64–66); during this process neither the appearance of peak III nor an increase in the half-width of peak II is observed [(64), E. Paleček, unpublished]. The changes observed in the height of peak II thus differ from those expected as a result of DNA melting. In the melting region, the peak height increases steeply with temperature; at lower melting temperatures, an increase of the half-width of the peak is observed; at higher temperatures, another peak appears (64). The steep increase in the peak height in the melting region agrees well with the optical density increase both at moderate (Fig. 10) and high ionic strengths (Fig. 2).

The appearance of peak II on a pulse polarogram of native DNA at room temperature has been explained by assuming that a small number of bases in DNA are not in the Watson–Crick structure, which predominates at room temperature, and that these are available for the electroreduction. It was further assumed that the conformation of DNA segments containing these bases differs from that of denatured DNA and that raising the temperature increases the number of bases available for the electrode process.

The possibility was also considered that, in the bulk of the solution,

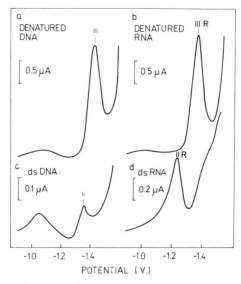

FIG. 9. Differential pulse polarograms of single-stranded and double-stranded DNA and RNA: (a) thermally denatured calf thymus DNA at a concentration of 50 μg/ml in 0.3 M ammonium formate with 0.1 M sodium phosphate, pH 6.8; (b) thermally denatured RNA of phage f2 at a concentration of 25 μg/ml in 0.3 M ammonium formate with 0.1 M sodium phosphate, pH 6.0; (c) native calf-thymus DNA at a concentration of 200 μg/ml in 0.3 M ammonium formate with 0.1 M sodium phosphate pH 6.8; (d) dsRNA of phage f2 at a concentration of 200 μg/ml in 0.3 M ammonium formate with sodium phosphate, pH 6.0. Measurements were performed on the PAR Polarographic Analyzer, Model 174; potentials were measured against a saturated calomel electrode.

all the bases of native DNA are included in the regular Watson–Crick double helix, and that the polarographic signal of native DNA appears in connection with the secondary unwinding of DNA on the electrode (53, 65). On the basis of experimental data summarized in Reference 65, this possibility was considered highly improbable. Recently, Flemming (67) measured, at various temperatures, the differential capacity of the double-layer of the mercury electrode immersed in a DNA solution and concluded that DNA premelting occurs in the bulk of the solution, not at the electrode surface. In our quite recent experiments (V. Brabec and E. Paleček, unpublished) with the large mercury-pool electrode, we found no sign of DNA unwinding even after more than 1 hour of interaction of dsDNA with the electrode charged to the potentials of peak II (in neutral 0.3 M CsCl). On the other hand, at more positive potentials, partial denaturation of DNA at the electrode surface was demonstrated (68, 69).

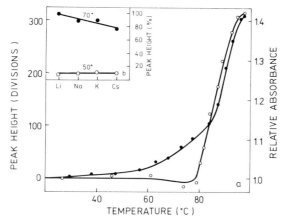

FIG. 10. (a) Thermal transition of calf thymus DNA followed by pulse-polarographic and spectrophotometric methods. DNA in 0.15 M ammonium formate with 0.025 M sodium phosphate, pH 6.5; ●——●, pulse polarography (100 μg of DNA/ml); ○——○, spectrophotometry. Both polarographic and spectrophotometric measurements were corrected for thermal expansion. Pulse-polarographic measurements were carried out on A 3100 Southern-Harwell Pulse Polarograph at an amplifier sensitivity of ⅛ or lower, the number of divisions being calculated for a sensitivity of 1/16. (b) Dependence of the height of the polarographic peak of calf thymus DNA on the nature of cation. DNA at a concentration of 200 μg/ml in 1 M Cl⁻ with 0.05 M Tris buffer pH 6.8; ●——●, at 70°C; ○——○, at 50°C. The height of the DNA peak in 1 M LiCl at 70°C was taken as 100%.

At moderate ionic strengths, the height of the pulse-polarographic peak II depends only slightly on salt concentration (Fig. 11). No marked dependence on the nature of the cation (in 1 M concentration) was observed at temperatures up to 50°C (Fig. 10b). However, at higher temperatures, the height of the peak decreased in the order Li > Na > K > Cs. If the molecular weight of DNA is decreased by sonication, a similar influence of the cation was observed even at room temperature. At salt concentrations higher than 1.5 M, the half-width and the peak height of native (unsonicated) DNA increase with salt concentration. At 2.8 M, the highest peak was observed in NaClO$_4$ (Fig. 11); smaller peaks appeared in CsCl and NaCl.

The premelting behavior of all DNA samples examined (from bacteriophages, bacteria and mammalian organs) is qualitatively similar (9, 61, 62); no simple quantitative correlation with the DNA (G + C)-content was found. Bacterial DNAs of the genus of *Bacillus* (ca. 43% G + C) yield almost identical temperature profiles (Fig. 12) if the samples come from genetically closely related organisms. The temperature profile of the DNA of the genetically distant *B. brevis* (27) differs

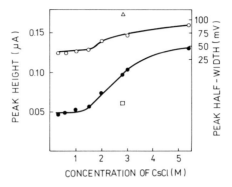

FIG. 11. Dependence of the height and half-width of the pulse-polarographic peak II of calf-thymus DNA on concentration of CsCl. DNA at a concentration of 300 μg/ml in 0.05 M sodium phosphate, pH 6.8, and CsCl in the concentration given in the graph. ●——●, peak height; ○——○; peak half-width; □, peak height in 2.8 M NaCl with 0.05 M Tris pH 6.8; △, peak height in 2.8 M NaClO₄ with 0.05 M sodium phosphate pH 6.8. Measurements were carried out on the PAR Polarographic Analyzer, Model 174 at 25°C.

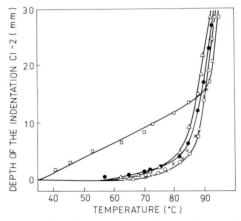

FIG. 12. Temperature dependence of the depth of the oscillopolarographic indentation of DNAs from bacteria of the genus *Bacillus*. DNA at a concentration of 100 μg/ml in 0.25 M ammonium formate with 0.025 M sodium phosphate (pH 7.0). ●——●, *B. subtilis* 168; ×——×, *B. natto*; ○——○, *B. subtilis* var. *niger*; △——△, *B. subtilis* var. *aterrimus*; □——□, *B. brevis*. After Paleček (62).

greatly from the others (Fig. 12). While both the (G + C)-contents and the molecular weights of *B. subtilis* and *B. brevis* DNAs are roughly the same, the difference in their polarographic behavior may be explained (62) by differences in nucleotide sequence. It was assumed that *B. subtilis* DNA contains a larger number of G · C pairs arranged in long blocks than does *B. brevis* DNA (62).

2. DOUBLE-STRANDED RNA

Pulse-polarographic behavior of dsRNA (the replicative form of phage f2) is similar to that of DNA (70). This dsRNA produced peak IIR (Fig. 9d) at potentials more positive than those of peak II of dsDNA (Fig. 9c); peak IIIR (Fig. 9b) of thermally denatured RNA appeared at more negative potentials than peak III of denatured DNA. Similarly to peak II of DNA (Fig. 11), peak IIR was almost independent of the salt concentration at moderate ionic strengths. In the region of high ionic strengths, the height of the peak of dsRNA increased with salt concentration more markedly than the height of the peak of DNA; the peak of dsRNA in 5.4 M CsCl was more than four times higher than in 1 M CsCl. An increase in temperature in the premelting zone resulted first in an increase of peak IIR (Fig. 13); however, a further increase in temperature

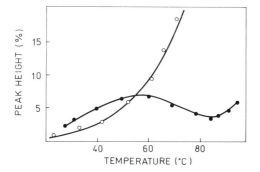

FIG. 13. Dependence of the height of pulse-polarographic peaks of double-stranded RNA and DNA on temperature. ●——●, Peak IIR of dsRNA at a concentration of 200 μg/ml in 0.5 M ammonium formate with 0.1 M sodium phosphate, pH 6.5; ○——○, peak II of calf thymus DNA at a concentration of 200 μg/ml in 0.5 M CsCl with 0.1 M sodium phosphate pH 6.5. The peak heights of dsRNA and DNA are expressed as percentages; the height of peak IIIR of thermally denatured RNA and of peak III of thermally denatured DNA measured at 40°C were taken as 100%. The measurements were performed on a Polarographic Analyzer PAR 174.

caused a decrease of this peak and, at temperatures close to the beginning of melting, peak IIR increased again. Even at high temperatures, the potential of peak IIR differed markedly from the potential of the well-separated peak IIIR; thus the increase of peak IIR at high temperatures (Fig. 13) could not be caused by a superposition of peak IIIR on peak IIR. At temperatures above 60°C, the extent of changes in the height of peak IIR of dsRNA was considerably smaller than that of peak II of calf-thymus DNA, in spite of the fact that the molecular weight of dsRNA was lower by an order of magnitude, and thus a greater influence of the ends of the molecule on the premelting might be expected (Section III, C).

3. POLYNUCLEOTIDES

The pulse-polarographic behavior of biosynthetic polynucleotides basically agrees with that of natural nucleic acids. The peaks of double-stranded polynucleotides are smaller and appear at more positive potentials than do those of the corresponding single-stranded polynucleotides (54, 71); in the case of poly(dG · dC) and poly(rG · rC), no peak appeared at room temperature and neutral pH [(71), E. Paleček and F. Jelen, unpublished]. A great difference in the dependence on temperature of G · C pairs contained in double helices and those with other base-pairs (A·T, A·U and I·C) was demonstrated. At temperatures close to 0°C, both the thermally more stable (according to t_m) poly(dG · dC) and the less stable poly(dA — dT) behave as almost nonreducible substances (Fig. 14). A rise in temperature results, however, in the appearance of a peak of poly(dA — dT), that increases almost linearly with temperature, starting from 25°C. On the contrary, poly(dG · dC) yields no peak even at temperatures above 80°C (Fig. 14). As with poly(rG · rC), an increase in temperature has no influence on the pulse polaro-

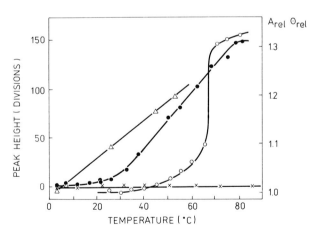

FIG. 14. Thermal transition of double-helical polydeoxynucleotides as measured by polarography, circular dichroism and spectrophotometry. Poly(dA-dT) in 0.15 M ammonium formate with 0.05 M Tris buffer, pH 6.7; ●——●, derivative pulse polarography; ○——○, optical density (261 nm). Poly(dA-dT) in 0.5 M NaCl with 7.5×10^{-3} M sodium phosphate with 1×10^{-3} M EDTA, pH 7: △——△, relative ellipticity at 262 nm. Poly(dG · dC) in 0.15 M ammonium formate with 0.05 M Tris buffer, pH 6.7; ×——×, derivative pulse polarography. For A_{rel} and θ_{rel} see Fig. 2.

gram of this polynucleotide complex. On the contrary, the peak of poly (rI · rC) increases linearly with temperature in the premelting zone. The results clearly show the nonreducibility of the cytosine residues arranged in long sequences of G · C pairs in the double-helical molecule. Changes in temperature in the premelting zone have practically no effect on this property of cytosine. On the other hand, a certain fraction of cytosine residues in poly(rI · rC) as well as a fraction of the adenine residues in poly(dA — dT) are polarographically reducible at room temperature, and the accessibility of these residues increases with temperature in the premelting zone.

D. Formaldehyde Reaction[3]

Interaction of formaldehyde with DNA results in hydroxymethylation of the basic exocyclic amino groups of cytosine, adenine and guanine, and of the acidic endocyclic imino groups of guanine and thymine (72–74). As with polarographic reduction (Section II, C), the hydroxymethylation of the bases in dsDNA does not take place unless the base-pairs are dissociated, and the bases that react with formaldehyde are not able to re-form pairs again. These two facts have been utilized by numerous workers in nucleic acid structure research (cited in References 72, 73). Recent studies of the kinetics of the reaction of dsDNA with formaldehyde indicate changes in DNA structure at premelting temperatures (72, 75–79). The reaction was studied mainly by spectrophotometry; the extents of the reaction and of denaturation estimated separately at two wavelengths (75, 79) indicate that the reacting base-pairs induced the denaturation of adjacent base-pairs (75). The induction effect and the parameter θ', which expresses the instantaneous degree of denaturation, increased with temperature (Fig. 15). θ' was the smallest for a relatively long duplex (phage T7 DNA) free of ss-breaks. The initial reaction rate increased with the concentration of the destabilizing salt ($NaClO_4$), i.e., with bringing the experimental conditions closer to the temperature of melting (79). Under certain conditions, a large number of dissociated but unreacted bases in the molecule was found (75, 80). The reaction between formaldehyde and bases contained in the double helix was explained by a "breathing" (temporary dissociation) of base-pairs (75). The dependence of CD on reaction time (M. Vorlíčková and E. Paleček, unpublished) suggests that formaldehyde-induced denaturation is not preceded by premelting (observed by means of CD in the absence of formaldehyde).

Experimental data from various laboratories as well as interpretations of these data do not always agree with one another. For instance,

---

[3] See article by Feldman in Volume 13 of this series (72).

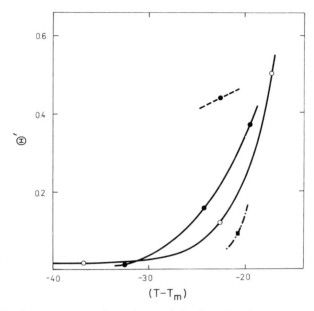

Fig. 15. The temperature dependence of the "breathing" parameter θ' of various double-helical polydeoxyribonucleotides. ●——●, poly(dA-dT) --●--, hydrolyzed poly(dA-dT); ○——○, poly(dA · dT); --■--, DNA of phage T7. After Utiyama and Doty (75).

Lazurkin et al. (76–78) and Utiyama and Doty (75) assumed that basepairs adjacent to duplex ends and ss-breaks are more reactive with formaldehyde than the base-pairs in the middle of the intact molecule. On the basis this assumption, a "kinetic formaldehyde method" was developed (76–78), and it was shown experimentally that this method can be used for the detection of "defects" (ss- and ds-breaks, base damage caused by irradiation) in the DNA secondary structure (Section III, C). Contrary to this is the belief (79) that most of the openings occur in the middle of the molecule and that the presence of "defects" in the double helix is insignificant for the course of formaldehyde reaction. This conclusion was based on the findings (79) that the initial rate of formaldehyde reaction with sonicated DNA (5×10^5 daltons) was only about twice the rate obtained with unsonicated DNA (5×10^6 daltons) and that no lag period was detected in the kinetic curve. However, such results might be conditioned by ss-breaks in the commercial calf-thymus DNA sample used in the experiments. Further discrepancies in the experimental data and in the views of different authors were recently discussed by Frank-Kamenetskii and Lazurkin (78); for a more definite explanation, further work will be necessary.

E. Enzymic Methods

It has long been recognized that information on the secondary structure of proteins and polypeptides can be obtained by using appropriate proteolytic enzymes as conformational probes (e.g., *81, 82*). The DNA double-helical structure was first investigated by means of an enzymic probe technique with micrococcal nuclease, EC 3.1.4.7 (*83*). Under certain conditions, this enzyme (Table I) attacks thermally denatured DNA much faster than it does native DNA. The site of the initial attack depends on DNA conformation; at 60°C denatured DNA is attacked essentially at random, whereas the initial attack on dsDNA occurs in (A + T)-rich regions. In spite of a marked difference between the initial rate of hydrolysis of native and denatured DNA, the activation energies for this process are practically the same for both substrates over the entire 30°–60°C temperature range. The estimated value of ΔH, -14 kcal/mol, is very similar to that obtained for the hydrolysis of randomly coiled proteins by proteolytic enzymes. It was therefore concluded that the rate-limiting step for the enzymic hydrolysis of both native and denatured DNA is the cleavage of phosphodiester bonds or some adsorption step that does not involve an enzyme-induced distortion of the DNA duplex. Further, the preference of the enzyme for (A + T)-rich sequences in dsDNA increases as the digestion temperature decreases (*84*). At 10°C, the initial acid-soluble product contained 88 mol-% of adenine and thymine, while at temperatures close to t_m the product composition corresponded nearly to that of undegraded DNA. Poly(dG · dC) is considerably more resistant to enzymic digestion than is poly(dA · dT) (*85*). It was concluded (*84*) that (A + T)-rich regions of dsDNA undergo local strand separation (or opening–closing reactions) to a greater extent (or at a greater rate) than do (G + C)-rich regions.

Johnson and Laskowski (*86*) isolated and characterized an enzyme called mung bean nuclease I (EC 3.1.4.9) with substantially greater preference for single-stranded polynucleotides (Table I), which is much better suited for the study of conformation of double-stranded polynucleotides. Incubation of poly(dG · dC) with a relatively high amount of this enzyme caused no change in the $s_{20,w}$. On the other hand, poly(dA-dT) is a good substrate. At low temperatures (17° and 27°C), a marked autoacceleration of the hydrolysis was observed (due to production of a more favorable single-stranded substrate during the course of the reaction), while at a higher temperature (37°C) normal kinetics appeared. It was apparent from the temperature coefficients ($k_0^{17}/k_0^7 = 8.1$; $k_0^{27}/k_0^{17} = 7.3$; $k_0^{37}/k_0^{27} = 23.5$) that a change from 27° to 37°C caused a considerably greater increase in reaction rate than could be expected

TABLE I
Some Nucleases with Preference for Single-Stranded Polynucleotides[a]

Nuclease	Mode of attack	Specificity toward sugar	Site of the cleavage	Base specificity	pH optimum	Activating ion	Note	References
Micrococcal, EC 3.1.4.7	Endo	Both	3' Former	None[c]	ca. 9	Ca^{2+}	Heat-stable	83–85
Mung bean I,[b] EC 3.1.4.9	Endo	Both	5' Former	Preference for A-pN, T-pN	5	None	3×10^4 higher initial reaction rate for denatured DNA than for native T4 phage DNA	86, 87
Aspergillus oryzae S_1[b] EC 3.1.4.21	Endo	Both	5' Former	None	4.3	Zn^{2+}, Ca^{2+}	Heat-stable	100, 101, 109
Neurospora crassa endo[b] EC 3.1.4.21	Endo	Both	5' Former	Preference for G-pN	7.5–8.5	None[d]	10^3 higher reaction rate for denatured DNA than for dsDNA, heat-stable	96, 97
Neurospora crassa exo[b] EC 3.1.4.25	Exo[e]	Both	5' Former	Preference for T-pN, U-pN	8.5	Mg^{2+}	Unstable	95, 98, 99
Escherichia coli I, EC 3.1.4.25	Exo	Deoxy	5' Former	3'-OH terminus	9.5	Mg^{2+}	4×10^4 higher reaction rate for denatured DNA than for dsDNA	91
Vipera lebetina venom, EC 3.1.4.25	Exo	Both	5' Former	3'-OH terminus	8.5	Mg^{2+}	—	92, 93, 94

[a] Other nucleases are listed only by name: endonuclease from dog fish liver (102, 103), lamb brain (104), sheep kidney (105) and yeast cells (106).
[b] Endonuclease with high specificity for polynucleotide conformation; may be considered as a region-specific nuclease.
[c] Virtually no base specificity observed with denatured DNA; preference for A+T-rich regions in native DNA.
[d] Hydrolysis of denatured DNA is stimulated about 2.5-fold by the addition of 0.01 M $MgCl_2$, $CaCl_2$ or $FeCl_2$; however, the same cations inhibit, RNA degradation by 40%.
[e] An extra endonucleolytic activity specific toward (A+T)-rich regions of dsDNA has been observed (95, 108).

from the enzyme thermal activation alone. This was ascribed to a premelting conformational transition. Optical density measurements at 260 nm at zero time at these temperatures showed little or no changes. The experiments with λ phage DNA indicate that a specific (A + T)-rich region of a dsDNA can be hydrolyzed without a significant digestion of the rest of the molecule. The center of the λ phage DNA molecule is rich in A and T, and the halves of the molecule obtained by shear degradation differ in base composition (88–90). Cesium density-gradient ultracentrifugation shows (86) that the enzymic digestion at 17°C produces halves of the molecule with base compositions nearly the same as those of the halves obtained by shear degradation. The central region deleted by the enzyme was shorter in length and richer in A and T than the central fragment obtained earlier (88) by another technique. However, at 37°C the hydrolysis went beyond the formation of two halves and a preferential digestion of the (A + T)-rich half took place.

The remarkable property of this enzyme to distinguish between "tight" and "loose" double-helical regions led Johnson and Laskowski (86) to propose the name "region-specific nucleases" for the new class of enzymes, the first representative of which was the mung bean nuclease I. The requirements for a "loose" region are not yet accurately known (e.g., what is the minimal length of A · T-pair blocks that makes a region susceptible to the enzyme?), but the significance of this enzyme for the study of the details of double-helical structures is obvious. An ability to digest certain regions in dsDNA has been found recently for other enzymes considered to be "single-strand specific," e.g., nucleases from *Neurospora crassa* and *Aspergillus oryzae* (Table I). These enzymes convert circular superhelical DNAs into their "relaxed" forms (Section III, D) and, at high enzyme concentrations, they produce more than one cut per supercoiled DNA molecule (108, 110–114). *Aspergillus* nuclease S_1 also introduced ss-breaks into the linear dsDNA (110, 111) of T7 phage (no ss-breaks were present in this DNA before digestion) and released acid-soluble products from dsDNA of calf thymus (107).

The cleavage of supercoiled DNA in solution depends on the concentration of NaCl (113, 114) and on temperature (114). Sedimentation of the product of cleavage of supercoiled DNA of simian virus 40 indicated that degradation of the linear molecules takes place in 0.01 M NaCl, whereas in 0.075 M and 0.25 M NaCl almost no degradation of linear DNA occurs. Conversion of supercoiled DNA to "nicked" (nonopposed single-strand cleavages) circles and linear DNA was more than five times slower in 0.25 M NaCl than in 0.075 M NaCl. Similar results were obtained with polyoma virus DNA (114); at high salt concentration (0.05 M NaCl) and at low temperature (35°C) digestion was considerably inhibited. At low ionic strength (0.005 M NaCl) and at high temperature (45°C), a wide range of DNA pieces smaller than 14 S was produced.

F. Other Techniques

1. PROTON NUCLEAR MAGNETIC RESONANCE

In Me_2SO, the resonances of the protons at N-1 of guanine and at N-3 of thymine resonate at 10.9 and 10.0 parts per million, respectively (115). These proton resonances are absent in aqueous solutions because of a rapid exchange with the solvent. Hydrogen-bond formation in the double helix shifts these resonances downfield (116). When the Watson–Crick hydrogen bonds are broken, they broaden and merge with the H_2O signal. Crothers et al. (117) studied the proton nuclear magnetic resonance spectra of the double-helical complex made of two complementary pentanucleotides d(A-A-C-A-A) and d(T-T-G-T-T). Well-resolved proton resonances of T-N3-H and G-N1-H were observed at 1°C, which became broader at 4°C and disappeared at 9°C. The t_m of the complex was 28°C and the beginning of the steep increase of optical density at 245 nm was around 9°C. Broadening and disappearing of the low-field resonances thus occurred in the premelting region. The results of a more detailed study by Patel et al. (118–119b) with the double-stranded oligonucleotides d(A-T-G-C-A-T), d(G-C)$_4$, d(T-A)$_3$, d(T-A)$_4$, d(T-A)$_5$, and d(A-G)$_4$ · d(C-T)$_4$ basically agreed with the observation made by Crothers et al. (117). Raising the temperature of d(A-T-G-C-A-T) to 25°C caused the three resonances observed to broaden at different temperatures (Fig. 16) (118, 119). This sequential broadening process was explained by fraying of the helix ends at premelting temperatures. It was assumed that the fraying process represents a rapid opening and closing of the hydrogen bonds and that it occurs much faster than does melting. For the given oligonucleotide, fraying would predominate at the terminal A · T pairs and proceed to the internal A · T pairs. Breakage of the central G · C pairs would constitute melting of the double helix. In d(T-A)$_{35}$ the resonances of the terminal A · T at 0°C were too broad to be observable; however, in longer sequences (octa- and decanucleotide), this resonance appeared. The spectra at 0°C of p(dA-dT)$_4$ and p(dT-dA)$_4$, i.e., of helices with the same length and nearest-neighbor sequences, were not identical. The differences in these spectra suggested that a terminal 5′-phosphate may influence the stability of the helix or the exchange with H_2O to different extents in the two oligonucleotides.

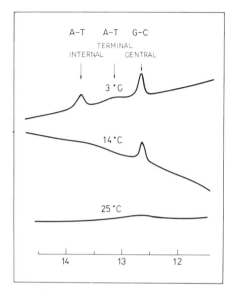

Fig. 16. Temperature dependence of the high-resolution proton nuclear magnetic resonance spectra of the double-helical oligonucleotide d(A-T-G-C-A-T) in H_2O. Adapted from Patel and Tonelli (119).

2. SMALL-ANGLE X-RAY SCATTERING

This technique can yield information about the shapes and dimensions of nucleic acid molecules in solution (120). At room temperature, DNA (in 0.15 M NaCl) is rodlike and the mass per unit length (μ) and the axial radius of gyration (R_c) are in agreement with the Watson–Crick model (121). An increase in the temperature at first causes no changes in the studied parameters, while heating to premelting temperatures above 60°C decreases μ even though the rodlike configuration is preserved.

3. UV-INDUCED CROSS-LINKS

The induction of pyrimidine dimers (cyclobutadipyrimidines) in DNA by UV light is dependent on the DNA conformation; thus dimer formation can serve as a photochemical probe of the DNA structure (122–124). The formation of cross-linking dimers is conditioned by local opening of the DNA double helix and the number of cross-links increases with temperature in the premelting zone (123, 125, 126, 128); the maximum number of cross-links is obtained when the denaturation approaches 20% (125) and the fraction of cross-linked molecules reaches zero at the temperature at which the denaturation becomes complete. The increase in DNA cross-linking appears at pH values preceding alkaline [(127), M. Vorlíčková and E. Paleček, unpublished] and acidic (125, 127) denaturation.

4. SEDIMENTATION AND VISCOSITY

The sedimentation coefficient of various phage DNAs increases slightly with temperature in a range of 5°–60°C and the persistence length decreases (129). In a temperature range of 25–50°C, only very small or no changes in viscosity of various DNA solutions are observed (11, 130, 131). At higher temperatures, decreases in the viscosities of various DNAs with increasing temperature are observed (11, 130–133). The intrinsic viscosity of phage T4 DNA in NaCl, KCl and LiCl of ionic strengths between 0.2 and 0.4 decreases by 6.3% (132) as the temperature is increased from 25° to 61°C (in 0.2 M NaCl). At high salt concentrations, the same viscosities are found at both temperatures. The viscosity of a sonicated calf-thymus DNA preparation in 0.2 M NaCl decreased (133) owing to temperature increase to the same extent as that of phage T4 DNA (132). In DNA fractions from a column of methylated albumin, an increase in temperature resulted in both decrease and increase of viscosity (131). The observed changes in viscosity were explained by various mechanisms, e.g., melting of A · T clusters (11), alterations of molecular dimensions (130), and changes in DNA aggregation (131).

Studies of the viscosity and optical density of poly(dA-dT) in combination with the temperature-jump method showed that at temperatures below the onset of melting the polynucleotide helix changes its conformation apparently by the formation of self-complementary branches (134, 135).

5. SEDIMENTATION OF SUPERHELICAL DNAs

The dependence of the sedimentation of enzymically closed and "nicked" circular DNAs on temperature suggests that an increase in temperature in the premelting region results in partial unwinding of the DNA molecule (136, 137). An increase of 1°C decreased the number of superhelical turns by about half a turn, and the average rotation of the DNA double helix by about 0.005 degree per base-pair (137). The changes were noncooperative. Addition of $NaClO_4$ caused a similar noncooperative unwinding of the DNA double helix (138). $Mg(ClO_4)_2$ was twice as effective in the induction of premelting unwinding of DNA (139). The temperature and salt effects were additive; the addition of 1 M $Mg(ClO_4)_2$ was equivalent to a temperature rise of 20°C. All these findings could be explained either as a uniform change in the angle between each and every pair of the adjacent bases in the duplex, or, at the other extreme, a change in the size of nonhelical (completely denatured) segments. Wang (137) considered the latter possibility to be physically unreasonable. Even at temperatures almost 100°C below the melting temperature of the given DNA, the temperature coefficient would require an increase of the denatured region by about 5 base-pairs for every degree centigrade; at 40°C, the denatured region would have to include 200 base-pairs, i.e., more than 0.5% of the total content of base-pairs. However, recent studies (Section III, D) show that superhelical turns in the DNA molecule cause local conformational changes similar but not identical to DNA denaturation.

III. DNA Conformation in Solution

A. Classical Forms of the Double-Helix

Prior to considering local irregularities in the DNA double helix, it will be useful to mention the regular geometries of the three classical DNA forms found experimentally in X-ray fiber studies (140, 141). At high humidities and in the presence of excess salt, DNA assumes the B configuration, which is characterized by furanose rings in a standard C3'-exo conformation (142), with their bases nearly perpendicular to the axis of the helix; the helix parameters are given in Table II. The fact that B-DNA is formed at high humidities regardless of the counterion nature suggests that this form corresponds to the DNA conformation in aqueous solution. This conclusion receives support from studies of DNA properties in solution by, e.g., small-angle X-ray scattering (143) (spacing per residue), birefringence (144) (base-pairs are nearly perpendicular to the axis of the helix), and CD (145) (the CD spectrum in solution is similar to that of a film of B-DNA). At low humidities, lithium DNA assumes the C form (Table II), which is closely related to

TABLE II
HELIX PARAMETERS OF THE CLASSICAL DNA CONFIGURATIONS[a]

Helix type	Pitch (Å)	Residues/ turn	Rotation/ residue (deg.)	Translation/ residue (Å)	Tilt (deg.)	Twist (deg.)
A-DNA	28.15	11	32.7	2.56	20	−8
B-DNA	33.7	10	36.0	3.37	−2	5
C-DNA	33.2	9.3	38.6	3.32	−6	5

[a] After Gennis and Cantor (14); data from Arnott (141).

the B form. A-DNA (Table II) is formed at low humidities with Na⁺ or K⁺ as counterion. It differs significantly from B- and C-DNA. A-DNA is characterized by furanose rings in the C3'-endo conformation (*141*), with bases substantially tilted from the axis of the helix. Compared to B-DNA, the helix of A-DNA has a larger diameter and the large groove is deeper and narrower. Both dsRNA and RNA · DNA hybrids assume structures very similar to A-DNA (*146*).

Until recently, it was generally accepted that DNA exists only in these forms and that its structure is independent of the base composition. In the last few years, an increasing body of evidence suggests that DNA secondary structure does depend on the base composition and that the A, B and C forms are not the only ones in which DNA can exist.

B. Dependence on the Nucleotide Sequence

Using wide-angle X-ray scattering, Bram demonstrated that the double-helical structures (in solution) of calf-thymus DNA (*147*) as well as of various DNAs with high and medium G + C (*148*) are similar to the B form, estimated on the basis of X-ray fiber diffraction (B type). On the other hand, DNAs very rich in A + T (*Clostridium perfringens, Bacillus cereus*) exhibit a marked difference in their X-ray patterns. A detailed analysis of the wide-angle scattering data obtained with various DNA samples disclosed that the sharpness of the two intensity maxima near 13 and 10 Å depended on the DNA base composition (*149*); the ratio of the 13 to the 10 Å maximum is an almost continuous function of the A + T content (Table III). The features of the (A + T)-rich DNA

TABLE III
INTENSITY RATIOS OF THE WIDE-ANGLE X-RAY SCATTERING
MAXIMA OF VARIOUS DNA SOLUTIONS[a]

DNA (Na salt)	A + T (%)	$I(10 \text{ Å})/I(13 \text{ Å})$ × 100
Poly[d(A − T)]	100	88
Clostridium perfringens	69	70
Bacillus cereus	65	69
Calf thymus	58	66
Escherichia coli	50	62
Micrococcus luteus	31	53

[a] The data are from (*148*). To a first approximation, these ratios are related to the relative maximum intensity on the third layer line to that of the second layer line, but contributions from other layer lines are expected to increase the intensity scattered near 10 Å from DNA solutions. After Bram (*149*).

scattering curve are similar to those of poly(dA-dT) and differ from those of the B type. The data indicate that the pitch of the (A + T)-rich structure is about 10% larger than that of the other DNAs tested. Poly(dI · dC), poly(dI-dC) and poly(dG · dC) (150) possess structures different from the A, B and C DNA forms, while that of poly(rI · rC) corresponds roughly to that of the A'-RNA form (twelve base-pairs per turn of 36 Å). While X-ray scattering measurements reflect only the overall secondary and tertiary structures of nucleic acids, not their chemical composition, the only explanation of this finding was that DNA secondary structure depends on base composition. However this method cannot determine the exact atomic structure of the nucleic acid. Therefore attempts have been made to find the forms of (A + T)-rich DNAs in fibers corresponding to those determined in dilute salt solutions by the X-ray scattering technique.

X-Ray fiber-diffraction studies at low humidities (44–66%) on (A + T)-rich DNAs from *C. perfringens, Cytophaga johnsonii* (151) and *S. aureus* (152) show that these DNAs can exist in hitherto unknown conformations; e.g., *C. perfringens* DNA (68% A + T) can adopt at least five different forms at the same water content (151, 153). At high LiCl concentrations and at low and intermediate relative humidities, the lithium salt of *C. perfringens* DNA gives a B-type X-ray diffraction pattern; in contrast to typical DNA, no transition to the C-form was observed at low humidities. *Cytophaga johnsonii* DNA (65% A + T) yields two distinct X-ray fiber diffraction patterns (J_1 and J_2), depending on the salt concentration, while *S. aureus* DNA (67% A + T) exists in yet another form (S) under similar conditions. The (A + T)-rich DNAs of *S. aureus* (152) and *C. johnsonii* (151) do not adopt the A-form under conditions yielding this conformation with DNA of intermediate A + T content.

High humidity X-ray fiber diagrams from DNAs with various base compositions (149) show the (B-type) meridional reflections (3.4 Å) and layer line spacing (34 Å) characteristic of the B-type structure, while the intensity distribution of the first three layer lines varies with the base composition (Table IV). In diagrams of DNAs very rich in A + T, the intensity of the first and third layer lines is about 3 times stronger than in the patterns of (G + C)-rich DNAs. The high-humidity X-ray fiber measurements thus agree with the X-ray scattering from DNA solutions (Table III). The inter-base-pair separation, pitch, and radius (148) do not vary significantly with the DNA base composition. Thus the only alterations that might be responsible for the diverse intensity distributions (Table IV) are in the angle between the bases in a pair or in the base-pair tilt (149). These could occur over relatively

TABLE IV
MAXIMUM RELATIVE INTENSITIES OF DNA X-RAY FIBER
PATTERNS AT HIGH HUMIDITY

DNA (Na salt)	A+T/G+C of DNA	Maximum relative intensity[a]	
		First layer line	Third layer line
Yeast mitochondria	4.6	105	85
Clostridium perfringens	2.3	85	80
Staphylococcus aureus	2.0	80	75
Cytophaga johnsonii	1.9	75	75
Calf thymus	1.3	65	60
Escherichia coli	1.0	55	55
Micrococcus luteus	0.45	30	35
M. luteus (Li salt)	0.45	30	35
Sarcina lutea	0.41	35	40

[a] Given as percentage of the maximum value of the X-ray intensity on the second layer line, which is equal to 100 for each DNA. After Bram (*149*).

large regions (clusters of a dozen or more base-pairs) or could be caused by a difference in the structure of each G + C and A + T pair or group of several pairs.

The results of X-ray diffraction analyses of fibers of biosynthetic double-stranded polydeoxynucleotides are in good agreement with those obtained from natural DNAs (*154–156*). Alternating copolymers poly-(dG-dC) and poly(dA-dT) (*155*) adopt helical structures with 8 base-pairs per helix turn, while poly(dA · dT) (*154*) exists in a structure with 10 base-pairs per turn. Neither poly(dA · dT) nor poly(dA-dT) adopt a stable A-type helix with C3′-endo furanose rings. In contrast, poly(dG · dC) shows a preference for the A-form (*156*).

According to Bram (*149*), the word "form," which has been used for a unique configuration, should be applied only to the type-A conformation, as this type appears independent of both the base composition (*157*) and the salt and water content (*158*). There are many B-type structures that depend upon the source of the DNA, and sodium DNA can exist at humidities as low as 44% in a B-like structure differing from that existing at higher humidities (*149, 152*).

X-Ray analysis yields very convincing evidence for the dependence of the double-helical structure on the nucleotide sequence. The results of the infrared spectroscopy of natural (*159–162*) and biosynthetic (*162*) polynucleotides, obtained approximately simultaneously, are in good agreement with the results of X-ray analysis. However, polarography demonstrated much earlier that the DNA structure must depend on the nucleotide sequence (Section II, C). A similar conclusion follows from more recent studies of various properties of DNA [electrophoretic mobility (163), certain optical properties (Section II, B), etc.]. The finding that DNA secondary structure depends on nucleotide sequence may represent a very important step in our understanding of the way in which DNA performs its complicated function. If each small group of base-pairs (or each base-pair) has a characteristic structure, manifested by a specific change in the local dimensions of the DNA grooves, the sequence of nucleotides in the double helix could easily be recognized without unwinding the strands (149) (Section IV).

C. Anomalies in the Primary Structure of Double-Helical DNA

In the preceding paragraphs, the structure of DNA was considered to be an infinitely long series of A · T and G · C pairs arranged in a double-helical structure. However, no mention was made of the facts that linear molecules of natural DNA have ends, that the sugar–phosphate backbone may be interrupted, that DNA may contain mismatched and rare bases, etc., and that, in the vicinity of these anomalies, the local DNA conformation may differ significantly from the structure of most of the molecule. This generally adopted approach to the DNA structure is understandable, considering that the structure was originally derived from X-ray diffraction studies, which determine only the *regularities* in the DNA secondary structure. Moreover, fiber-diffraction patterns of DNA contain a relatively small number of diffractions (140, 141) (compared with the data obtainable from single crystals), and research on DNA structure depends heavily on model-building procedures. However, anomalies in the DNA primary structure undoubtedly do exist (whether they be natural or artificial), and it is therefore necessary to try to obtain information on their influence on the DNA secondary structure, using indirect methods with a sufficient degree of sensitivity.

Particularly important among these methods are CD, polarography, studies of DNA reactivity with formaldehyde, nuclear magnetic resonance, and enzymic methods, i.e., methods that have supplied important information on DNA premelting conformational changes (Section II). We shall now attempt to summarize the data obtained in studies of the anomalies in DNA structure.

1. CHAIN INTERRUPTIONS

The introduction of ss-breaks (by the action of ionizing radiation or DNase I) or ds-breaks (shearing in a capillary, sonication) leads to an increase in the initial rate of DNA unwinding under the action of formaldehyde (*164–167*), to an increase in polarographic reducibility (*60, 64, 168–170*) of dsDNA (Fig. 17b), and to increased cross-linking induced by UV light (*169*). The "single-strand specific" nuclease S_1 (Section II, E) splits the polyoma virus DNA (*114*) and that of simian virus 40 (*113, 171*) opposite ss-breaks (introduced by γ-radiation, ^{32}P decay, or DNase I). Acid or base titrations of γ- and X-irradiated DNA solutions indicate the rupture of a large number of hydrogen bonds (*172–175*). Rupture of hydrogen bonds in terminal base-pairs at premelting temperatures was observed in double-stranded oligodeoxynucleotides (*119*) using proton nuclear magnetic resonance (Section II, F, 1). Neither the introduction of ss-breaks into DNA by the action of DNase I nor decreasing its size by sonication (M. Vorlíčková and E. Paleček, unpublished) yield any changes in the CD. The degradation of DNA by moderate doses of γ-radiation (Fig. 18) produces a slight increase in the positive CD band (*176*). However, conformational changes in the vicinity of bases damaged by radiation could be responsible for these CD

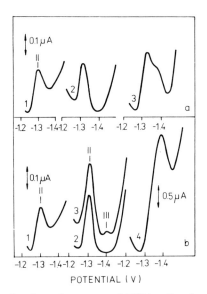

Fig. 17. Differential pulse polarograms of UV-irradiated and γ-irradiated DNAs. (a) UV radiation: DNA was irradiated in the concentration of 460 μg/ml in 0.0075 M sodium phosphate with 0.01 M EDTA, pH 6.9. Curve 1, control; 2, 2.1×10^4 erg/mm^2; 3, 6×10^5 erg/mm^2. (b) γ-Radiation: DNA was irradiated under the conditions given in (a). Curve 1, control; 2, 10^4 rads; 3, the sample 2 heated 6 minutes at 50°C and quickly cooled; 4, 6×10^5 rads. The differential pulse polarograms were measured at a Polarographic Analyzer PAR 174 in 0.3 M ammonium formate with 0.1 M sodium phosphate, pH 6.9, at a DNA concentration of 400 μg/ml.

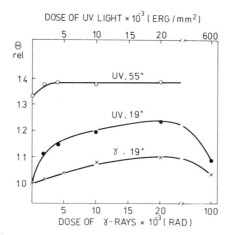

FIG. 18. Dependence of the relative ellipticity at 275 nm on the dose of UV and γ-radiation. DNA in a concentration of 80 μg/ml was irradiated in the medium given in Fig. 17. UV radiation: ●——●, CD was measured at 19.5°C; ○——○, at 55.5°C. CD of γ-irradiated samples, ×——×, measured at 19.5°C. θ_{rel}, the ratio between the ellipticity (275 nm) of the given sample and of the unirradiated sample at 19.5°C. The measurements were performed on Roussel–Jouan Dichrograph, Model 185.

changes. High radiation doses (around 10^6 rads) cause a decrease in the positive CD band (Fig. 18) (*176, 177*), the pulse polarographic peak shows a marked increase (Fig. 17b), and its potential shifts to negative values closer to the potential of peak III (*176*). These results have been explained by structural changes, brought about by partial denaturation. It should be emphasized that, with high radiation doses to DNA, large decreases in viscosity, changes in hypochromicity, and decreases in t_m occur (*172–175, 178*). In contrast, no detectable changes in t_m and the hypochromicity and only small changes in the DNA viscosity due to ds-breaks (*169, 172–175*) occur with doses of up to 10^4 rads.

The data suggest that a few ss-breaks in DNA do not lead to local denaturation, including unstacking. It might be presumed that, if the bases on either side of the chain interruption unstack, a region of flexibility will result. However, interruption of 0.1 to 0.2% of the phosphodiester bonds by DNase I produces no change in viscosity (*179, 180*). Sedimentation-viscosity measurements of "nicked" DNA at various temperatures (*181*) showed no increased flexibility at room temperature, but, near the beginning of melting, the viscosity of "nicked" DNA was lower than that of control DNA. This decrease in the viscosity was explained by formation of short ss-fragments splitting off between two closely spaced ss-breaks. Formation of such fragments has been demonstrated experimentally (*169, 176, 182*) using pulse polarography (Fig. 17b).

2. BASE MODIFICATION

Ultraviolet radiation induces formation of pyrimidine dimers in DNA (*183*); other types of damage (pyrimidine hydrates, ss- and ds-breaks, etc.) are formed to a substantially lesser degree. The irradiation of DNA by high radiation doses (10^5 to 10^6 erg/mm^2) leads to changes in a number of DNA properties that indicate partial denaturation (*183, 184*). Such doses causes, among other things, increased height and broadening of the pulse-polarographic peak II (Fig. 17a) or the appearance of peak III and a decrease in the positive CD band (*176, 185*). The t_m value decreases, and a broadening of the melting curve and a large increase in the rate of formaldehyde reaction occur (*183, 184*).

In contrast, small and moderate doses (10^3 to 10^4 erg/mm^2) produced some qualitatively different changes. The dependence of the positive CD band on the dose of 254 nm UV (Fig. 18) (*176, 185*), or of 313 nm UV (*186*) in the presence of acetophenone as a sensitizer (to introduce selectively thymine dimers into DNA) increases. The steepness of the dependence of the ellipticity on the radiation dose (254 nm) decreases (Fig. 18) as the temperature increases (*176*); at 55°C the amplitude of the positive CD band is almost independent of the dose. Further, a slight but significant increase in the height of pulse polarographic peak II is observed, while peak III is completely absent (Fig. 17a). Other changes are in qualitative agreement with those observed for high doses: the initial rate of unwinding under the action of formaldehyde increases with thymine dimer formation (*183, 187*). However, the concentration of defects detected by the kinetic formaldehyde method was much less than the thymine dimer concentration. Irradiation of superhelical and nonsuperhelical duplex DNAs made these DNAs more sensitive to treatment with "region-specific" nucleases (*111, 187a, 188*). The extent of "nicking" by low enzyme concentration was directly proportional to the UV dose. At much higher enzyme concentrations, in addition to ss-breaks, ds-breaks also occur in the irradiated DNA.

Sodium bisulfite selectively deaminates the cytosine residues in ssDNA and does not cause strand breaks (*189*).[4] In dsDNA, with the amount of modified residues below 1%, no changes in the CD spectra were observed, while the pulse-polarographic peak II decreased (*190*). This decrease was ascribed to selective modification of the cytosine residues (to polarographically nonreducible uracil) contained in open, polaro-

---

[4] See review by Hayatsu in Volume 16 of this series.

graphically reducible regions. This result also showed that peak II cannot be caused by the reduction of cytosine residues released randomly from the double-helical structure due to "breathing." If modification or polarographic reduction of the bases in dsDNA occurred at random, the decrease in the number of bases available for the polarographic reduction would be so low that no lowering of peak II would be observed.

Covalent binding of the carcinogen N-acetoxy-2-acetylaminofluorene to 1.7% of the base residues (guanine and perhaps adenine) of dsDNA gave an increase in the initial rate of formaldehyde action compared to unmodified DNA (*80, 191*), while almost no changes in the positive CD band were observed. It would be interesting to know how other types of chemical modification (e.g., methylation) of isolated nucleotide residues influence the local structure of the double helix. While many papers have been devoted to studies of DNAs containing greater proportions of bases modified with various agents, further data on the properties of DNAs containing less than 1% of modified nucleotide residues are still missing.

3. THE RELATIONSHIP BETWEEN THE NATURE OF LOCAL CONFORMATIONAL CHANGES AND THE NATURE OF ANOMALIES IN THE DNA PRIMARY STRUCTURE

The evidence summarized in the preceding paragraphs indicates that anomalies in the primary structure may cause local conformational changes in the DNA double helix that are not identical to denaturation. Lazurkin et al. (*76–78, 164, 165*), who studied differently damaged samples of DNA by the kinetic formaldehyde method, assumed that, in the vicinity of breaks, thymine dimers, and other anomalies of DNA primary structure, "defects" arise in the secondary structure whose character does not depend on the chemical nature of the anomaly. For the anomaly-affected places, the authors (*78*) suggested a very high probability (close to 1) of fluctuating opening of base-pairs (the size of the opening region was assumed to be 1 to 2 base-pairs) including rupture of hydrogen bonds and unstacking of bases. In other words, the authors assumed that a defect corresponds only to local denaturation of DNA.

However, the results summarized in the foregoing paragraphs suggest also that various types of DNA damage may produce various conformational changes. For example, dimerization of 1% of the thymine (ca. 0.29% of all bases) in DNA led to a significant increase in the positive CD band (ca. 10%) and to a relatively small increase in pulse-polarographic peak II (ca. 5%) (*176*). In contrast, the cleavage of about 0.2% of the phosphodiester bonds by DNase I led to a marked increase in peak II (50%) (*168, 170*), while the CD remained virtually unchanged (M. Vorlíčková and E. Paleček, unpublished observations); changes in the CD did not appear even when the number of bonds broken was several times higher. The velocity of DNA unwinding caused by formaldehyde was the same for defects associated with breaks (*165*); however, it was seven to eight times slower at pyrimidine-dimer sites than at chain-break sites (*192*). DNA treated with DNase I reacted with formaldehyde at about the same rate as gamma-irradiated DNA containing the same amount of ss-breaks (*167*). This finding suggests that unwinding might also proceed substantially more slowly at defects associated with bases damaged by gamma radiation than at chain breaks.

Thus it may be concluded that the character of a local structural change depends on the chemical nature of the anomaly. If the fraction of anomalies is very high and the probability of formation of clusters greatly increases, partial denaturation may prevail. If, however, the fraction of anomalies is very low (not exceeding tenths of a percent), local structural changes similar to those of premelting occur. The essence of these structural changes is discussed in Section III, E, 3 in connection with the nature of premelting.

D. Superhelical Turns

It has been demonstrated in recent years that the presence of superhelical turns in natural covalently closed circular dsDNAs (*193*) may induce local conformational changes in the DNA double helix. It is interesting that, similarly to the conformational changes mentioned in Section III, C, conformational changes in superhelical regions do not seem to be identical with denaturation and that information on these changes was obtained mostly using methods that are useful in premelting studies.

Measurements of changes in buoyant density during the alkaline titration of superhelical DNA revealed the early melting of about 3% of the closed molecule (*194*). On the other hand, no early melting was observed in studies of thermally induced helix-coil transition in 7.2 M $NaClO_4$ as followed by optical density measurements (*138*). Simultaneous optical density and sedimentation measurements of superhelical and relaxed forms of phage φX174 and PM2 DNAs in the presence of formaldehyde (*195*) revealed a marked difference in the reactivity of these two forms: the superhelical form started to react with formaldehyde at temperatures 15°–20°C lower than those at which the "nicked," relaxed circles reacted. At 30°C (when only the superhelical DNA reacted with formaldehyde), exponential kinetics was observed, which gave rise to pseudo first-order plots (this type of kinetics is typical of tRNA containing unpaired bases, while the dependence for native duplex DNA is sigmoidal); the rate constant was lower by a factor of two than that for the ss-form. The initial formaldehyde thermal transition was accompanied by about a 15% increase in the number of superhelices as compared to the number of superhelices in the native molecule (*193, 195*). Superhelical DNA of phage PM2 reacts with methyl mercuric hydroxide (a chemical probe for unpaired bases) more rapidly (*196, 197*) than does the nicked DNA. Early binding of CH_3HgOH, detected by an increase of sedimentation velocity, occurred only with DNA molecules having a superhelix density greater than −0.025 (*197*). On the basis of a comparison of the HCHO and CH_3HgOH reactions, it was concluded that the agents react with (A + T)-rich regions in the superhelical DNA (*196*). Modification of the DNA of simian virus 40 with another probe for unpaired bases, i.e., N-cyclohexyl-N'-2-(4-methylmorpholinium)-ethylcarbodiimide resulted in an increase in the sedimentation velocity of this superhelical DNA (*197*), while no change in $s_{20,w}$ was observed with the "nicked" form. It was estimated that 108 mol of the diimide were bound per mole of this superhelical DNA.

The observed enhanced reactivity of the superhelical DNAs (compared to "nicked" DNAs) could reflect either alterations in their secondary structure or an increase in their structural motility or "breathing" rate (Section III, D). Increased conformational motility of superhelical DNA might result from bending or torsional stresses in the DNA secondary structure (*198*). To distinguish between two possible explanations of the observed enhanced reactivity of the superhelical DNA, hydrogen-exchange analysis was performed and the results were compared with those of the formaldehyde reaction kinetics. Comparison of K_{conf} (characterizing the extent of a transiently open conformation) for various DNAs estimated from the formaldehyde reaction kinetics (Table V) showed great differences between superhelical and nonsuperhelical DNAs. On the other hand, the kinetics of the hydrogen-exchange process (at 0°C) for superhelical and "nicked" forms were very similar. The secondary structure of both forms was thus characterized by the same "breathing" rate for tritium-hydrogen exchange.

Electron microscopic examination of a superhelical DNA treated with formaldehyde revealed no denatured regions (*199*). However, the alkaline-denaturation mapping of the formaldehyde-treated DNA revealed six loops at low denaturation conditions where no denaturation was visualized for the "nicked" DNA. Introduction of superhelical turns into lambda phage DNA, E. coli 15 plasmid DNA (*200*) and in φX dsDNA (*201*) resulted in an increase of the positive CD band (Section II, B, 1), while the negative CD band was little affected. The changes in the

TABLE V

COMPARISON OF THE CONFORMATIONAL CONSTANTS OF LINEAR DUPLEX
DNAs AND SUPERHELICAL DNAs, USING HCHO AS A CHEMICAL PROBE[a]

DNA	Temperature (°C)	$K_{conf.}$	References
1. Calf thymus	60	0.01	79
2. T7	58	0.093	75
3. "Nicked" PM2 and φX-RF	30	NM[d]	
4. φX-RF Superhelical	30	0.61	195
5. Calf thymus	30	1×10^{-7}	Extrapolated data 79
6. PM2[b] superhelical	30	0.56	198
7. φX-RF[c] single stranded	30	1.0	195

[a] After Jacob, Lebowitz and Printz (198).
[b] Pseudo first (4% HCHO) order $K_{obs} = 1.29 \times 10^{-2}$.
[c] Pseudo first (4% HCHO) order $K_3 = 2.30 \times 10^{-2}$.
[d] NM, not measurable.

positive CD band were proportional to the degree of superhelicity (200), and were more pronounced at higher ionic strengths [about 2 M (200) and 4 M Na⁺ (201) compared to 0.1 M (200) and 0.3 M Na⁺ (201)].

Various nucleases considered to be "single-strand specific" or "region specific" (Section II, E) digest preferentially the superhelical DNA as compared to its "nicked," relaxed form (110–114, 201a, 202). With increasing concentration of the enzyme, increased conversion of superhelical DNA to its relaxed form (with one ss-break per molecule) was observed (112, 201a). The digestion rate depended on temperature and the ionic strength (Section II, E). At higher superhelical densities (roughly above 0.03), the rate of introduction of the first ss-break with the endonuclease from *Neurospora crassa* and with mung bean nuclease I increased sharply with the degree of superhelicity (197, 202). For nuclease S1 (113, 114), mung bean nuclease I (202), and *Neurospora crassa* endonuclease (197), positional specificity of superhelical DNA cleavage was demonstrated. Superhelical simian virus 40 DNAs, both modified (by the diimide) and unmodified, were digested to fragments of the same size (197); binding of diimide to certain DNA fragments corresponded to the sites attacked by nuclease S1. The superhelical DNA treated with formaldehyde was sensitive to lower concentrations of the endonuclease of *Neurospora crassa*, while no sensitization of the "nicked" form was found. The sensitization of superhelical DNA increased as the time of formaldehyde reaction was prolonged. As the temperature of the formaldehyde reaction was raised, both DNA forms became increasingly sensitized to the enzyme. All superhelical molecules were cleaved following formaldehyde treatment at 36°C. On the other hand, temperatures of 42°–58°C were necessary to sensitize the "nicked" DNA.

The evidence summarized in this section clearly shows that the introduction of superhelical turns into the DNA molecule results in changes in its secondary structure. These changes occur at a small number of sites rich in A + T (*196*) and affect only several percent of the whole molecule. The extent of these changes depends on the degree of superhelicity. The altered regions can be characterized as more open (cleavage by "region-specific" nucleases, reaction with formaldehyde, CH_3HgOH and carbodiimide) compared to the rest of the molecule; they are, however, not identical with single-stranded (melted) regions. This dissimilarity with ssDNA can be documented by the following evidence. (a) For the "region-specific" endonucleases, the initial ss-cleavage (nicking) rate on superhelical DNA is much lower than on ssDNA; in the case of mung bean nuclease I, it is only 1/450 of that on ssDNA (*202*). The reaction rate of formaldehyde with superhelical DNA is also lower than with ssDNA (*195*). (b) Sedimentation velocity changes caused by modification of superhelical DNAs with CH_3HgOH and carbodiimide (*196, 197*) can hardly be explained by reaction with preexisting denatured regions. It is more probable that the chemical agents disrupt labilized double-helical regions (*197*), which results in frictional changes manifested by changes in sedimentation. (c) The positive CD band increases with the degree of superhelicity (*200*), while it decreases as a result of DNA melting (Section II, B, 1). (d) No DNA synthesis by DNA polymerase I was detected in the presence of oligodeoxynucleotides on the superhelical DNA (*202*); however, synthesis was readily initiated on ssDNA.

We can thus see that the characteristics of the structural changes induced in closed-circular DNA by the introduction of superhelical turns are similar to those of premelting conformational changes (Sections II and III, E) as well as of local structural changes due to UV-induced thymine dimers (Section III, C). Further discussion of the nature of altered regions in superhelical DNA is given in Section III, E.

E. Nature of DNA Premelting

As shown in Section II, evidence for the existence of premelting conformational changes in DNA has come from basically different methods. The authors have generally endeavored to offer explanations of premelting that agree with their own experimental data, without treating the results obtained by other techniques in greater detail. In this section, we try to discuss the nature of premelting using data obtained by various methods. However, we first want to show that premelting is not identical with melting nor with a B-to-A transition, and probably not even with "breathing."

1. STRUCTURAL TRANSITIONS THAT CAN BE RULED OUT AS THE SOLE EXPLANATION OF PREMELTING

a. Melting. At first sight, the most simple explanation seems to be that premelting is qualitatively consistent with melting, but with the proviso that it affects only a small number of base pairs, a number that falls below the sensitivity limit of the usual spectrophotometric measurements at 260 nm. The results of pulse-polarographic and CD measurements are, however, at variance with this explanation: (i) No polarographic peak III, typical for denatured DNA, appears as a result of a temperature increase in the premelting zone (formation of about 0.5% of denatured DNA due to a temperature increase from 20° to 30°C can be excluded). (ii) Peak II, typical for dsDNA, gradually increases in height with increasing temperature (Figs. 10 and 13) showing that the extent of premelting changes is not below the sensitivity limit of the polarographic method. (iii) Relatively large changes in the positive CD band are the opposite of those caused by melting (Figs. 2 and 3). Thus it can be concluded that premelting is qualitatively different from melting.

b. B-to-A Transition. The increase in the positive CD band with increasing temperature (Section II, B) could be taken as evidence for the transition of B-DNA to the A form. However, a very large drop in the ellipticity at 210 nm is equally characteristic of this transition (*203*). On raising the temperature no marked drop at 210 nm is observed (*14*). Besides, the DNA of phage T2r⁺, which is glucosylated and does not assume the A form, gave changes in the CD with temperature similar to those of other DNAs (*14*). DNAs with high A + T content (Fig. 5), poly(dA − dT) and poly(dA · dT) (Fig. 4), which are equally incapable of adopting the A-type helix (Section III, A), behave similarly. In addition, recent studies (*39, 41*) of the B-to-A transition in ethanolic solutions have shown that this transition is independent of the temperature. Similarly, comparison of the Raman spectra of DNA at various temperatures with those of known backbone conformations (*49–51*), along with the results of polarographic measurements, exclude the possibility of a B-to-A transition [in dsRNA in the A′ form, the bases are mostly inaccessible for electroreduction at premelting temperatures (Fig. 13)]. Thus it seems clear that premelting is not identical with transition to the classical A form of DNA.

2. "Breathing"

In some papers, especially in those studying the reactivity of DNA with formaldehyde (Section II, D), the behavior of DNA in the premelting region is explained by a frequent opening and closing, or "breathing," of the base pairs. According to von Hippel et al. (79, 207), random thermal motions occasionally twist one segment of the DNA helix relative to another, resulting in unwinding and twisting of the intervening helical sections. The existence of local structural fluctuations at temperatures below melting is considered in many theories of DNA unwinding, and it is assumed that, in a large DNA molecule, these local fluctuations comprise the nuclei from which the melted region grows as the temperature is increased. Direct evidence for DNA "breathing" comes from hydrogen-exchange analysis (204). To estimate the extent to which DNA "breathing" could be responsible for DNA premelting behavior, a concise summary of the results of studies of DNA using this method follows.

Hydrogen Exchange Analysis. It is known from protein hydrogen-exchange experiments that hydrogens involved in intramolecular hydrogen-bonding exchange more slowly with the solvent than do other exchangeable hydrogens. Reasoning from protein studies, hydrogen-exchange analysis should reflect any transconformation reaction in DNA that involves changes in the interchain hydrogen-bonding. Experiments with double-stranded polynucleotides (204–211) show that the protons involved in hydrogen-bonding (and thus buried in the interior of the double helices) do exchange with solvent protons. Based on this fact, it has been suggested that the double helix must participate in some kind of reversible opening process. On the other hand, hydrogen-exchange analysis has yielded certain unexpected results. The rate of exchange of the hydrogens involved in hydrogen-bonding is independent of the DNA base composition (207) and of the presence of agents destabilizing the double helix. To explain these facts, McConnell and von Hippel (207) assumed that at least two locally open conformations occur in native DNA at temperatures below t_m. One is formed noncooperatively without base unstacking and with a low activation energy [$\Delta H^\pm = 18$ kcal/mol (205) or 24 kcal/mol (211)] and prevails at temperatures far below t_m. The other form is cooperative, with a high activation energy, characterized by both hydrogen-bond breakage and unstacking of adjacent bases, and dominates exchange at temperatures close to t_m. It was suggested (207) that destabilizing salts influence only the stacking–destacking equilibrium, and that these agents have little or no effect on the exchange proceeding via the low energy conformation, which does not involve unstacking. On the other hand, increasing the temperature (at constant t_m) markedly increases the exchange rate (*211*).

The numbers of slowly exchangeable hydrogens per nucleotide pair quoted in earlier papers (*206, 207*) are very close to the number expected for the internucleotide hydrogen-bonded hydrogens. However, recent improvements in experimental techniques (*208*) disclose the inaccuracy of these data. It has been shown that all amino hydrogens of the DNA bases participate in the slow exchange (*208, 209*). In agreement with this finding, poly(rG · rC) and poly(dG · dC) yielded 5 slowly exchanging hydrogens per base-pair (*210*) while for poly(rA · rU), poly(rI · rC) and poly(dA — dT), 2 or 3 hydrogens were found (*208, 210*).

Quite recent work (*212, 213*) has raised doubts about some aspects of the earlier models for DNA hydrogen-exchange, in particular the identification of the hydrogen atoms involved in the overall exchange process. From studies of the influence of H_3O^+, OH^-, and general acid and base catalysts on the different exchangeable hydrogens in adenine- and cytosine-containing double helices, it appears that the relative rates of the hydrogens are determined not by their "exposure" in the native state, but by their intrinsic exchange chemistry. However, if the chemical exchange mechanism is correctly known, the parameters of opening of the double helix can be computed. The equilibrium opening constant, the opening rate constant (0.06 s^{-1}) and the closing rate constant (about 6 s^{-1}) of poly(dA — dT) and poly(rA — rU) show that, at 0°C, the average base-pair is open about 1% of the time (*212*). The opening rate constant for poly(rG · rC) is 0.01 s^{-1} and the closing rate constant is about 25 s^{-1} (*213*). However, uncertainties in the cytosine amino group analyses do not allow us to estimate the latter two numbers accurately; they could be in error by as much as a factor of ten.

The slow opening and closing rates and the large equilibrium opening constants indicate that opening of the double helix is a cooperative process involving relatively long polynucleotide segments (one or more helical turns) and that stacking of bases in the open segments is not substantially influenced. It is supposed (*212*) that open long-lived loops migrate by a twisting random-walk process along the helical length.

Hydrogen-exchange analysis shows that local structural fluctuations do occur in dsDNA. However, can premelting be explained solely in terms of these fluctuations? If so, these fluctuations would have to exhibit a significant dependence on the sequence of nucleotides and its anomalies, would have to be sensitive to the presence of destabilizing agents, etc. None of these properties has so far been proved by hydrogen analysis. Can this be attributed to insufficient elaboration of the method? or to the fact that structural fluctuations are not identical with premelting and can, when some other methods are used, represent only a kind of "background," independent of the factors mentioned above? These questions cannot yet be answered with certainty. Hydrogen-exchange analysis works usually with degraded DNA at temperatures near 0°C, which makes it difficult to compare the results of this method with those of other methods. Thus we can conclude only that premelting cannot be explained solely on the basis of essentially random structural fluctuations. However, if we take into account the fact that polymorphic dsDNA with anomalies in its molecule contains segments that differ considerably in their stability at premelting temperatures, we can imagine that thermal motions might result in a highly selective "breathing" of DNA segments. Unequivocal experimental evidence supporting this concept is lacking. The results of the study of DNA reactivity with formaldehyde, as well as of polarographic measurements, can be explained, not only in terms of "breathing," but also by the presence of permanently open regions in the double helix. For the sake of simplicity, "breathing" will not be considered in the discussion of the nature of DNA premelting. However, this aspect of DNA structure dynamics undoubtedly deserves further investigation.

3. Terminology

We have demonstrated that the presence of superhelical turns or of isolated anomalies in the DNA primary structure (Section III, D, C) produces changes in the secondary structure that differ from those of denatured DNA and resemble those of premelting conformational changes (Section II). Thus it does not seem reasonable to term the regions in the vicinity of isolated thymine dimers, molecule ends, superhelical turns, etc., as locally "melted" or "single-stranded," because these terms can lead to confusion. The same is true of regions whose structure has been changed by temperature increase in the premelting region. It is probable that further advances of knowledge in this field will lead to precise terminology distinguishing among various types of local structural changes in the double helix. For the present, we can probably manage with terms such as locally "open," "distorted" or "premelted" ds-regions. "Open" ds-region refers to a region characterized by an increased ability of its bases to react with their environment (including weakening or rupture of hydrogen bonds), while the terms "distorted" and "premelted" regions are more general and may include various changes in interatomic angles and distances in the double-stranded region. The terms "melted" and "single-stranded" regions should be reserved for segments where the ds-structure is absent; in the "melted" segments (besides the absence of the inter-base hydrogen bonds), complete unstacking of bases is assumed. The term "intermediary form," devised originally for premelting structural changes, was based on the assumption of a very limited number of structural forms in which DNA can exist; this term does not fully correspond to present knowledge.

4. COMPLEXITY OF PREMELTING

The results summarized in Section II suggest that premelting conformational changes are of a very complex nature, and that a major proportion of them are reversible. While the results of CD measurements could be explained both by local structural changes and by uniform changes in the average parameters of the double helix, the results of polarographic (Sections II, C and III, C, 2), formaldehyde, and enzymic methods (Sections II, D and E) are hardly compatible with the latter explanation. Similarly the nonmelting structural changes caused by isolated anomalies and by superhelical turns exhibit a pronounced local character (Sections III, C and D). It can be thus concluded that the premelting is connected with local structural changes; however, participation of changes in the average helix parameters cannot be excluded. It seems that the local opening of the double helix related to weakening or rupture of hydrogen bonds as well as base tilting dominate over the complex of premelting changes.

a. Local Opening of the Double Helix, Rupture or Weakening of Hydrogen Bonds. At elevated premelting temperatures, the formaldehyde and polarographic methods (Sections II, D and C) reflected the increased ability of a portion of the bases in the double helix to react with their environment. This phenomenon can be explained through the existence of opened regions resembling denatured DNA in some of their properties. This resemblance probably consists of rupture or weakening of the hydrogen bonds. Raman spectroscopy supplies evidence supporting the rupture of hydrogen bonds (48, 49), and proton nuclear magnetic resonance (117–119) demonstrates hydrogen bond rupture in double-stranded oligodeoxynucleotides. Changes in DNA susceptibility to region-specific nucleases (Section II, E) and the cross-link induction by UV light in the premelting region (125, 127) (Section II, F) also support local opening of the double helix.

However, the polarographic reducibility of adenine in an A·T pair might also be conditioned by, e.g., Hoogsteen pairing (140) with the hydrogen bonds intact. In a Hoogsteen A·T pair, the reduction site of adenine is outside the hydrogen-bonding system (Fig. 19a). The reducibility of adenine contained in the double helix has been observed in protonated poly(adenylic acid) (214) in which the reduction sites are not involved in hydrogen-bonding (Fig. 19b). However, no evidence supporting the existence of Hoogsteen A·T pairing in DNA yet exists.

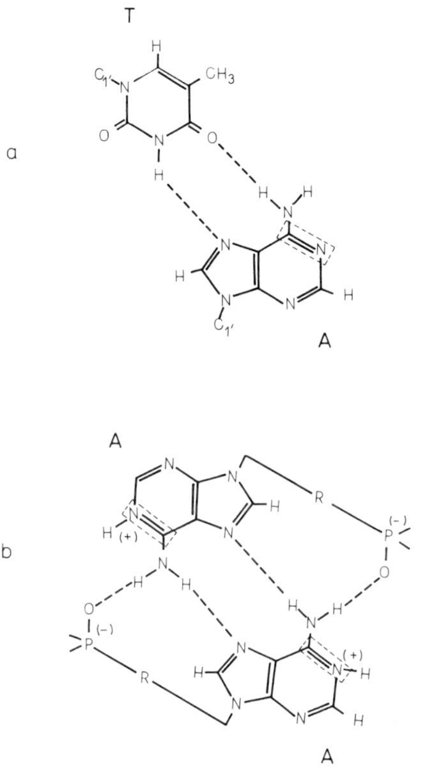

FIG. 19. Schematic presentation of hydrogen bonding and the primary site of the polarographic reduction. (a) In the Hoogsteen A · T pair; (b) in the protonated, double-stranded poly(adenylic acid). The reduction sites of adenine are indicated by rectangles (interrupted line).

b. Base Tilting, Changes in Angles, and Distances between Adjacent Bases. The premelting changes in CD spectra could be explained especially by changes in base tilting and by an increase in the distance of the bases from the helix axis. These changes most strongly affect the rotational strength of the DNA-positive CD band (215–217) and, considering the local character of premelting, they appear to be the most probable conformational changes responsible for the relatively large changes in CD (Section II, B, 1). An increase in the distance of the bases from the helix axis might be expected in the open ds-regions. Evidence in support of this possibility is provided by the qualitative agreement between CD results and polarographic measurements (i.e., a small influence of the ionic strength on premelting at moderate ionic strengths, dependence on the cation nature, no or relatively small premelting changes in polydeoxyribonucleotides containing G · C pairs only, but large changes in A · T polynucleotides, substantially smaller changes in dsRNA than in DNA, etc.) (Sections II, B, 1 and II, C); this agreement is understandable assuming that the distorted regions in DNA reflected by the two methods overlap to a certain extent.

Evidence supporting base tilting is that poly(dG · dC), which shows a preference for the A form in fibers (where bases are tilted) (Section III, B), exhibits almost no changes in its CD spectra due to temperature changes in the premelting region (Fig. 4), and that the changes in the CD spectra of dsRNA are minimal over the premelting region and are the opposite of the CD changes observed in DNA. Changes in the base-pair tilt or in the angle between the bases in a pair appear to be the only possible explanation of the observed dependence of the X-ray diffraction patterns on the DNA (A + T)-content at room temperature (Section III, B).

c. *Backbone Conformation.* Two related methods could indicate changes in the deoxyribose–phosphate backbone conformation. Infrared spectroscopy indicates small changes in the backbone conformation, whereas Raman spectroscopy suggests that the geometry of the backbone remains intact up to the melting point (Section II, B, 2). The dependence of the sedimentation of superhelical DNA on temperature indicates unwinding of the double helix at higher temperatures (Section II, F), which should result in backbone conformation changes. However, Wang's assumption (*137*) of a uniform change occurring in the angle between each pair of adjacent bases in the double helix is open to question. The results summarized in Section III, D suggest that, in superhelical DNA, local changes in the secondary structure occur that could influence Wang's data (*136, 137*) on the change in the average rotation of the helix with temperature (Section II, F) to a considerable degree. Thus it can be concluded that unwinding does occur, but most probably not as a uniform change in the angle between each and every pair of adjacent bases, but rather as changes with local character.

d. *Stacking.* The question of whether unstacking of bases participates in the premelting changes is not easy to answer. Recent measurements of difference absorption spectra (Section II, B, 3) indicate that unstacking might occur in a small number of bases. However, most of these changes are irreversible. Raman spectra results could be explained in terms of reversible changes in base stacking (Section II, B, 2). Since the changes in the UV absorption spectra are slight, it is necessary to assume either that unstacking involves only a small number of bases (corresponding to the results of UV absorption spectroscopy), or that the Raman spectra actually reflect changes in the base arrangement and not complete unstacking, which occurs during melting and is reflected by changes in the hypochromicity.

If unstacking occurs to a substantially lesser degree than the rupture or weakening of hydrogen bonds, it is necessary to suppose that the segments of the double helix with non-hydrogen-bonded (or weakly hydrogen-bonded) but stacked bases are stable under the given conditions. It has been theoretically derived that stacking forces suffice to maintain helix stability even in the absence of hydrogen bonding (1). The neutral ss-forms of poly(C) and poly(A) contain helical regions with stacked non-hydrogen-bonded bases (218). However, Von Hippel and Wong (79) suggest that stacking as manifested by hypochromism occurs in DNA only in double-helical regions containing hydrogen-bonded base-pairs. They support their view by the observation that the fraction of bases that remained stacked, as measured by hypochromism after thermal denaturation of DNA, is approximately equal to the fraction of the double helix remaining as determined by hydrogen-exchange analysis. If it is borne in mind that the hydrogen-exchange analysis was carried out in an arrangement where a considerable portion of the slowly exchanging hydrogens could not be detected, and that the chemical exchange mechanism was not correctly known, then this argument (79) tends to become irrelevant. Recent nuclear magnetic resonance studies (Section II, F) show that elevating the temperature in the premelting region ruptures hydrogen bonds in double-stranded oligonucleotides (117–119) without simultaneous changes in the hypochromism. Thus it can be expected that, in dsDNA, also, rupture or weakening of hydrogen bonds at premelting temperatures need not result in complete unstacking of the bases.

e. Nucleotide Sequence. The CD, polarographic and enzymic methods (Sections II, B, C and E) show that premelting occurs preferentially in (A + T)-rich regions. It is probable that there is a direct connection between this phenomenon and the specific structure of the (A + T)-rich DNAs established by X-ray analysis (Section III, B). Polarography and the enzymic methods suggest *opening* of the structure of (A + T)-rich DNAs at premelting temperatures, which probably includes breakage or weakening of the hydrogen bonds. The two methods indicate consistently that open (A + T)-rich ds-regions may already exist close to room temperature. It then becomes important to decide whether X-ray analysis reflects these open ds-regions or corresponds only to the structure of the regions that are "closed" at room temperature. For the time being, only a single piece of evidence is available, obtained from the study of lambda phage DNA cleavage by "region-specific" nuclease (87), from which it follows that the structure of an (A + T)-rich ds-region can exhibit an open character at room temperature even when located in the center of the intact molecule. The length of this region in lambda phage DNA represented about one-tenth of the entire molecule, i.e., roughly 4000 base-pairs.

In addition to nucleotide sequences rich in A and T, the presence of various repetitions of nucleotides [found in DNAs from various sources (e.g., *219–221*)] might have a certain importance for the premelting conformational changes. A special role might be played by inverted repetitions, characterized as "*palindromes*" (*220–224*). If palindromes are present in a single-stranded DNA, they reassociate by an intramolecular reaction to form "hairpins." It is possible that some of the open regions formed in dsDNA by the increase in thermal energy in the premelting region could migrate until they reach a segment with intrastrand complementarity and form a hairpin structure. The possibility of such a mechanism is suggested by the premelting structural changes in poly-(dA − dT) (*134, 135*) observed by means of viscosimetry and absorption spectrophotometry (Section II, F) and by the greater degree of the premelting polarographic changes in poly(dA − dT) than in poly(dA · dT) (M. Vorlíčková and E. Paleček, unpublished observations).

f. Hydration. The existence of a hydration shell surrounding the DNA molecule has been demonstrated in numerous experiments (*225–233*). However, knowledge of the detailed structure of this hydration shell, which probably includes water molecules bound to DNA with different affinities, is missing. Eisenberg (*225*) concluded in his review that two water molecules could be firmly associated with each phosphate group. Up to six molecules of water could be bound less firmly in the grooves of the double helix. Further molecules could be only loosely connected to DNA. Lewin (*226*) assumes that the DNA double-helical structure is buttressed by hydrogen-bonded water-bridging and cation water-bridging. A series of papers (*228–233*) indicates that elevating the temperature leads to DNA dehydration. The nuclear magnetic resonance spin-echo technique (*229*) shows that, after a gradual decrease in the DNA hydration in the premelting region, a rapid drop in the melting region follows. On the other hand, dilatometry (*232*) shows no deviations from a monotonic temperature dependence in the melting zone. The difference in these results could be connected with the different sensitivities of the methods to the water molecules bound to DNA. It is probable that changes in the hydration shell with increasing temperature in the premelting region lead to destabilization of the double helix and facilitate DNA premelting. This assumption is supported by a correlation between the heat of hydration of the cation and the magnitude of the CD (*22*) and polarographic changes (Fig. 10) at elevated temperatures.

High concentrations of salts in solution of DNA may cause partial dehydration of the molecule (226). Even then (as with premelting), structural changes occur in DNA that result in almost no changes in the hypochromism (22). It has been concluded from sedimentation measurements on superhelical DNAs (Section II, F), that increasing salt concentrations ($NaClO_4$ and $MgClO_4$) and increasing temperatures have an additive effect (138, 139) and lead to unwinding of the double helix. On the contrary, in the CD spectra the same effect (decreases of the positive band) was produced by increasing the salt concentration (Fig. 6a) and decreasing the temperature (22, 33, 234). An increase in salt concentration, like an increase in temperature, leads to an increase in the pulse-polarographic peak (Fig. 11). However, in contrast to the effect of temperature changes, increasing salt concentration still causes a marked widening of the peak. In dsRNA, increasing salt concentration (CsCl or $NaClO_4$) produces a greater increase in the pulse-polarographic peak than in DNA, but it produces almost no changes in the CD spectrum of dsRNA (E. Paleček, unpublished). All these facts suggest that premelting structural changes in DNA produced by temperature changes and by changes in the salt concentration cannot be identical, even though these two types of changes are accompanied by dehydration. Thus changes in the DNA hydration may affect the premelting conformational changes; however, they are probably not the only cause.

g. *Extent of Premelting.* It is obvious that none of the methods so far applied to the study of premelting can reflect the whole complex of premelting structural changes. Thus quantitative determination of the extent of all the changes, e.g., due to a change in the temperature by 10°C, is difficult. However, knowledge of the precise extent of the partial changes detected by each method would be useful. Unfortunately, these methods do not permit even such a determination, be it either in principle or because they have not yet been sufficiently developed. Polarography seems to come the closest to this aim, but for purposes of accurate determination, some information is lacking on the mechanism of the electrode process that dsDNA undergoes. So far, only a rough estimation is possible, from which it follows that the fraction of bases available for the electroreduction in native calf-thymus DNA varies around one to a few percent at room temperature (at moderate ionic strength and neutral pH); at 65°C, the fraction is approximately ten times higher.

5. STRUCTURAL CHANGES IN THE VICINITY OF ANOMALIES AND SUPERHELICAL TURNS

The data given in Section III, C and summed up in Table VI suggest different local conformational changes in the vicinity of isolated thymine

TABLE VI

SUMMARY OF THE RESULTS OF STUDIES OF THE LOCAL STRUCTURAL CHANGES IN DNA BY MEANS OF VARIOUS TECHNIQUES

	Increase in the positive CD band	Increase of the polarographic signal	Increase of the initial rate of HCHO reaction	Digestion with "region-specific" nucleases	Changes in hypochromicity at 260 nm
Thermally induced premelting	Yes	Yes	Yes	Yes	Below 1%
Pyrimidine dimers induced by UV-light	Yes[a]	Yes	Yes	Yes	No[a]
Chain breaks	No	Yes	Yes	Yes	No
Superhelical turns	Yes	ND[b]	Yes	Yes	No

[a] After low and moderate doses of UV radiation.
[b] No data available.

dimers and chain breaks. Introduction of the dimer is likely to lead to a certain strain in the double-helical structure; this strain may result in local structural changes including base tilt, and/or changes in the distances of the bases from the helix axis (CD) as well as helix opening (formaldehyde reaction, polarography, nucleases). It seems, however, that the open regions lie only in the closest vicinity of the dimer and are shorter than the open ds-regions in the vicinity of chain breaks. Absence of changes in CD spectra on introduction of chain breaks into DNA indicates that, in the vicinity of the break, no homogeneous changes occur in the distance and/or tilting of bases over a range comparable with those arising in the vicinity of dimers. It is possible that the region in the neighborhood of a chain break is less organized, and that the structural changes to which CD is sensitive are mutually compensated.

It is very difficult to determine the size of a structurally changed region in the vicinity of an anomaly. The difficulties arising here are caused by factors similar to those encountered in determining the extent of temperature-induced premelting, for example, data on the number of broken hydrogen bonds per ss-break induced by ionizing radiation range within the limits of about 3 to 60 (236, 237). The results are influenced by the method employed and by the fact that experimental data were frequently obtained at high radiation doses and extrapolated to lower values, assuming that qualitatively the same conformational changes occur after low radiation doses. From a study of the conformational changes in the vicinity of the pyrimidine dimer in the decanucleotide $(dT)_4dT<>dT(dT)_4$, which formed a duplex with poly(dA), Hayes et al. (238) concluded that about four of the hydrogen-bonded base-pairs in the duplex were disrupted by the presence of the dimer. This could imply that, in addition to the hydrogen bonds broken in the formation of the photodimer, only the hydrogen bonds in the nearest adjacent base-pairs on each side of the dimer were broken. Unfortunately, CD measurements, which could assist in a more accurate characterization of the structural changes in the vicinity of the dimer—not confined in DNA to the breakage of hydrogen bonds alone (176, 185, 186)—were not performed.

Superhelical turns in DNA produce changes in CD similar to those produced by thymine dimers (Table VI). In superhelical DNAs, these changes were attributed to tilting of the bases (200); however, the authors could not distinguish whether each base-pair in the double helix undergoes a small perturbation or whether local structural changes occur. Subsequent studies (Section III, D) render the latter explanation more probable. Moreover, the regions of superhelical turns are "open" (formaldehyde and other chemical reactions, nucleases). Lebowitz et al. explained their experimental data (195–199) (Section III, D) by the formation of hairpin structures. They assumed that ss-regions produced by superhelical torsional forces can migrate until they reach regions where sufficient intrastrand complementarity occurs. However, since the extent of conformational changes depends upon the superhelix density, it would be necessary to assume a considerable reserve of intrastrand complementarity (palindromes) in the circular DNAs studied. This assumption could be verified experimentally. On the other hand, it is possible that hairpin structures do not represent the only possible changes in the secondary structure of superhelical DNAs.

6. POLYMORPHY OF DNA SECONDARY STRUCTURE

On the basis of the preceding discussion, a schematic picture of the structure of natural linear DNA in solution under physiological conditions (e.g., at 36°C, moderate ionic strength, and pH 7) can be drawn. We can assume that the double-helical structure of the very long $(A + T)$-rich regions differs from the structure of the major part of the molecule and that some of the $(A + T)$-rich segments are open (Fig. 20). An open ds-structure can be assumed in the region of chain termini and/or in the vicinity of ss-breaks and other anomalies in the DNA primary structure. The exact changes in the open ds-regions will depend on the nucleotide

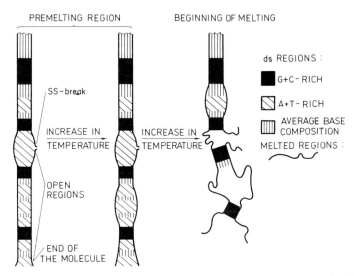

Fig. 20. Schematic presentation of premelting structural changes in double-helical DNA.

sequence as well as on the chemical nature of the anomaly. Most of the molecule will exhibit an average Watson–Crick B-structure with local deviations given by the nucleotide sequence. Elevating the temperature in the premelting region (Fig. 20) is likely to lead to the opening of other regions and, eventually, to expansion of the existing distorted ds-regions and to further structural changes. Thus the course of the conformational changes as a function of temperature (premelting) will be determined by the distribution of the nucleotide sequences and anomalies in the primary structure, and may have an almost continuous character.

Consequently, even if we do not consider "breathing," not only the architecture of a DNA double-helical molecule, but also its mechanics or dynamics can be taken into account.

To determine whether, e.g., only the (A + T)-rich molecule ends will be open at a certain temperature or also long A + T regions in the center of the molecule, further experimental research with better-defined samples of viral and synthetic nucleic acids will be necessary. Further work will undoubtedly provide new information on the details of the local arrangement of nucleotide residues in the double helix, as well as on DNA conformational motility. Thus a more accurate picture of DNA structure will emerge, whose characteristic feature will be polymorphy of the double helix, in contrast to the classical, highly regular DNA structure models.

IV. Relation between DNA Conformation and Function

In realizing its function, DNA interacts specifically with various substances, especially with the proteins. It is necessary that the interacting protein recognize, in the double-helical DNA, a specific site, determined by the nucleotide sequence, that represents but a very small fraction of the entire chromosome (e.g., about 10^{-5} of the *E. coli* genome). The question arises how a protein molecule can recognize a specific sequence on the double helix in which the bases are hidden inside. This problem has been considered in numerous reviews (e.g., 239–241). Here we confine ourselves to discussing this problem in connection with the premelting conformational changes.

Of the proteins binding specifically to DNA, the bacterial and viral repressors (*lac* and *lambda*) and the bacterial RNA polymerase (239–241) have been studied to the greatest extent. These proteins undergo recognition reactions with the operator and promoter regions, respectively; both reactions are extremely specific and rapid. Three major mechanisms of recognition have recently been considered (241): (i) recognition of a unique tertiary structure (e.g., hairpin loop); (ii) recognition of a unique DNA secondary structure (e.g., C-DNA); and (iii) recognition of a unique nucleotide sequence involving functional groups in the grooves of the double helix. Of these, the third is, according to Chamberlin (241), heavily favored in the repressor-operator interaction, while the first two can probably be ruled out. The third mechanism is based on the fact that access to the edges of the stacked bases is available from the grooves (240) of the native double helix. Thus a protein of the proper geometry (and charge) to make it possible to enter the grooves should be able to recognize A · T and G · C pairs from the grooves. However, the steric and other requirements of such recognition are very strict. The protein must possess a binding surface fitting into the grooves over a distance of at least 45 Å. Furthermore, the presence of a single incorrectly located functional group in the binding region would reduce the affinity of the protein for this region by several orders of magnitude, etc. The available physicochemical data make it unlikely that, e.g., *lac* and *lambda* repressors possess a structure capable to assume the necessary shape.

In this paper, the data summarized show that the DNA local structure depends on the primary structure and its anomalies, and that even in the premelting region, it depends on environmental conditions, especially on temperature. Such polymorphy of the DNA structure may facilitate recognition of the specific base sequence in the double helix. Local differences in base tilting, the existence of open ds-regions, differences in base twisting and stacking, changes in the helix pitch, hairpin formation, etc., offer a great variety of possibilities that can be utilized in the process of recognition.

Until recently, the existence of small distorted regions in the DNA double helix was practically not considered in the literature. It is therefore not surprising that no systematic studies of the relationship between the content of distorted (premelted) ds regions in DNA and the function of these regions have been undertaken. Nevertheless, certain data concerning this relationship have been obtained in connection with the studies of other problems, e.g., *in vitro* transcription studies with DNA-dependent RNA polymerase. It was shown that the introduction of ss-breaks into DNA markedly stimulates RNA polymerase activity (*242, 243*). This stimulation, observed with core RNA polymerase, was explained by formation of new sites for the initiation of RNA synthesis; a local conformational change was presumed in the vicinity of the DNA strand break (*243*). The holoenzyme of RNA polymerase binds firmly to DNA in the region of ss-breaks and at the termini formed by DNA degradation. RNA polymerase exhibited a preference for $(A + T)$-rich DNA segments (*244, 245*), and poly$(dA - dT)$ was the best template for the core enzyme. Still more markedly than the strand breaks, transcription was stimiulated by superhelical turns in the circular replicative form of phage ϕX174 DNA (*246*) and in circular DNAs of PM2 phage (*247*) and of simian virus 40 (*248–251*). The finding of 2-fold symmetry in the nucleotide sequence of the sites binding RNA polymerase (*253–255*) suggest that formation of hairpin loops may play an important role in the transcription process. It was further shown that the transcription of phage DNAs depends on ionic strength and temperature in the premelting zone (*241*) and that the optimum conditions depend on DNA source (*241, 252*). It may be presumed that temperature influences the DNA local structure and, in this way, the interaction of DNA with the RNA polymerase. It can be expected that further research will show the importance of local distortions in the DNA structure for the realization of their known and, as yet, unknown functions.

V. Concluding Remarks

DNA premelting has been demonstrated under conditions preceding thermal, acidic and alkaline denaturation of aqueous DNA solutions. Certain signs of premelting changes have also been shown to arise from the addition of organic solvent to DNA solution (*256*). It thus appears that premelting might be as general a phenomenon as DNA melting itself. However, whereas melting is an event related rather to the entire DNA molecule, premelting is more limited to the small portions of the DNA molecule that differ from the predominating part of the molecule. It is therefore not surprising that the premelting is more influenced by the DNA primary structure and its anomalies than is the melting. The t_m value in natural DNAs is practically independent of the nucleotide sequence (*3*), the presence of isolated single-strand breaks (*178*), molecular weight (*3, 257*) (down to about 6×10^5 daltons), etc. However, a dependence of t_m on nucleotide sequence also appears in synthetic polynucleotides with repeating (alternating) nucleotide sequences (*258, 259*), i.e., in cases where the nucleotide sequence characterizes the entire molecule rather than only local regions. Similarly, the presence of free molecule-ends decrease the t_m value in shorter dsDNA molecules (*257*) or in dA — dT oligomers (*260*), where the end regions do not represent a negligible fraction of the molecule.

In conclusion, it should be emphasized that the structure of double-stranded nucleic acids is so complicated and the possibilities of structural changes so great that at the present time no substantial progress can be made on the basis of the results obtained with only a single experimental technique. It has been shown in this review that while one technique shows no changes in DNA structure, another can indicate a certain type of structural change, and a third another type of change. For further achievements in this field it will be necessary to study well-defined nucleic acid samples using a number of various methods differing in principle. Besides the studies of nucleic acids in solution, it could be useful to extend research to conditions simulating in some respects conditions existing *in vivo*, e.g., to study the structure and properties of the DNA during its adsorption on various electrically charged surfaces.

This article summarizes the results of DNA studies in a region to which, until recently, relatively little attention has been devoted. If this article stimulates discussions and research into local nonmelting structural changes in double-helical nucleic acids, its main purpose will be fulfilled.

ACKNOWLEDGMENTS

I should like to express my deep gratitude to the Corresponding Member of the Czechoslovak Academy of Sciences Dr. J. Říman for his support and stimulation which helped me to finish this work. I am indebted to Drs. M. Daune, W. Guschlbauer and J. Lebowitz who provided me with unpublished information. I am also grateful to Drs. V. Kleinwächter, Z. Pechan, J. Šponar and M. Vorlíčková for critical reading of the manuscript.

References

1. J. Josse and J. Eigner, *ARB* **35**, 789 (1966).
2. G. Felsenfeld and H. T. Miles, *ARB* **36**, 407 (1967).
3. J. Marmur, C. L. Schildkraut and B. Rownd, This series. **1**, 231 (1963).
4. W. Szybalski, in "Thermobiology" (A. H. Rose, ed.), p. 73. Academic Press, New York, 1967.
5. K. W. Kohn, O. L. Spears and P. Doty, *JMB* **19**, 287 (1966).
6. P. Doty, H. Boedtker, J. R. Fresco, R. Haselkorn and M. Litt, *PNAS* **45**, 482 (1958).
7. P. O. P. Ts'o and G. Helmkamp, *Tetrahedron* **13**, 198 (1961).
8. J. R. Fresco, *Tetrahedron* **13**, 185 (1961).
9. E. Paleček, *Z. Chem.* **2**, 260a (1962); *Abhandl. Deut. Akad. Wiss. Berlin, Kl. Med.* p. 270 (1964).
10. V. R. Glišin and P. Doty, *BBA* **61**, 458, 1000 (1962).
11. A. M. Freund and G. Bernardi, *Nature (London)* **200**, 1318 (1963).
12. E. Paleček, *JMB* **11**, 839 (1965).
13. E. Paleček, *JMB* **20**, 263 (1966).
14. R. B. Gennis and C. R. Cantor, *JMB* **65**, 381 (1972).
15. T. Samejima and J. T. Yang, *JBC* **240**, 2094 (1965).
16. T. Samejima and J. T. Yang, This series **9**, 224 (1969).
17. C. A. Busch and J. Brahms, in "Physico-chemical Properties of Nucleic Acids" (J. Duchesne, ed.), Vol. 2, p. 147. Academic Press, New York, 1973.
18. C. A. Busch, in "Basic Principles of Nucleic Acid Chemistry" (P. O. P. T'so, ed.), Vol. 2, p. 91. Academic Press, New York, 1974.
19. J. Brahms and W. H. F. M. Mommaerts, *JMB* **10**, 73 (1964).
20. G. Luck, C. Zimmer, G. Snatzke and G. Söndgerath, *EJB* **17**, 514 (1970).
21. E. Paleček and I. Frič, *BBRC* **47**, 1262 (1972).
22. D. S. Studdert, M. Patroni and R. C. Davis, *Biopolymers* **11**, 761 (1972).
23. E. J. Ramm, V. J. Vorobev, T. M. Birshtein, I. A. Bolotina and M. V. Volkenshtein, *EJB* **25**, 245 (1972).
24. M. T. Sarocchi and W. Guschlbauer, *EJB* **34**, 232 (1973).
25. A. F. Usatyi and L. S. Shlyakhtenko, *Biopolymers* **12**, 45 (1973).
26. D. M. Gray and F. J. Bollum, *Biopolymers* **13**, 2087 (1974).
27. J. Marmur, E. Seaman and J. Levine, *J. Bact.* **85**, 461 (1963).
28. P. Henson and I. O. Walker, *EJB* **16**, 524 (1970).
29. F. X. Wilhelm, D. M. de Murcia and M. P. Daune, *NARes* **1**, 1043 (1974).
30. R. Mandel and G. D. Fasman, *BBRC* **59**, 672 (1974).
31. T. Samejima, H. Hashizume, K. Imahozi, I. Fujii and K. Miura, *JMB* **34**, 39 (1968).
32. P. N. Borer, O. C. Uhlenbeck, B. Dengler and I. Tinoco, *JMB* **80**, 759 (1973).
33. V. I. Ivanov, L. E. Minchenkova, A. K. Shyolkina and A. I. Poletayev, *Biopolymers* **12**, 89 (1973).
34. E. P. Geiduschek and I. Gray, *JACS* **78**, 879 (1956).
35. J. H. Coates and D. O. Jordan, *BBA* **43**, 214 (1960).
36. T. T. Herskowitz, S. J. Singer and E. P. Geiduschek, *ABB* **94**, 99 (1961).
37. D. Lang, *JMB* **78**, 247 (1973).
38. J. C. Girod, W. C. Johnson, Jr., S. K. Huntigton and M. F. Maestre, *Bchem* **12**, 5092 (1973).

39. V. I. Ivanov, L. E. Minchenkova, E. E. Minyat, M. D. Frank-Kamenetski and A. K. Shyolkina, *JMB* **87**, 817 (1974).
40. D. M. Gray and R. L. Ratliff, *Biopolymers* **14**, 487 (1975).
41. A. Usatyi and L. S. Shlyakhtenko, *Bipolymers* **13**, 2435 (1974).
42. M. Tsuboi, S. Takahasi and I. Harada, in "Physico-chemical Properties of Nucleic Acids" (J. Duchesne, ed.), Vol. 2, p. 91. Academic Press, New York, 1973.
43. M. Shie, I. G. Kharitonenkov, T. I. Tikhonenko and Yu. N. Chirgadze, *Nature* (*London*) **235**, 386 (1972).
44. Yu. N. Chirgadze, M. Shie and I. G. Kharitonenkov, *Dokl. Akad. Nauk USSR* **203**, 959 (1972) (in Russian).
45. E. W. Small and W. L. Peticolas, *Biopolymers* **10**, 68 (1971).
46. E. W. Small and W. L. Peticolas, *Biopolymers* **10**, 1377 (1971).
47. K. Morikawa, M. Tsuboi, S. Takabashi, Y. Kyogoku, Y. Mitsui, Y. Iitaka and G. J. Thomas Jr., *Biopolymers* **12**, 799 (1973).
48. L. Rimai, V. M. Maher, D. Gill, I. Salmeen and J. J. McGormick, *BBA* **361**, 155 (1974).
49. S. C. Erfurth and W. L. Peticolas, *Biopolymers* **14**, 247 (1975).
50. S. C. Erfurth, P. J. Bond and W. L. Peticolas, *Biopolymers* **14**, 1245 (1975).
51. E. B. Brown and W. L. Peticolas, *Biopolymers* **14**, 1259 (1975).
52. M. T. Sarocchi and W. Guschlbauer, in preparation.
53. E. Paleček, This series **9**, 31 (1969).
54. E. Paleček, in "Methods in Enzymology," Vol. 21: Nucleic Acids, Part D (L. Grossman and K. Moldave, eds.), p. 3. Academic Press, New York, 1971.
55. E. Paleček, *J. Electroanal. Chem.* **22**, 347 (1969).
56. V. Brabec and E. Paleček, *Biophysik* **6**, 290 (1970).
57. B. Janík and P. J. Elving, *Chem. Rev.* **68**, 295 (1968).
58. J. W. Webb, B. Janík and P. J. Elving, *JACS* **95**, 8495 (1973).
59. E. Paleček and V. Vetterl, *Biopolymers* **6**, 917 (1968).
60. M. Vorlíčková and E. Paleček, *FEBS Lett.* **7**, 38 (1970).
61. E. Paleček, *JMB* **11**, 839 (1965).
62. E. Paleček, *JMB* **20**, 263 (1966).
63. H. Berg and H. Bär, *Monatsber. DAW* (*Berlin*) **7**, 210 (1965).
64. E. Paleček, *ABB* **125**, 142 (1968).
65. E. Paleček and V. Brabec, *BBA* **262**, 125 (1972).
66. E. Paleček, *Studia Biophys.* **42**, 59 (1974).
67. J. Flemming, *Studia Biophys.* **45**, 21 (1974).
68. E. Paleček, *Collect. Czech. Chem. Commun.* **39**, 3449 (1974).
69. V. Brabec and E. Paleček, *Biophys. Chem.* **4**, 79 (1976).
70. E. Paleček and J. Doskočil, *Anal. Biochem.* **60**, 518 (1974).
71. A. Bezděková and E. Paleček, *Studia Biophys.* **34**, 141 (1972).
72. M. Ya. Feldman, This series **13**, 1 (1973).
73. J. D. McGhee and P. H. von Hippel, *Bchem* **14**, 1281 (1975).
74. J. D. McGhee and P. H. von Hippel, *Bchem* **14**, 1297 (1975).
75. H. Utiyama and P. Doty, *Bchem* **10**, 1254 (1971).
76. E. N. Trifonov, Yu. S. Lazurkin and M. D. Frank-Kamenetskii, *Mol. Biol.* **1**, 164 (1967) (in Russian).
77. Yu. S. Lazurkin, M. D. Frank-Kamenetskii and E. F. Trifonov, *Biopolymers* **9**, 1253 (1970).
78. M. D. Frank-Kamenetskii and Yu. S. Lazurkin, *Annu. Rev. Biophys. Bioeng.* **3**, 127 (1974).

79. P. H. von Hippel and K. Y. Wong, *JMB* **61**, 587 (1971).
80. R. P. P. Fuchs and M. P. Daune, *Bchem* **13**, 4435 (1974).
81. W. F. Harrington, P. H. von Hippel and E. Mihalyi, *BBA* **32**, 32 (1959).
82. W. G. Miller, *JACS* **83**, 259 (1961).
83. P. H. von Hippel and G. Felsenfeld, *Bchem* **3**, 27 (1964).
84. L. Wingert and P. H. von Hippel, *BBA* **157**, 114 (1968).
85. E. Sulkowski and M. Laskowski, Sr., *JBC* **244**, 3818 (1969).
86. P. H. Johnson and M. Laskowski, Sr., *JBC* **243**, 3421 (1968).
87. P. H. Johnson and M. Laskowski, Sr., *JBC* **245**, 891 (1970).
88. A. Skalka, E. Burgi and A. D. Hershey, *JMB* **34**, 1 (1968).
89. D. D. Hogness and J. R. Simmons, *JMB* **9**, 411 (1964).
90. J. C. Wang, U. S. Nandi, D. S. Hogness and N. Davidson, *Bchem* **4**, 1697 (1965).
91. I. R. Lehman and A. L. Nussbaum, *JBC* **239**, 2628 (1964).
92. L. Rejthar and E. Paleček, *Collect. Czech. Chem. Commun.* **32**, 2687 (1967).
93. I. I. Nikolskaya, N. M. Shalina and T. I. Tikhonenko, *BBA* **91**, 354 (1964).
94. I. I. Nikolskaya, O. S. Kislina, N. M. Shalina and T. I. Tikhonenko, *Biokhymiya* **30**, 1245 (1965) (in Russian).
95. E. Z. Rabin, H. Tenenhouse and M. J. Fraser, *BBA* **259**, 50 (1972).
96. S. Linn and I. R. Lehman, *JBC* **240**, 1287 (1965).
97. S. Linn, in "Methods in Enzymology," Vol. 12: Nucleic Acids, Part A (L. Grossman and K. Moldave, eds.), p. 247. Academic Press, New York, 1967.
98. E. Z. Rabin, M. Mustard and M. J. Fraser, *Can. J. Biochem.* **46**, 1285 (1968).
99. E. J. Rabin and J. Fraser, *Can. J. Biochem.* **48**, 389 (1970).
100. T. Ando, *BBA* **114**, 158 (1966).
101. W. D. Sutton, *BBA* **240**, 525 (1971).
102. H. Ashe, E. Seaman, H. Van Vunakis and L. Levine, *BBA* **99**, 298 (1965).
103. R. Jayzaman and E. P. Goldberg, *PNAS* **64**, 198 (1969).
104. J. W. Healey, D. Stollar, M. Y. Simon and L. Levine, *ABB* **103**, 461 (1963).
105. K. Kasai and M. Grunberg-Manago, *EJB* **1**, 152 (1967).
106. S. Y. Lee, Y. Nakao and R. M. Boek, *BBA* **151**, 126 (1968).
107. T. St. John, J. D. Johnson and J. Bonner, *BBRC* **57**, 240 (1974).
108. G. Roizes, *NARes* **1**, 443 (1974).
109. V. M. Vogt, *EJB* **33**, 192 (1973).
110. G. N. Godson, *BBA* **308**, 59 (1973).
111. K. Shishido and T. Ando, *BBRC* **59**, 1380 (1974).
112. I. M. Mechali, A. M. de Recondo and M. Girard, *BBRC* **54**, 1306 (1973).
113. P. Beard, J. F. Morrow and P. Berg, *J. Virol.* **12**, 1303 (1973).
114. J. E. Germond, V. M. Vogt and B. Hirt, *EJB* **43**, 591 (1974).
115. L. Katz and S. Penman, *JMB* **15**, 220 (1966).
116. D. R. Kearns, D. Patel, R. G. Shulman and T. Yamane, *JMB* **61**, 271 (1971).
117. D. M. Crothers, C. W. Hilbers and R. G. Shulman, *PNAS* **70**, 2899 (1973).
118. D. J. Patel and A. E. Tonelli, *PNAS* **71**, 1945 (1974).
119. D. J. Patel and A. C. Tonelli, *Biopolymers* **13**, 1943 (1974).
119a. D. J. Patel and C. W. Hilbers, *Bchem* **14**, 2651 (1975).
119b. C. W. Hilbers and D. J. Patel, *Bchem* **14**, 2656 (1975).
120. O. Kratky, *Progr. Biophys.* **13**, 107 (1963).
121. V. Luzzati, A. Mathis, F. Mason and J. Witz, *JMB* **10**, 28 (1964).
122. K. C. Smith and P. C. Hanawalt, "Molecular Photobiology," p. 57. Academic Press, New York, 1969.

123. Z. Tramer, K. L. Wierzchowski and D. Shugar, *Acta Biochim. Polon.* **16**, 83 (1969).
124. R. B. Setlow and J. K. Setlow, *Annu. Rev. Biochem. Bioeng.* **1**, 293 (1972).
125. V. R. Glišin and P. Doty, *BBA* **142**, 314 (1967).
126. R. O. Rahn, J. K. Setlow and J. L. Hosszu, *Biophys. J.* **9**, 510 (1969).
127. E. Paleček, E. Lukášová, M. Vorlíčková and H. Ambrová, *Studia Biophys.* **24/25**, 123 (1970).
128. J. L. Hosszu, and R. O. Rahn, *BBRC* **29**, 327 (1967).
129. H. B. Gray Jr. and J. E. Hearst, *JMB* **35**, 111 (1968).
130. T. I. Tikhonenko, G. A. Perevertajlo and E. N. Dobrov, *BBA* **68**, 500 (1963).
131. M. Boublík, L. Pivec, J. Šponar and Z. Šormová, *Collect. Czech. Chem. Commun.* **30**, 2645 (1965).
132. P. D. Ross and R. L. Scruggs, *Biopolymers* **6**, 1005 (1968).
133. G. Cohen and H. Eisenberg, *Biopolymers* **4**, 429 (1966).
134. H. Ch. Spatz and R. L. Baldwin, *JMB* **11**, 213 (1965).
135. R. L. Baldwin, in "Molecular Associations in Biology" (B. Pullman, ed.), p. 145. Academic Press, New York, 1968.
136. J. C. Wang, D. Baumgarten and B. M. Olivera, *PNAS* **58**, 1852 (1967).
137. J. C. Wang, *JMB* **43**, 25 (1969).
138. W. R. Bauer, *JMB* **67**, 183 (1972).
139. W. R. Bauer, *Bchem* **11**, 2915 (1972).
140. D. R. Davies, *ARB* **36**, 321 (1967).
141. S. Arnott, *Progr. Biophys. Mol. Biol.* **21**, 265 (1970).
142. S. Arnott and D. W. L. Hukins, *JMB* **81**, 93 (1973).
143. H. Eisenberg and G. Cohen, *JMB* **37**, 355 (1968).
144. M. F. Maestre and R. Kilkson, *Biophys. J.* **5**, 275 (1965).
145. M. J. B. Tunis-Schneider and M. F. Maestre, *JMB* **25**, 521 (1970).
146. G. Milman, R. Langridge and M. J. Chamberlin, *PNAS* **57**, 1804 (1967).
147. S. Bram, *JMB* **58**, 277 (1971).
148. S. Bram, *Nature NB* **232**, 174 (1971).
149. S. Bram, *PNAS* **70**, 2167 (1973).
150. S. Bram, *Nature NB* **233**, 161 (1971).
151. S. Bram and P. Tougard, *Nature NB* **239**, 128 (1972).
152. S. Bram, *BBRC* **48**, 1088 (1972).
153. S. Bram and P. Baudy, *Nature (London)* **250**, 414 (1974).
154. S. Arnott and E. Selsing, *JMB* **88**, 509 (1974).
155. S. Arnott, R. Chandrasekaran, D. W. L. Hukins, P. J. C. Smith and L. Watts, *JMB* **88**, 523 (1974).
156. S. Arnott and E. Selsing, *JMB* **88**, 551 (1974).
157. L. D. Hamilton, R. K. Barclay, M. H. F. Wilkins, G. L. Brown, H. K. Wilson, D. A. Marvin, H. Ephrussi-Taylor and N. S. Simmons, *Biophys. Biochem. Cytol.* **5**, 397 (1959).
158. P. J. Cooper and L. D. Hamilton, *JMB* **16**, 562 (1966).
159. J. Pilet and J. Brahms, *Nature NB* **236**, 99 (1972).
160. J. Pilet and J. Brahms, *Biopolymers* **12**, 387 (1973).
161. J. Brahms, J. Pilet, Tran-Thi Phuong Lan and L. R. Hill, *PNAS* **70**, 3352 (1973).
162. J. Pilet, J. Blicharski and J. Brahms, *Bchem* **14**, 1869 (1975).
163. R. S. Zeiger, R. Salomon, C. W. Dingman and A. C. Peacock, *Nature NB* **238**, 65 (1972).

164. E. N. Trifonov, N. H. Shafranovskaya, M. D. Frank-Kamenetskii and Yu. S. Lazurkin, *Mol. Biol.* 2, 887 (1968) (in Russian).
165. Yu. A. Bannikov and E. N. Trifonov, *Mol. Biol.* 5, 734 (1970) (in Russian).
166. G. S. Komolova, E. N. Trifonov and I. A. Egorov, *Dokl. Akad. Nauk USSR* 207, 222 (1972) (in Russian).
167. A. M. Poverennyi, N. I. Ryabchenko, Yu. I. Gamov, B. P. Ivannik and V. V. Simonov, *Mol. Biol.* 6, 524 (1972) (in Russian)
168. E. Paleček, *BBA* 145, 410 (1967).
169. E. Lukášová and E. Paleček, *Radiat. Res.* 47, 51 (1971).
170. V. Brabec and E. Paleček, *Biopolymers* 11, 2577 (1972).
171. K. Chowdhury, P. Gruss, W. Waldeck and G. Sauer, *BBRC* 64, 709 (1975).
172. G. Scholes, *Progr. Biophys. Mol. Biol.* 13, 59 (1963).
173. J. J. Weiss, *This series* 3, 103 (1964).
174. W. Ginoza, *Annu. Rev. Nucl. Sci.* 17, 469 (1967).
175. D. T. Kanazir, *This series* 9, 117 (1969).
176. M. Vorlíčková and E. Paleček, *Int. J. Radiat. Biol.* 26, 363 (1974).
177. R. Uliana, P. V. Creac'h and A. Ducastaing, *Biochimie* 53, 461 (1971).
178. J. Boháček and G. Blažíček, *Biophysik* 2, 233 (1965).
179. V. N. Shumaker, E. G. Richards and H. K. Schachman, *JACS* 78, 4230 (1956).
180. C. A. Thomas, Jr., *JACS* 78, 1861 (1956).
181. J. B. Hays and B. H. Zimm, *JMB* 48, 297 (1970).
182. E. Lukášová and E. Paleček, *Folia Biol.* 18, 307 (1972).
183. R. O. Rahn, in "Photophysiology" (A. Giese, ed.), Vol. 8, p. 231. Academic Press, New York, 1973.
184. J. Marmur, W. F. Anderson, L. Mathews, K. Berns, E. Gajevska and P. Doty, *J. Cell Comp. Physiol.* 58, Suppl. 1, 33 (1961).
185. H. Lang, and G. Luck, *Photochem. Photobiol.* 17, 387 (1973).
186. H. Lang, *NARes* 2, 179 (1975).
187. N. N. Shafranovskaya, E. N. Trifonov, Yu. S. Lazurkin and M. D. Frank-Kamenetskii, *Nature NB* 241, 58 (1973).
187a. A. C. Kato and M. J. Fraser, *BBA* 312, 645 (1973).
188. R. Das Gupta and S. Mitra, *BBA* 374, 145 (1974).
189. R. Shapiro, B. Bravermann, J. B. Louis and R. E. Servis, *JBC* 248, 4060 (1973).
190. E. Lukášová and E. Paleček, *Studia Biophys.* in press.
191. R. Fuchs and M. Daune, *Bchem* 11, 2659 (1972).
192. R. O. Rahn and R. S. Stafford, *Nature (London)* 248, 52 (1974).
193. W. Bauer and J. Vinograd, in "Basic Principles in Nucleic Acid Chemistry" (P.O.P. Ts'o, ed.), Vol. 2, p. 265. Academic Press, New York, 1974.
194. J. Vinograd, J. Lebowitz and R. Watson, *JMB* 33, 173 (1968).
195. W. W. Dean and J. Lebowitz, *Nature NB* 231, 5 (1971).
196. T. A. Beerman and J. Lebowitz, *JMB* 79, 451 (1973).
197. N. P. Salaman, J. Lebowitz, M. Chen, E. Sebring and C. F. Garon, *CSHSQB* 39, 209 (1974).
198. R. J. Jacob, J. Lebowitz and M. P. Printz, *NARes* 1, 549 (1974).
199. R. J. Jacob, J. Lebowitz and A. K. Kleinschmidt, *J. Virol.* 13, 1176 (1974).
200. M. F. Maestre and J. C. Wang, *Biopolymers* 10, 1021 (1971).
201. A. M. Campbell and D. S. Lochhead, *BJ* 12, 661 (1971).
201a. A. C. Kato, K. Bartok, M. J. Fraser and D. T. Denhardt, *BBA* 308, 68 (1973).
202. J. C. Wang, *JMB* 87, 797 (1974).
203. M. F. Maestre, *JMB* 52, 453 (1970).

204. P. H. von Hippel and M. P. Printz, *FP* **24**, 1458 (1965).
205. C. W. Lees and P. H. von Hippel, *Biochemistry* **7**, 2480 (1968).
206. B. McConnell and P. H. von Hippel, *JMB* **50**, 297 (1970).
207. B. McConnell and P. H. von Hippel, *JMB* **50**, 317 (1970).
208. C. V. Hanson, *JMB* **58**, 847 (1971).
209. J. J. Englander and P. H. von Hippel, *JMB* **63**, 171 (1972).
210. J. J. Englander, N. R. Kallenbach and S. W. Englander, *JMB* **63**, 153 (1972).
211. R. E. Bird, K. G. Lark, B. Curnutte and J. E. Maxfield, *Nature* (London) **225**, 1043 (1970).
212. H. Teitelbaum and S. W. Englander, *JMB* **92**, 55 (1975).
213. H. Teitelbaum and S. W. Englander, *JMB* **92**, 79 (1975).
214. E. Paleček, V. Vetterl and J. Šponar, *NARes* **1**, 427 (1974).
215. D. S. Studdert and R. C. Davis, *Biopolymers* **13**, 1377 (1974).
216. D. S. Studdert and R. C. Davis, *Biopolymers* **13**, 1391 (1974).
217. D. S. Studdert and R. C. Davis, *Biopolymers* **13**, 1404 (1974).
218. A. M. Michelson, J. Massoulié and W. Guschlbauer, This series **6**, 83 (1967).
219. C. A. Thomas, Jr., B. A. Hamkalo, D. N. Misra and C. S. Lee, *JMB* **51**, 621 (1970).
220. H. Schaller, H. Voss and S. Gucker, *JMB* **44**, 445 (1969).
221. D. A. Wilson and C. A. Thomas Jr., *JMB* **84**, 115 (1974).
222. A. B. Forsheit and D. S. Ray, *PNAS* **67**, 1534 (1970).
223. R. Barzilai and C. A. Thomas, Jr., *JMB* **51**, 145 (1970).
224. D. A. Wilson and C. A. Thomas, Jr., *BBA* **331**, 333 (1973).
225. H. Eisenberg, in "Basic Principles in Nucleic Acid Chemistry" (P.O.P. Ts'o, ed.), Vol. 2, p. 171. Academic Press, New York, 1974.
226. S. Lewin, "Displacement of Water and Its Control of Biochemical Reactions." Academic Press, New York, 1974.
227. P. J. Killion and L. H. Reyerson, *J. Colloid. Interfac. Sci.* **22**, 582 (1966).
228. B. Lubas, T. Wilczok and O. K. Daszkiewicz, *Biopolymers* **5**, 967 (1967).
229. B. Lubas and T. Wilczok, *BBA* **224**, 1 (1970).
230. B. Lubas and T. Wilczok, *Biopolymers* **10**, 1267 (1971).
231. B. Lubas and T. Wilczok, *Acta Biochim. Polon.* **19**, 161 (1972).
232. R. E. Chapman, Jr. and J. M. Sturtevant, *Biopolymers* **7**, 527 (1969).
233. J. Vinograd, R. Gruenwald and J. E. Hearst, *Biopolymers* **3**, 109 (1965).
234. C. Zimmer and G. Luck, *BBA* **361**, 11 (1974).
235. B. Wolf and S. Hanlon, *Bchem* **14**, 1661 (1975).
236. B. Collyns, S. Okada, G. Scholes, J. J. Weiss and C. M. Wheeler, *Radiat. Res.* **25**, 526 (1965).
237. R. Frey and U. Hagen, *Rad. Environm. Biophys.* **12**, 111 (1975).
238. F. N. Hayes, D. L. Williams, R. L. Ratliff, A. J. Varghese and C. S. Rupert, *JACS* **93**, 4940 (1971).
239. J. P. Richardson, This series **9**, 75 (1969).
240. P. H. von Hippel and J. D. McGhee, *ARB* **41**, 231 (1972).
241. M. J. Chamberlin, *ARB* **43**, 721 (1974).
242. D. C. Hinkle, J. Ring and M. J. Chamberlin, *JMB* **70**, 197 (1972).
243. J. P. Dausse, A. Sentanac and P. Fromageot, *EJB* **31**, 394 (1972).
244. J. Y. Le Talaer and Ph. Jeanteur, *FEBS Lett.* **12**, 253 (1971).
245. K. Shishido and Y. Ikeda, *BBRC* **44**, 1420 (1971).
246. Y. Hayashi and M. Hayashi, *Bchem* **10**, 4212 (1971).
247. J. P. Richardson, *Bchem* **13**, 3164 (1974).

248. P. Hossenlopp, P. Oudet and P. Chambon, *EJB* **41**, 397 (1974).
249. J. L. Mandel and P. Chambon, *EJB* **41**, 367 (1974).
250. J. L. Mandel and P. Chambon, *TJB* **41**, 379 (1974).
251. J. J. Champoux and B. L. McConaughy, *Bchem* **14**, 307 (1975).
252. C. Escarmis, E. Domingo and R. C. Warner, *BBA* **402**, 261 (1975).
253. T. Manialis, M. Plashne, B. G. Barrel, and J. Davelson, *Nature* (*London*) **250**, 394 (1974).
254. T. Sekiya and H. G. Khorana, *PNAS* **71**, 2978 (1974).
255. K. Sugimoto, T. Okamoto, H. Sugisaki and M. Takanami, *Nature* **253**, 410 (1975).
256. G. Bressan, R. Rampone, E. Bianchi and A. Ciferri, *Biopolymers* **13**, 2227 (1974).
257. Y. Miyzava and C. A. Thomas, Jr., *JMB* **11**, 223 (1965).
258. R. D. Wells, J. E. Larson, R. C. Grant, B. E. Shortle and R. C. Cantor, *JMB* **54**, 465 (1970).
259. P. N. Borer, B. Dengler, I. Tinoco, Jr. and O. C. Uhlenbeck, *JMB* **86**, 843 (1974).
260. I. E. Scheffler, E. L. Elson and R. L. Baldwin, *JMB* **48**, 145 (1970).

Addendum

Since the completion of this review, several papers have been noted. Measurements of dielectric conductivity indicate changes in the DNA ionic atmosphere at premelting temperatures (*261*). The fluorescence depolarization of ethidium bromide (*262*) and of proflavin (*263*) show that the bases may oscillate around the helix axis in a nanosecond time range. Fluorescence studies of the binding of oligopeptides containing lysine and tryptophane residues to DNA reveal the ability of the peptide to "recognize" distorted DNA regions containing pyrimidine dimers (*264*).

It has been shown unequivocally that ethidium bromide unwinds the duplex (*265*) but that the unwinding angle (and consequently the superhelical densities) are between two and three times the previously reported values. This fact considerably influences earlier data (Section II, F) concerning the unwinding of the helix due to temperature and ionic strength changes. New calculations based on the earlier results (*137*) as well as on recent electrophoretic studies of superhelical DNA's (*266*, *267*) show values around 0.010–0.012 degrees/°C. The recent data agree much better with the results of CD (*33*, *265*, *268*). It has been further shown that treatment of "nicked" circular DNA with ligase or "nicking-closing" enzyme yields a population of covalently closed circles heterogeneous in the topological winding number (*266*, *267*). This fact has been explained by thermal fluctuations of the DNA conformation during DNA closure. The possibility of various local conformational changes in the vicinity of single-strand breaks (which can be formed at various sites differing in nucleotide sequences) was not taken into consideration.

The conception of polymorphy of the double helix (depending on the nucleotide sequence) has gained support on theoretical grounds (*269*); contrary to Bram's findings (Section III, B) further x-ray diffraction study (*270*) showed no noticeable difference between three DNA samples with various (G + C)-content.

In an attempt to explain how DNA is folded in chromatin, Crick and Klug (*271*) have recently introduced a new concept of a "kinky helix." They have assumed that at the kink, one base pair is completely unstacked from the adjacent one while the parts of DNA on each side of the kink are straight and remain in the normal B form. It is assumed that every kink is exactly the same. The kink induces a small negative twist to the DNA. The authors believe essentially that kinking may be a way of partly exposing a small group of bases without any great expenditure of energy. However, they point out that there is no compelling evidence that DNA in chromatin and/or in solution is kinked. Taking into consideration that data on DNA nonmelting conformational changes summarized in this review, we may say that the results of certain techniques are compatible with the idea of kinking (e.g., polarography, formaldehyde reaction—local exposure of bases; CD—helix unwinding). On the other hand, the presumption that every kink is exactly the same excludes the possibility that kinking is the only local non-melting change in DNA conformation. Thus the kink, if it does occur in the DNA helix, may represent only one of the types of local nonmelting changes in DNA conformation.

References to Addenum

261. H. Grassi, M.-A. Rix-Montel, H. Kranck and D. Vasilescu, *Biopolymers* **14**, 2525 (1975).
262. P. Wahl, J. Paoletti and J. B. LePecq, *PNAS* **65**, 417 (1970).
263. S. Georghiou, *Photochem. Photobiol.* **22**, 103 (1975).
264. J. J. Toulmé, M. Charlier and C. Hélène, *PNAS* **71**, 3185 (1974).
265. D. E. Pulleyblank and A. R. Morgan, *JMB* **91**, 1 (1975).
266. R. E. Depew and J. C. Wang, *PNAS* **72**, 4275 (1975).
267. D. E. Pulleyblank, M. Shure, D. Tank, J. Vinograd and H. P. Vorsberg, *PNAS* **72**, 4280 (1975).
268. D. M. Hinton and V. C. Bode, *JBC* **250**, 1071 (1975).
269. R. V. Polozov, V. I. Poltev and B. I. Sukhorukov, *J. Theor. Biol.* **55**, 491 (1975).
270. S. Premilat and G. Albiser, *JMB* **99**, 27 (1975).
271. F. H. C. Crick and A. Klug, *Nature (London)* **255**, 530 (1975).

Quantum-Mechanical Studies on the Conformation of Nucleic Acids and Their Constituents

BERNARD PULLMAN AND
ANIL SARAN

*Institut de Biologie
Physico-Chimique
Laboratoire de Biochimie
Théorique associé au C.N.R.S.
Paris, France*

I. Introduction	216
II. The Quantum-Mechanical Methods	217
A. The Basic Idea of the Method of Molecular Orbitals	217
B. The Self-Consistent Field Method	218
C. The Classical Hückel Approximation	222
D. The Extended Hückel Theory (EHT)	222
E. The Zero-Differential Overlap (ZDO) Approximation	223
F. The Complete Neglect of Differential Overlap (CNDO) Method	224
G. Configuration Interaction (CI)	226
H. Perturbative Configuration Interaction (PCI)	227
I. Localized Orbitals	228
J. The PCI over Localized Orbitals (PCILO) Method	228
III. Types of Torsion Angles and Definitions	230
A. Torsion Angles in Polynucleotides	231
B. Torsion Angles in Mononucleotides	233
C. Torsion Angles in Nucleosides	234
IV. The Glycosyl Torsion Angle χ_{CN}	236
A. The *syn–anti* Equilibrium in Nucleosides	236
B. The Rigidity and Flexibility of 5'-Nucleotides	251
V. Conformation about the Exocyclic C4'—C5' Bond	255
A. β-Nucleosides	256
B. α-Nucleosides	270
C. 5'-β-Nucleotides	272
VI. The Backbone Structure of Di- and Polynucleotides	275
A. Conformational Energy Maps about Two Consecutive Torsion Angles	276
B. The Geometry of the Phosphate Group: Key to the Conformation of Polynucleotides?	291
VII. Conformation of the Sugar Ring: The Pseudorotational Representation	301
A. Pseudorotational Parameters, P and τ_M	302
B. Conformational Properties of Ribose and Deoxyribose	303
C. Relation between P and χ_{CN}	304
VIII. Related Subjects	308
IX. Concluding Remarks	310
References	313

I. Introduction

The conformational properties of nucleic acids, polynucleotides and their constituents have recently aroused a tremendous amount of attention, and investigations in this field are being carried out in a number of laboratories all over the world. These investigations involve both experimental studies by a large variety of physicochemical and biophysical techniques, and theoretical investigations by different computational methods. Broadly speaking the theoretical ones are of two types:

a. The so-called "empirical" procedures, which consist of partitioning the potential energy of the system into several discrete contributions—such as nonbonded interactions, electrostatic interactions, barriers to internal rotations, hydrogen-bonding—which are then evaluated with the help of empirical formulas deduced from studies on model compounds of small molecular weight. In the simplest approximation of these procedures (the "hard sphere" approximation), the problem is limited to the evaluation of allowed or forbidden contacts, with the help of van der Waals (or similar) radii.

b. The quantum-mechanical methods. In a fundamental difference to the preceding methods, these aim at a direct evaluation of the total molecular energy associated with the different atomic configurations of the system, and thus at a direct prediction of the preferred molecular conformations. The practical realization of this scheme became possible only recently owing to the development of the all-valence-electrons and all-electron-molecular-orbital methods.

Among these procedures, it is the quantum-mechanical ones that have been most extensively used in the field of the conformational properties of the nucleic acid and their constituents, and that yielded the most complete and, as will be seen, also the most satisfactory description of the subject.

The hard-sphere approximation obviously takes cognizance of short-range repulsions only and disregards completely attractive forces between atoms. Moreover, it operates at an all-or-nothing level. While indicating *grosso modo* the allowed and disallowed conformational zones, it gives no information about the positions of the energy minima, i.e., about the nature of the preferred conformations. It was the goal of the partitioned-potential-energy approximation to refine this situation by evaluating the interatomic interactions in a more precise way. Unfortunately, these procedures suffer from a number of defects, the principal ones being a certain arbitrariness and frequent incompleteness in the energy partition and the diversity of the empirical formulas used by different investigators. For these reasons, they appear to have been rather strikingly unsuccessful for a number of important aspects of the conformational properties of the nucleic acids and their constituents.

The quantum-mechanical methods, all being approximative methods, also have their difficulties, although altogether their performance is better. Different methods of this kind have been applied to the problem, and the quality of their results depends upon the refinement of their techniques. Among these methods, the most prominent applied to the study of polynucleotides and their constituents are the Extended Hückel method the CNDO/2 method, the PCILO method and the so-called *ab initio* procedure. By far the most complete and abundant computations in this field have been carried out by the PCILO method.

The aim of this review is the presentation of the results of the quantum-mechanical computations of the conformational properties of nucleic acids and their constituents, comparing the results obtained by different methods among themselves with the results of the empirical computations and with the available experimental data as obtained by different techniques, both in the solid state and in solution. The theoretical computations refer essentially, at least in the present stage of their development in this field, to free molecules, and the comparison of their results with experimental data enables one to estimate to what extent the observed conformations correspond to the intrinsically preferred ones and to what extent they are influenced by environmental forces. At the same time, the computations furnish for these molecules an amount of information that, from some point of view, is larger than what is available experimentally in the sense that it yields conformational energy *maps* instead of conformations preferred in a given situation. It is thus obvious that the field is an excellent one for a very fruitful collaboration between theory and experiment. A somewhat similar comparative study has been presented recently for the conformational properties of proteins, polypeptides and their constituents (1).

We start this review by presenting a brief general summary of the basic principles of the molecular orbital procedure and of the above-mentioned methods. Interested readers may find more details in the references indicated.

II. The Quantum-Mechanical Methods

A. The Basic Idea of the Method of Molecular Orbitals

The basic idea that lies at the foundations of *the method of molecular orbitals* is a very general one, first used in the quantum-mechanical description of polyelectronic atoms. It consists of *constructing the wave function of a polyelectronic system as a suitable combination of individual one-electron wave functions.*

The most "suitable" combination has the general form

$$\begin{vmatrix} a(1) & b(1) & c(1) & \cdots \\ a(2) & b(2) & c(2) & \cdots \\ \cdot & \cdot & \cdot \\ \cdot & \cdot & \cdot \\ \cdot & \cdot & \cdot \\ a(n) & b(n) & c(n) & \cdots \end{vmatrix} \qquad (1)$$

a notation that stands for the determinant built on the n individual wave functions a, b, c, etc. Since each of those is a product of an *"orbital"* part (function of the space coordinates only)

$$\varphi(x,y,z) \qquad (2)$$

and a spin function α or β, the total wave function for an even number of electrons is written

$$\Psi = (n!)^{-\frac{1}{2}} \begin{vmatrix} \varphi_1(1) & \alpha(1) & \varphi_1(1) & \beta(1) & \varphi_2(1) & \alpha(1) & \varphi_2(1) & \beta(1) & \cdots \\ \varphi_1(2) & \alpha(2) & \varphi_1(2) & \beta(2) & \varphi_2(2) & \alpha(2) & \varphi_2(2) & \beta(2) & \cdots \\ \cdot & & \cdot & & \cdot & & \cdot \\ \cdot & & \cdot & & \cdot & & \cdot \\ \cdot & & \cdot & & \cdot & & \cdot \\ \varphi_1(n) & \alpha(n) & \varphi_1(n) & \beta(n) & \varphi_2(n) & \alpha(n) & \varphi_2(n) & \beta(n) & \cdots \end{vmatrix}$$

(3)

Such a "Slater determinant," as it is often called, would in fact be the correct wave function for a system of noninteracting electrons. However, electrons do interact in real molecular systems. Thus, in order to obtain a satisfactory representation preserving the convenient one-electron model, one tries to determine *the individual orbitals φ so as to take into account the presence of the other electrons.*

B. The Self-Consistent Field Method

The best procedure, which allows the determination of the individual molecular orbitals, is the "self-consistent field" method, the main features of which are as follows:

a. One writes the exact total Hamiltonian for the system with explicit inclusion of electron interactions:

$$H = \sum_\nu H(\nu) + \sum_{\mu<\nu} \frac{1}{r_{\mu\nu}} \qquad (4)$$

where $H(\nu)$ is the Hamiltonian for one electron ν in the field of all the bare nuclei.

b. One expresses the total energy of the system by the standard quantum mechanical expression

$$E = \frac{\int \Psi^* H \Psi d\tau}{\int \Psi^* \Psi d\tau} \qquad (5)$$

in terms of the individual orbitals φ, by using the determinantal expression of Ψ.

c. One satisfies the variation principle for the energy. This is a standard procedure based on a fundamental theorem of quantum mechanics, namely, that the energy calculated by the above expression using an approximate wave function always lies higher than the exact energy. Thus, if one uses an approximate wave function expressed in terms of certain parameters, minimization of the energy with respect to these parameters will yield the best possible energy value attainable with this form of Ψ. Carrying out this program yields the general "Fock" equations, one for each individual orbital φ:

$$F\varphi_i = \epsilon_i \varphi_i \qquad (6)$$

where F is an operator playing the role of an individual Hamiltonian and ϵ_i is the individual energy of one electron occupying the orbital φ_i.

An essential characteristic of the Fock equations resides in the fact that each individual operator F depends on all the orbitals that are occupied in the system (on account of the explicit inclusion of the interaction terms): thus each φ is given by an equation that depends on all the φ's. The way out of this difficulty is to choose arbitrarily a starting set of φ's, calculate the $F(\nu)$'s, solve the series of equations for a new set of φ's, and go over the same series of operations again and again until the pth set of φ's reproduces the $(p-1)$th set to a good accuracy. Hence the name "self-consistent" given to the procedure. The orbitals obtained in this fashion are, in principle, the best possible orbitals compatible with a determinantal Ψ.

However, one restriction must be placed on this last statement: a choice must be made of a starting set of φ's. Since it is impossible to guess *ab initio* the appropriate analytical form of a molecular orbital, one must rely on a "reasonable" possibility. Thus the final orbitals are the best possible orbitals of the form chosen.

The classical choice of the starting orbitals is based on the following idea: suppose that we deal with a chemical bond formed between two monovalent atoms A and B by the pairing of their valence electrons, one on A, the other on B. It is natural to assume that when one electron in the molecule is close to nucleus A, its molecular orbital will resemble the atomic orbital that it would occupy in A, and a similar situation would occur in the vicinity of B. This leads to the idea that *the molecular orbital may be approximated by a linear combination*

$$\varphi = c_1\chi_A + c_2\chi_B \qquad (7)$$

where the χ's are the atomic orbitals.

The idea can be extended to a polyatomic molecule and generalized so that *each molecular orbital in a molecule is a linear combination of all the atomic orbitals occupied by the electrons in the constituent atoms:*

$$\varphi = \sum_r c_r \chi_r \qquad (8)$$

This is the classical and general LCAO approximation (linear combination of atomic orbitals) of the molecular orbital method.

Given this form of molecular orbitals, the Fock equations yield a system of homogeneous linear equations in the c's, the Roothaan equations (2)

$$\sum_q c_{iq}(F_{pq} - \epsilon S_{pq}) = 0 \qquad (9)$$

where

$$S_{pq} = \int \chi_p^*(\nu)\chi_q(\nu)d\tau_\nu \qquad (10)$$

is the overlap integral, and F_{pq} is the pq matrix element of the Fock operator:

$$F_{pq} = \int \chi_p^*(\nu)F\chi_q(\nu)d\tau_\nu \qquad (11)$$

All the necessary elements can be calculated in terms of integrals over the atomic orbitals χ, integrals that involve either the nuclear attraction operators in $H(\nu)$ or the interelectronic repulsions $1/r_{\mu\nu}$.

As is well known, a set of equations like Eq. 9 has a nontrivial solution only if

$$|F_{pq} - \epsilon S_{pq}| = 0 \qquad (12)$$

Solution of Eq. 12 yields the energy parameters ϵ, and their replacement in the system of Eq. 9 yields the c_{ir}'s. But the elements F_{pq} depend on the coefficients; indeed, writing

$$F_{pq} = H_{pq} + G_{pq} \qquad (13)$$

yields

$$H_{pq} = \int \chi_p^*(\nu)H(\nu)\chi_q(\nu)d\tau_\nu \qquad (14)$$

and

$$G_{pq} = \sum_{jr,s} c_{jr}c_{js}[2(rp,qs) - (rq,ps)] \qquad (15)$$

with

$$(pq,rs) = \iint \chi_p^*(\mu)\chi_q(\mu)\frac{1}{r_{\mu\nu}}\chi_r^*(\nu)\chi_s(\nu)d\tau_\mu d\tau_\nu \qquad (16)$$

Thus, the practical way of solving the Roothaan equations is to choose an initial set of c_r^0's, calculate the F_{pq}'s, solve the equations for a new set, and iterate again until consistancy is attained.

In principle, this kind of scheme may be carried out for any molecule, with any number of electrons and any number of atomic orbitals χ in the LCAO basis set. The practical calculation, however, involves the tedious evaluation of a large number of integrals, a number that increases so rapidly with the number of electrons that, for large molecules, complete self-consistent field (SCF) calculations are not easily feasible on a large scale. Nevertheless, in recent years, substantial technical progress has been made in this field to the point that such computations have been carried out for some aspects of the conformational problems related to the nucleic acids and their constituents. They are referred to in the literature as SCF *ab initio* computations.

In such computation all the electrons of the system are considered in the field of all the nuclei. All electron–electron interactions are included, and the equations are solved exactly for the molecular orbitals expressed as linear combination of a preselected set of atomic orbitals. It is the choice of this atomic "basis set" that thus fixes the limits to the accuracy of the results, and it is generally a compromise between accuracy and computational feasibility. Although presently the practical computations are made through the mediation of gaussian functions (see below), one may still characterize the most currently utilized basis sets as follows: suppose that the molecular orbitals are expressed as linear combinations of the type of Eq. 8, and the χ's are Slater-type atomic orbitals for each set of quantum numbers n, l, m,

$$\chi_{n,l,m} = r^{n-1} e^{-\zeta r} Y_{lm}(\theta, \varphi)$$

a. *Minimal basis sets* are those using only one χ function per orbital of quantum number n, l, m filled in each atom (e.g., one for each $1s$, $2s$, $2p_x$, $2p_y$, $2p_z$, etc. . . . orbital). This still leaves the choice of the ζ exponents. In the early days of SCF computations, it was customary to use Slater exponents; later, the use of "best-atom" exponents, optimized to give the best possible atomic energies, was advocated. More recently, the utilization of "molecule-optimized exponents" has developed. In this particular case, the optimal ζ-value should be different according to the molecule considered, but in fact average standard values may be found.

b. *Double-ζ basis sets* correspond to the utilization of two exponential functions with different values of ζ for each atomic orbital; this doubles the number of unknown coefficients in the expansion and augments considerably the flexibility, and hence the possible accuracy, of the results. A simplified variant of this is the "split-valence shell" basis which uses double-ζ functions only for the valence shell of atoms in molecules.

Triple-ζ, quadruple-ζ basis are obvious extensions.

c. *Sets including "polarization functions"* are either kind of basis sets augmented by atomic orbitals of higher symmetry than that of the occupied orbitals in the atom: that is, p functions for hydrogen, and d-type functions for atoms containing normally s and p electrons, etc. Again, this increases the flexibility in the molecular linear combination.

d. *The Hartree–Fock limit* corresponds to the best possible value of the total energy obtained by extending the basis set further and further until no more improvement can be obtained in an SCF computation. Understandably, very few computations have indeed reached this limit for molecular systems, and still fewer for supermolecular systems, but a few extrapolations to the limit are available as guidelines.

One more explanatory word is needed concerning the nomenclature: as already stated, the computations are in fact performed in terms of linear combinations of atom-centered gaussian functions $e^{-\alpha r^2}$ instead of the exponentials e^{-r}. These functions[1] are called the primitive, or uncontracted, gaussians. In a very common notation, $(7s, 3p/2s)$ indicates that $7s$ and $3p_x$, $3p_y$, $3p_z$ primitives have been used for each "heavy" atom, with $2s$ primitives for each hydrogen.

In the practical computations, these primitives are combined into small subsets of the type:

$$a_1 e^{-\alpha_1 r^2} + a_2 e^{-\alpha_2 r^2} + a_3 e^{-\alpha_3 r^2}$$

with preselected fixed coefficients, each of these *contracted* gaussians corresponding to one exponential of the previous development, so that the nomenclature minimal basis, double ζ, etc. may be kept as a reference.

---

[1] More precisely, for all ns functions, $e^{-\alpha r^2}$ replaces the polynomial-exponential product; for all np functions, $re^{-\alpha r^2}$ replaces the polynomial-exponential product.

For technical reasons, the basis sets utilized in various studies vary according to the group of molecule considered. The choice(s) made in each case must thus be specified, and the possible consequences affecting the accuracy of the conclusions must be borne in mind.

C. The Classical Hückel Approximation

There is another approach to the determination of individual molecular orbitals that differs in spirit from the preceding one in that it forgets about the apparent rigor of the self-consistent formalism. In this approach, instead of trying to determine the best molecular orbitals of an LCAO form that minimize the energy corresponding to a determinantal wave function, one looks for *approximate* molecular orbitals (always of an LCAO form). The method has been used for a long time, in particular for the study of π-electronic systems: one considers that each π-electron of the system moves in an *"effective" field* resulting from the field of the σ-core including the nuclei and the averaged repulsions of the other π-electrons. Defining the corresponding individual *"effective Hamiltonian,"* H_{eff}, the solving of an individual Schrödinger equation:

$$H_{eff}\varphi = \epsilon\varphi \qquad (17)$$

yields the individual energy ϵ and orbital φ.

If one looks for a molecular orbital of an LCAO form, one is led to the equations:

$$\sum_s c_s(H_{rs} - \epsilon S_{rs}) = 0 \qquad (18)$$

with the definitions:

$$H_{rs} = \int \chi_r^* H \chi_s d\tau \qquad (19)$$
$$S_{rs} = \int \chi_r^* \chi_s d\tau \qquad (20)$$

where H is the *individual* effective Hamiltonian. The individual energies are solutions of the equation:

$$|H_{rs} - \epsilon S_{rs}| = 0 \qquad (21)$$

Formally, these equations are similar to the SCF–LCAO–MO equations. Thus, a good definition of H_{eff} may yield satisfactory solutions without the tedious iteration procedure of the SCF method. In practice, two calculation procedures have evolved from these equations, the Hückel approximation and the Wheland–Mulliken approximation. Neither of them specifies the analytical form of H, but instead treats some of the matrix elements H_{rs} as adjustable parameters. Their main difference resides in the neglect (Hückel) or nonneglect (Wheland–Mulliken) of overlap. Their common feature is the tight-binding approximation, namely, the neglect of all matrix elements involving nonbonded atoms.

D. The Extended Hückel Theory (EHT)

The extended Hückel method is essentially due to Hoffman (3, 4) and is a pioneering contribution in the field of computations including all *valence* electrons in large molecules. In Hoffmann's method, the molecular orbitals φ_i are built up as a linear combination of Slater atomic orbitals χ_r:

$$\varphi_i = \sum_r c_{ir}\chi_r \qquad (22)$$

the coefficients and the orbital energies being obtained by solving the set of secular equations:

$$\sum_s c_{is}(H_{rs} - ES_{rs}) = 0 \qquad i = 1, 2, \ldots n \qquad (23)$$

where H is an effective Hamiltonian.

The atomic basis set is made up of one $1s$ orbital for each hydrogen atom, and one $2s$ and three $2p$ orbitals for each carbon and each heteroatom (with Slater orbital exponents). The diagonal matrix elements H_{rr} are the atomic valence state ionization potentials for each orbital, and the nondiagonal matrix elements H_{rs} are approximated by the Wolfsberg–Helmholtz formula:

$$H_{rs} = 0.5K(H_{rr} + H_{ss})S_{rs} \qquad (24)$$

with $K = 1.75$ everywhere.

This method is obviously an extension of the Hückel approximation for π-electrons in the sense that the molecular orbitals are obtained as the eigenvalues of an effective Hamiltonian H that is not made explicit, the matrix elements of H being treated as empirical input characteristics of the atoms involved. For these reasons the procedure is called the *Extended Hückel Theory* although it differs from the Hückel set of hypotheses in that it includes overlap as well as all nondiagonal elements in the secular determinant.

The usual set of input parameters of the Extended Hückel Theory is listed in Table I.

E. The Zero-Differential Overlap (ZDO) Approximation

Another way of getting around the difficulties of practical computations on large molecules and avoiding the use of a totally empirical effective Hamiltonian is to

TABLE I
VALUES OF THE DIAGONAL ELEMENTS OF
THE HAMILTONIAN USED IN THE
EXTENDED HÜCKEL THEORY FOR
s AND p ATOMIC ORBITALS (eV)

Atom	s	p
H	13.6	—
C	21.4	11.4
N	26.0	13.4
O	35.3	17.76

simplify the self-consistent method in various ways, conserving, however, its main formalism. The bottleneck of rigorous SCF calculations is the difficulty of calculating the integrals over the atomic orbitals. Thus, reducing their number has been an imperative requirement and has led to the fundamental zero-differential overlap approximation (ZDO), which assumes

$$\chi_p(\mu)\chi_q(\mu) \equiv 0 \tag{25}$$

for $p \neq q$. This hypothesis was initiated by Pariser and Parr (5) in the case of π-electrons and was later generalized by Pople et al. (6) for any pair of valence orbitals. It is easily seen that such an assumption simplifies considerably both the SCF equations and the calculation of the matrix elements involved in them, since all overlap integrals vanish, and, moreover, among the (pq,rs) integrals, only those of the (pp,pp) or (pp,qq) type remain.

The Roothaan LCAO–SCF equations in the ZDO approximation are thus simplified to

$$\sum_q c_{iq}(F_{pq} - \epsilon\delta_{pq}) = 0 \tag{26}$$

where δ_{pq} is the Kronecker symbol and the F_{pq}'s involve interaction elements G_{pq}, which reduce to

$$G_{pq} = -\sum_j c_{jp}c_{jq}(pp,qq) \tag{27}$$

and

$$G_{pp} = \sum_i \sum_r 2c_{ir}^2[(rr,pp)] \tag{28}$$

F. The Complete Neglect of Differential Overlap (CNDO) Method

In this approximation, all the valence electrons of a molecule are treated explicitly in the LCAO–SCF framework previously described (Eqs. 9–12). The inner-shell electrons are included in the core, and all the products

$$\chi_p(\mu)\chi_q(\mu) \tag{29}$$

between valence orbitals, whether on the same atom or on different atoms, are neglected. Thus, the equations to be solved are those of the ZDO approximation (Eq. 26) and the matrix elements G_{pq} and G_{pp} are given by Eqs. 27 and 28, which we write now for convenience in Pople's usual notations; the Greek-letter indices refer to atomic orbitals, the capital-letter indices to atoms. Thus, Eq. 26 becomes

$$\sum_\nu c_{i\nu}(F_{\mu\nu} - \epsilon\delta_{\mu\nu}) = 0 \tag{30}$$

Using the definition of the charge-density and bond order matrix:

$$P_{\mu\nu} = 2\sum_i c_{i\mu}c_{i\nu} \tag{31}$$

where i's are the occupied molecular orbitals, one obtains:

$$G_{\mu\mu} = \tfrac{1}{2} P_{\mu\mu}\gamma_{\mu\mu} + \sum_{\lambda \neq \mu} P_{\lambda\lambda}\gamma_{\lambda\mu} \tag{32}$$

$$G_{\mu\nu} = -\tfrac{1}{2} P_{\mu\nu}\gamma_{\mu\nu} \tag{33}$$

where the notation $\gamma_{\mu\nu}$ denotes the integral $(\mu\mu,\nu\nu)$.

These are the equations of the CNDO procedure, where CNDO stands for "complete neglect of differential overlap" (all products, Eq. 29, neglected). Other possibilities exist using ZDO in less complete fashion, letting some of the most important terms remain in the equations: this yields the INDO procedure (intermediate neglect of differential overlap) which retains all one-center exchange integrals or the NDDO method (neglect of diatomic differential overlap) (6, 7).

As to the CNDO method, it adds to the ZDO hypothesis two other fundamental approximations: the first one concerns the values of the Coulomb integrals $\gamma_{\mu\nu}$, which are all approximated as Coulomb integrals over s Slater atomic orbitals. If μ and ν are on the same second-row atom A:

$$\gamma_{\nu\mu} = \gamma_{2s_A 2s_A} = \gamma_{AA} \tag{34}$$

If not:

$$\gamma_{\mu\nu} = \gamma_{2s_A 2s_B} = \gamma_{AB} \tag{35}$$

This is a sort of mean value between $\gamma_{\sigma\sigma}$ and $\gamma_{\pi\pi}$, and this approximation ensures the invariance with respect to a rotation of the molecular axes (6), an important requirement if the theory is to be correct. When hydrogen atoms are involved, 1s atomic orbitals with $\zeta = 1.2$ are used in the calculations of the corresponding γ values.

With this approximation, and putting

$$P_{AA} = \sum_{\mu A} P_{\mu\mu} \tag{36}$$

the $G_{\mu\mu}$ matrix elements become

$$G_{\mu\mu} = (P_{AA} - \tfrac{1}{2}P_{\mu\mu})\gamma_{AA} + \sum_{B \neq A} P_{BB}\gamma_{AB} \tag{37}$$

The last fundamental hypothesis of the CNDO procedure concerns the core matrix elements $H_{\mu\nu}$. These correspond to the α and β parameters of the Pariser–Parr–Pople method, the core including here only the 1s electrons and the nuclei.

Using a partition of the core Hamiltonian into atom-centered fractions, the one center–one orbital core integral may be written:

$$\begin{aligned} H_{\mu_A\mu_A} &= U_{\mu_A\mu_A} + \sum_{B \neq A} (\mu_A|V_B|\mu_A) \\ &= U_{\mu_A\mu_A} - \sum_{B \neq A} V_{BA} \end{aligned} \tag{38}$$

where V_B is the potential due to the Bth core.

The values of all $H_{\mu_A \nu_A}$ for two different orbitals on the same atom are neglected. Concerning the $H_{\mu_A \nu_B}$ values, they are *assumed* to be expressible as

$$H_{\mu_A \nu_B} = \tfrac{1}{2}(\beta_A^0 + \beta_B^0) S_{\mu\nu} \tag{39}$$

where $S_{\mu\nu}$ is the overlap integral between orbitals μ and ν, which is calculated using the appropriate atomic Slater orbitals. The proper parameters of the procedure are then the values of the $U_{\mu\mu}$'s and the β^0's characteristic of each atom. A discussion of this choice is detailed in the original papers (6, 8, 9). In the version called CNDO/2, one approximates the $U_{\mu\mu}$'s as:

$$U_{\mu\mu} = -\tfrac{1}{2}(I_\mu + A_\mu) - (Z_A - \tfrac{1}{2})\gamma_{AA} \tag{40}$$

and the V_{AB}'s as:

$$V_{AB} = Z_B \gamma_{AB} \tag{41}$$

where I_μ and A_μ are the orbital ionization potential and electron affinity, and Z_B the Bth core charge.

Inside this framework, the β^0 values have been chosen so that the results fit as well as possible with those of nonempirical calculations for small molecules. Table II summarizes the numerical values used in CNDO/2.

G. Configuration Interaction (CI)

Let us come back to the purely theoretical SCF one-electron model described earlier. Even carried out exactly, this method yields a molecular wave function that contains a "built-in" error, the *correlation error* due to the model itself, in which two electrons of different spins may occupy the same molecular orbital. This allows for the possibility of finding two electrons in the same place at the same time, in fundamental contradiction with the fact that similarly charged particles avoid each other as much as possible because of their Coulomb repulsion.

One classical device which permits correction of this error is the following: when the SCF–LCAO equations are solved, one obtains more molecular orbitals then are needed to build the determinant Ψ (which is constructed on the orbitals of lowest energies). The remaining ones (the *virtual* orbitals) can be used for constructing

TABLE II
ATOMIC PARAMETERS FOR CNDO/2 (eV)[a]

Atom	$\tfrac{1}{2}(I + A)$		β^0
	s	p	
H	7.175	—	−9
C	14.051	5.572	−21
N	19.316	7.275	−25
O	25.390	9.111	−31

[a] CNDO = Complete Neglect of Differential Overlap.

other determinants Ψ_k of the same type, in which one of the initial φ's is replaced by one virtual orbital. These determinants correspond to "excited" configurations of the system, in the sense that one electron has been "excited" to an orbital of higher energy. Configurations may be singly, doubly, triply, etc., excited. Clearly, mixing all these configurations in an appropriate fashion is a way of correcting for the correlation error inserted in a single-determinantal wave function. The appropriate fashion is to compute the best possible linear combination of all the Ψ_k's:

$$\theta = \Sigma d_k \Psi_k \quad (42)$$

This is called "configuration mixing" or "configuration interaction" (CI). The classical CI technique uses a variational procedure for obtaining the coefficients d_k, which minimize the total energy corresponding to θ. However, this way of improving over the SCF results presents two major difficulties that are intertwined: obviously, in order to introduce a significant amount of the correlation correction, the number of configurations must be as large as possible, but this entails the solving of a very large configuration interaction matrix for obtaining the d's. The number of possible "excitations" increases so rapidly with the enlargement of the system that drastic arbitrary truncation of the CI matrix is imposed for practically every molecule of reasonable interest.

H. Perturbative Configuration Interaction (PCI)

Since one is looking for a correction to a relatively good representation, it appears natural to turn to perturbation techniques for solving the configuration interaction problem. The single configuration determinants Ψ_k may be considered as the eigenvectors of an unperturbed Hamiltonian H_0, which differs from the exact Hamiltonian $\mathcal{H}$ (including correlation) by a small perturbation term λV:

$$\mathcal{H} = H_0 + \lambda V \quad (43)$$

with

$$(H_0)_{kk} = E_k \quad (44)$$

The wave function and energy corresponding to the exact solution may thus be developed into Taylor expansions in the neighborhood of $\lambda = 0$:

$$\epsilon = E_k + \lambda E'_1 + \lambda^2 E'_2 + \cdots \quad (45)$$

$$\theta = \Psi_k + \lambda \Phi'_1 + \lambda^2 \Phi'_2 + \cdots \quad (46)$$

for a given k. Taking $k = 0$, the first-, second-, third-order, etc., corrections E_i' and Φ_i' to E_0 and Ψ_0 are given by the classical formulas of the Rayleigh–Schrödinger perturbation theory in terms of the matrix elements of V and of the eigenvalues of H_0. For instance[2]

$$E_1' = V_{00} \quad (47)$$

$$E_2' = \sum_{k \neq 0} \frac{|V_{0k}|^2}{E_0 - E_k} \quad (48)$$

$$E'_3 = \sum_{\substack{k \ l \\ \neq 0}} \frac{V_{k0} V_{l0} V_{0k}}{(E_0 - E_k)(E_0 - E_l)} \quad (49)$$

[2] λ is now included in the perturbation for convenience.

The definition of the perturbation potential may be done in different fashions. The classical definition identifies H_0 with the SCF Hamiltonian, but this entails a slow convergency of the perturbation process, thus the necessity of computing a large number of terms in the series. A better convergency may be obtained if one adopts a definition of H_0 and V that does not require the exploitation of the operators themselves, but only of their matrix elements over the Ψ_k's (10, 11). Thus one imposes

$$\mathcal{H}_{IJ} = V_{IJ'} \quad \text{for } I \neq J \quad (50)$$

$$\mathcal{H}_{II} = (H_0)_{II'} \quad \text{for all I's} \quad (51)$$

where $\mathcal{H}$ is the known exact Hamiltonian. In this way, the E_k values of the denominators in Eqs. 47–49 are the energies of the single determinants computed with $\mathcal{H}$, all V_{II}'s are zero [so that the first-order correction is zero (Eq. 47)], and the other necessary matrix elements of V are the usual elements of the CI matrix computed with $\mathcal{H}$. Note that in the CI matrix, only single- and double-excitation terms interact with Ψ_0, since $\mathcal{H}$ contains only one- and two-electron operators. Thus the perturbation up to second-order requires only the double-excitation terms.

I. Localized Orbitals

It is well known (12) that the relatively delocalized canonical SCF molecular orbitals can be transformed through a matrix transformation into an equivalent set of orthogonal *localized orbitals*—localized, for instance, over the chemical bonds, leaving the SCF energy invariant.

Aside from the fact that such localized orbitals appeal to chemists, they have the important property of being advantageous for performing configuration interaction: it has been shown that the configurations contributing most strongly to the correlation energy are those constructed from orbitals as similar as possible to the occupied orbitals that they replace in the reference state (13–16). This condition is automatically fulfilled by localizing the SCF orbitals on the chemical bonds and the virtual orbitals into "antibonding" orbitals in the corresponding regions (17).

This important property of localized orbitals may be utilized inside a perturbative configuration treatment like that outlined in Section II, H. *Perturbative configuration interaction over localized orbitals* is the general framework of the PCILO method.

J. The PCI over Localized Orbitals (PCILO) Method

The originality of the PCILO method proper resides in the particular choice of the localized orbitals (18, 19). In the perturbative scheme developed above, nothing imposes a requirement to start with the SCF molecular orbitals (or their localized equivalents) for building the zero-order determinant. The sole restriction is that of using an orthonormal set of orbitals. On the other hand, best convergency is to be expected from the use of occupied and virtual orbitals localized in the same spatial region: hence the idea of using a set of bond orbitals and their antibonding counterparts defined *a priori* on the chemical bonds. (Such orbitals are distinct from the localized SCF orbitals, which are not fully localized over the bonds but have tails in other parts of the molecules. Nevertheless, the similarity is sufficient to ensure a reasonable zero-order approximation.)

A PCILO procedure using this idea can be developed in an *ab initio* fashion (see, e.g., 20). However, the method usually called PCILO designates an all-valence electrons version of the method, which introduces the following simplifications: (a) the use of the full ZDO hypothesis, (b) the adoption of the CNDO/2 integral approximations, both for the Coulomb integrals and for the core matrix elements.

The starting molecular orbitals are built as bond orbitals and lone-pair orbitals according to the picture provided by the chemical formula, as two-by-two combinations of suitable atomic *hybrids*:

$$i = \alpha_{\chi 1} + \beta_{\chi 2} \qquad (52)$$

An antibonding orbital is made to correspond to each bond orbital:

$$i^* = -\beta_{\chi 1} + \alpha_{\chi 2} \qquad (53)$$

(for the lone-pairs χ_2 is rejected to infinity: β and $i° = 0$).

The appropriate hybrids are constructed from the usual Slater atomic orbitals so as to ensure maximum overlap on the chemical bonds and orthogonality between bonds by a procedure due to Del Re (21).

The ZDO condition has the considerable advantage of making the starting set of bond orbitals automatically orthogonal. The zero-order ground-state determinant is built over all the i's, and all possible "excited" determinants are built using the $i°$'s. This and the total Hamiltonian define entirely both the E_k and the necessary perturbation matrix elements V_{ok} according to the formulas of Section II, H. The advantage of using bond orbitals is both to simplify the computation of the elements and to allow a simple physical interpretation of the various terms (19).

Thus, the second-order correction to the energy comprises (a) single excitations $i \rightarrow i°$ (polarization term) or $i \rightarrow j°$ (delocalization term); (b) double excitations reduced by ZDO to: $ii \rightarrow i°i°$ (intrabond "correlation"), $ij \rightarrow i°j°$ (interbond "correlation").

The third-order terms (Eq. 49) involve interaction between the previous configurations, namely: single excitations with single excitations; single excitations with double excitations; double excitations with double excitations. The first combination is made of polarization and/or delocalization corrections; the second and third involve again "correlation."

An important point must be noted so as to keep clear the relation between PCILO and other procedures: in the PCILO method, the starting orbitals not being self-consistent, the single excitations terms do not cancel out as in a conventional CI over an SCF determinant. The SCF level lies somewhere in the neighborhood of the "zero-order + single excitations correction" level *without being identical to it* (see Table III for an example). The further addition of the double excitations terms in PCILO thus corresponds roughly to what would be obtained by adding double excitations by perturbation to an SCF determinant, but again is not identical to it.

In practice, the currently available PCILO program[3] goes to the third order in the energy. The second-order correction is a summation over only negative terms (Eq. 48) and is large; thus it descends below the correct energy that would be obtained by full CI; the third-order corrections are appreciably smaller. They result in a global positive contribution, thus bringing the energy closer to the exact value. Table III reproduces, as an example, the energy output for the formaldehyde molecule. In that case the CNDO/2 SCF energy is $-16,838$ kcal/mol. CI performed

---

[3] Which may be obtained from Q.C.P.E. (Quantum Chemistry Program Exchange) at the Department of Chemistry, University of Indiana, Bloomington, Indiana.

TABLE III
A Typical PCILO[a] Computer Output: Case of Formaldehyde[b]

Term	Energies (kcal/mol)	
Energy of the fully localized determinant		−0.16799055E 05
Polarization energy	−0.28453348E−03	
Delocalization energy	−0.29723450E 02	
Intrabond correlation energy	−0.33245941E 02	
Interbond correlation energy	−0.30136505E 02	
Second-order energy correction		−0.93106171E 02
Total energy after the second order		−0.16892160E 05
One-bond correlation–polarization interaction	−0.44230334E−02	
Polarization–polarization interaction	−0.69539863E−04	
Polarization–delocalization interaction	−0.89407538E−02	
Polarization–2-bond correlation interaction	−0.23979165E−02	
Delocalization–2-bond correlation interaction	−0.20620240E−01	
One-bond correlation–1-bond correlation interaction	0.13736535E 02	
Delocalization–delocalization interaction	−0.62138920E 01	
Two-bond correlation–2-bond correlation interaction	0.49700222E 01	
Third-order energy correction		0.12484096E 02
Total energy after the third order		−0.16879676E 05
The perturbation process has required		0 MN 0 S 12

[a] PCILO = Perturbative Configuration Interaction over Localized Orbitals.
[b] All the contributions involving polarization are made very small by an optimization of the bond polarities at zero-order.

on this starting point lowers this value by 57 kcal/mol (22). The third-order PCILO value is 42 kcal/mol below the SCF level, thus yielding over 70% of the correlation energy.

III. Types of Torsion Angles and Definitions

Let us recall first that a torsion angle of the bonded atoms A—B—C—D is the angle between the planes formed by atoms A,B,C and B,C,D. The torsion angle is considered positive (0°–360°) for a right-handed rotation: when looking along the bond B—C, the far bond C—D rotates clockwise with respect to the near bond A—B. The zero value of the torsion angle corresponds to a *cis*-planar arrangement of the bond A—B and C—D.

The conformational properties of nucleic acids and their constituents depend upon a series of torsion angles that can be divided into three groups: (i) the glycosyl torsion angle, which defines the relative orientation of the pyrimidine and purine bases with respect to the sugar; (ii) the torsion angles along the backbone of the chain; and (iii) the torsion angles about the bonds of the sugar that determine the puckering of the sugar.

A. Torsion Angles in Polynucleotides

The schematic diagram of a nucleotide unit in polynucleotide chains is shown in Fig. 1. In that figure are also indicated the notations we use for the different torsion angles throughout this paper. More precisely, they are defined as:

$$\chi_{CN} = O1'\text{—}C1'\text{—}N1\text{—}C6 \quad \text{for pyrimidine nucleosides}$$
$$\chi_{CN} = O1'\text{—}C1'\text{—}N9\text{—}C8 \quad \text{for purine nucleosides}$$

$$\Phi_{P-O3'} = O5'\text{—}P\text{—}O3'\text{—}C3'$$
$$\Phi_{O3'-C3'} = P\text{—}O3'\text{—}C3'\text{—}C4'$$
$$\Phi_{C3'-C4'} = O3'\text{—}C3'\text{—}C4'\text{—}C5'$$
$$\Phi_{C4'-C5'} = C3'\text{—}C4'\text{—}C5'\text{—}O5'$$
$$\Phi_{C5'-O5'} = C4'\text{—}C5'\text{—}O5'\text{—}P$$
$$\Phi_{O5'-P} = C5'\text{—}O5'\text{—}P\text{—}O3'$$

$$\tau_0 = C4'\text{—}O1'\text{—}C1'\text{—}C2'$$
$$\tau_1 = O1'\text{—}C1'\text{—}C2'\text{—}C3'$$
$$\tau_2 = C1'\text{—}C2'\text{—}C3'\text{—}C4'$$
$$\tau_3 = C2'\text{—}C3'\text{—}C4'\text{—}O1'$$
$$\tau_4 = C3'\text{—}C4'\text{—}O1'\text{—}C1'$$

Unfortunately, no generally adopted nomenclature now exists in this field, and different groups of authors use different notations. An *ad hoc* International Committee composed of Drs. B. Pullman, W. Saenger, V. Sasisekharan, M. Sundaralingam and H. R. Wilson under the chairmanship of Dr. Sundaralingam was formed at the Jerusalem Symposium on Conformation of Biological Molecules and Polymers (held in April 1972) and made a proposal for a universal nomenclature (23). However, this proposal has not yet been officially accepted. Although we favor the Jerusalem proposals, we adopt in this review a more "neutral" nomenclature so as to accommodate those who use other notations.

The problem concerns essentially the torsion angles of the backbone. Although the definition of these angles is the same in all proposals and the same as given above, and corresponds to the original proposal of Sundaralingam (24, 25), different authors (24–29) have used different letters for the description of these angles (Table IV). This situation makes it difficult to compare results obtained by different workers. [It may also be added that, in distinction from all other authors who have used a *cis*-planar arrangement of the terminal bonds as being the zero value of the torsion angle, this value corresponds in Olson and Flory's notations (29) to the *trans*-planar arrangement of these bonds.] For that reason, we have simply designated here the various torsion angles

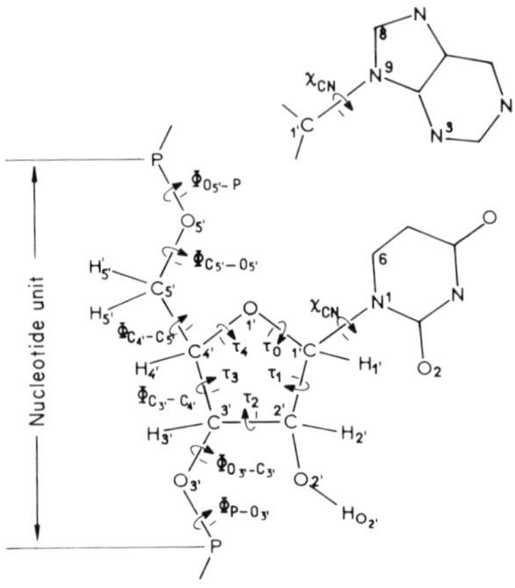

FIG. 1. The torsion angles of the nucleotide unit.
FIG. 2. Schematic diagram of the puckerings of the ribose ring.

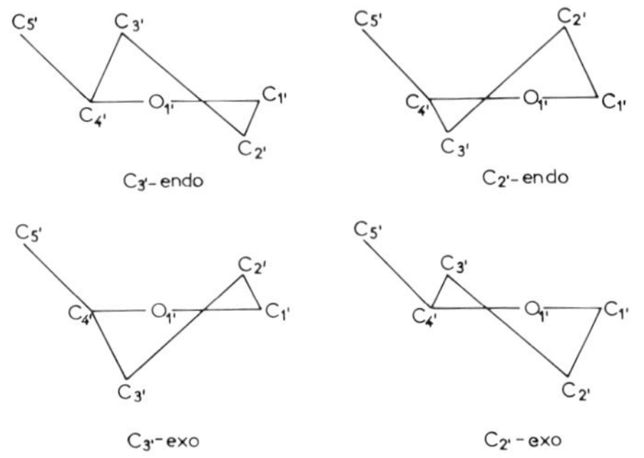

TABLE IV

Various Notations Proposed by Different Authors for the Same Backbone Torsion Angles

This study	Sundaralingam (24, 25)	Arnott (26)	Lakshminarayan and Sasisekharan (27), Sasisekharan (28)	Olson and Flory (29)
$\Phi_{P-O3'}$	ω'	Φ	Φ	Ψ'
$\Phi_{O3'-C3'}$	Φ'	ω	$\theta 3$	ω''
$\Phi_{C3'-C4'}$	Ψ'	σ	σ	ω'
$\Phi_{C4'-C5'}$	Ψ	ξ	$\theta 2$	Φ''
$\Phi_{C5'-O5'}$	Φ	θ	$\theta 1$	Φ'
$\Phi_{O5'-P}$	ω	Ψ	Ψ	Ψ''

by Φ's, using subscripts to indicate the bond around which the rotation is carried out. As the numbering of the atoms in a polynucleotide chain is universally accepted, this procedure avoids all misunderstandings and has the obvious advantage of immediate comprehension.

On the other hand, we use the symbol χ_{CN}, introduced by Sundaralingam (24, 25), for the glycosyl torsion angle, as it seems to be quite universally adopted today. It is closely related to the glycosyl torsion angle Φ_{CN}, first proposed by Donohue and Truebood (30), by the equation $\chi_{CN} \approx -\Phi_{CN}$. However, it must be added that various other definitions for χ_{CN} have also been used, depending upon the choice of the atoms appended to the glycosyl bond (26, 31, 32; for a review see ref. 25). Following a widely accepted definition, we denote by the terms *anti* and *syn* the ranges of the glycosyl torsion angle χ_{CN}, corresponding, respectively, to $\chi_{CN} = 0° \pm 90°$ and $\chi_{CN} = 180° \pm 90°$. The problem of the *syn-anti* equilibrium of nucleosides and nucleotides is among the most debated ones in conformational studies on this type of molecules.

The conformation of the sugar ring (Fig. 1) may be described by five torsion angles τ_0, τ_1, τ_2, τ_3 and τ_4, corresponding, respectively, to the rotations about the O1'—C1', C1'—C2', C2'—C3', C3'—C4' and C4'—O1' bonds. Currently, four major classical conformations (Fig. 2) are considered, corresponding to the so-called C3'-endo, C2'-endo, C3'-exo and C2'-exo puckers (envelope conformations). The key to this nomenclature is the position of the C3' or C2' atom with respect to the C5' atom and the mean plane passing through the other four atoms of the ring (24, 25, 33, 34). For example, the conformation of the sugar ring is said to be C3'-endo (notation ^3E) when the atom C3' is displaced from this plane on the same side as the C5' atom; on the contrary, in a C3'-exo conformation (notation $_3E$), the atom C3' is displaced on the opposite side of C5' from the same plane. Subsequently, in a more realistic representation, Sundaralingam (35) has completed the description of the conformations of the sugar rings by introducing the twist (T) forms. In such forms, no four atoms of the furanose ring lie in a plane, and the two atoms that show the largest deviations from the five-atom plane are considered to be the puckered atoms. The displacements of these two atoms from the remaining three-atom plane are again described with reference to C5'. For example $^2T_3$ means C2'-endo–C3'-exo. The number that precedes the letter T denotes the atom that shows the major puckering, and the number that follows this letter denotes the secondary puckering. Other notations have also been utilized to describe the conformations of the sugar ring in the pseudorotational concept (28, 36–38), and they are presented in Section VII, where the conformation of the sugar ring is described in more detail. Besides these, Arnott (26) has designated the torsion angles of the sugar ring by different notations.

In the following sections, dealing with the glycosyl and principal torsion angles of the backbone, we generally use the simple envelope notation for designing the sugars possibly involved.

B. Torsion Angles in Mononucleotides

Figure 3 shows the schematic diagram of a 5'-nucleotide (39). The glycosyl torsion angle χ_{CN}, the torsion angles τ_0 to τ_4 of the sugar and the torsion angles $\Phi_{C3'-C4'}$, $\Phi_{C4'-C5'}$ and $\Phi_{C5'-O5'}$ of the backbone remain unchanged. The remaining torsion angles are defined as

$$\Phi_{O5'-P} = C5'-O5'-P-OIII$$
$$\Phi_{P-OIII} = O5'-P-OIII-HOIII$$
$$\Phi_{C2'-O2'} = C1'-C2'-O2'-HO2'$$
$$\Phi_{C3'-O3'} = C2'-C3'-O3'-HO3'$$

It should be noticed that the torsion angle $\Phi_{C3'-O3'}$ defined for a 5'-nucleotide is different from $\Phi_{O3'-C3'}$ defined for the polynucleotide chain. The two torsion angles differ in the choice of the terminal atoms.

In the case of a 3'-nucleotide, the phosphate group attached to O5' is replaced by HO5' and HO3' attached to O3' is replaced by the phosphate group; some torsion angles are thus redefined as:

$$\Phi_{C5'-O5'} = C4'-C5'-O5'-HO5'$$
$$\Phi_{O3'-C3'} = P-O3'-C3'-C4'$$
$$\Phi_{P'-O3'} = OIII-P-O3'-C4'$$
$$\Phi_{OIII-P} = HOIII-OIII-P-O3'$$

In a 2'-nucleotide the phosphate group attached to O5' is replaced by HO5', and HO2' attached to O2' is replaced by the phosphate group. Some torsion angles have to be redefined; they are:

$$\Phi_{C5'-O5'} = C4'-C5'-O5'-HO5'$$
$$\Phi_{C2'-O2'} = C1'-C2'-O2'-P$$
$$\Phi_{O2'-P} = C2'-O2'-P-OIII$$
$$\Phi_{P-OIII} = O2'-P-OIII-HOIII$$

C. Torsion Angles in Nucleosides

The schematic diagram of a nucleoside is shown in Fig. 4. It differs from that of a 5'-nucleotides (Fig. 3) in having a hydrogen atom HO5' attached to O5' instead of the phosphate group and consequently, $\Phi_{C5'-O5'}$ is defined as (39):

$$\Phi_{C5'-O5'} = C4'-C5'-O5'-HO5'$$

All the other torsion angles remain the same as for the 5'-nucleotide. It is relevant here to point out that Hart and co-workers (40–42) use a different convention to describe the glycosyl torsion angle in their solution studies on nucleosides by nuclear Overhauser effect. Their torsion angle γ is related to χ_{CN} and Φ_{CN} by the simple relations:

$$\gamma = -\chi_{CN} - 120°$$
$$\gamma = \Phi_{CN} - 120°$$

The conformation about the exocyclic C4'—C5' bond is described by the relative orientation of C5'—O5' bond with respect to the C4'—O1' and C4'—C3' bonds. In a widely used notation, Shefter and Trueblood (43) have designated the angle between the projected C5'—O5' and C4'—O1' bonds along the O4'—C5' bond as Φ_{OO} and the angle between the projected C5'—O5' and C4'—C3' bonds along the C4'—C5' bond as Φ_{OC}. Three stable orientations of C5'—O5' bond have been observed in crystal structures, and these are denoted as *gauche-gauche* ($\Phi_{OO} \approx 60°$, $\Phi_{OC} \approx 60°$), *gauche-trans* ($\Phi_{OO} \approx 60°$, $\Phi_{OC} \approx 180°$) and *trans-gauche* ($\Phi_{OO} \approx 180°$, $\Phi_{OC} \approx 60°$) In the *gauche-gauche* (*gg*) conformation, the C5'—O5' bond is in between the C4'—O1' and C4'—C3' bonds and above the sugar ring; in the *gauche-trans* (*gt*) conformation, the C5'—O5' bond

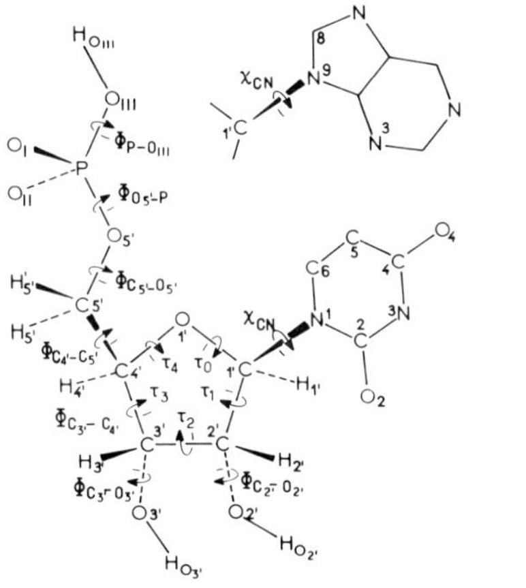

FIG. 3. 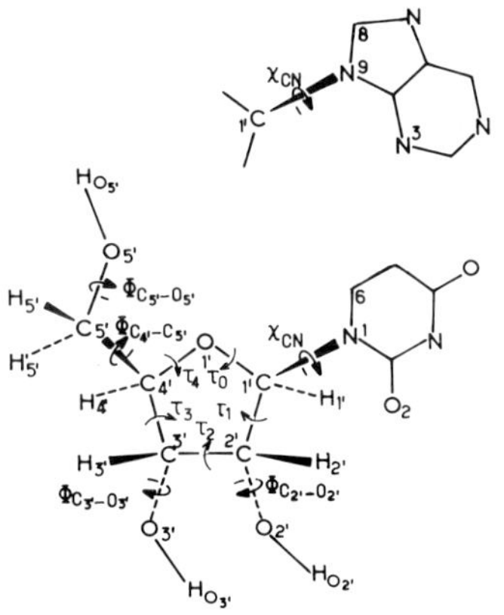   FIG. 4.

FIG. 3. The torsion angles in 5'-β-nucleotides.

FIG. 4. The torsion angles in β-nucleosides.

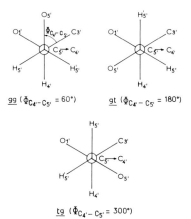

gg ($\Phi_{C_{4'}-C_{5'}} = 60°$)   gt ($\Phi_{C_{4'}-C_{5'}} = 180°$)

tg ($\Phi_{C_{4'}-C_{5'}} = 300°$)

FIG. 5. The *gg*, *gt* and *tg* conformations about the exocyclic C4'—C5' bond.

is outside the ring at the side of C4′—O1′ bond and opposite to the C4′—C3′ bond; in the *trans–gauche* (*tg*) conformation, the C5′—O5′ is also outside the sugar ring at the side of C4′—C3′ bond and opposite to the C4′—O1′ bond (Fig. 5). The torsion angles Φ_{OO} and Φ_{OC}, as defined in ref. 43, do not carry a sign. However, more recently, a number of authors have attributed a sign to these angles and denote the *gauche–gauche* conformation by $\Phi_{OO} \approx -60°$ $\Phi_{OC} \approx 60°$; the *gauche–trans* conformation by $\Phi_{OO} \approx 60°$, $\Phi_{OC} \approx 180°$ and the *trans–gauche* conformation by $\Phi_{OO} \approx 180°$, $\Phi_{OC} \approx -60°$.

In the present study, the conformation about the exocyclic C4′—C5′ bond is defined by the torsion angle

$$\Phi_{C4'-C5'} = C3'-C4'-C5'-O5'$$

similar to the definition used for polynucleotides and nucleotides. By analogy with Φ_{OC} and Φ_{OO} notations of Shefter and Trueblood (*43*), we may also add the definition:

$$\Phi_{C4-C5'} = O1'-C4'-C5'-O5'$$

In our discussion, the *gg*, *gt* and *tg* conformations about the exocyclic C4′—C5′ bond correspond to the values of $\Phi_{C4'-C5'}$ approximately equal to 60°, 180° and 300°, respectively.

IV. The Glycosyl Torsion Angle χ_{CN}

The study of the glycosyl torsion angle χ_{CN} in the constituents of nucleic acids has been divided into two parts, concerned respectively with nucleosides and nucleotides.

A. The *syn–anti* Equilibrium in Nucleosides

It is a current belief among many that the *anti* conformers of purine and pyrimidine nucleosides are *intrinsically more stable* than the *syn* conformers. This belief originates mainly from the observation that the majority of the crystal structures of nucleosides (and nucleotides) determined by X-ray are of the *anti* type. It is still supported by the observation that the nucleosides that occur in the *syn* conformation frequently (but not always) have relatively bulky substituents at a position of the base crucial for the *syn–anti* arrangement (e.g., a bromine atom at position 8 of guanosine or adenosine, or a CH_3 group at position 6 of uridine). This suggests that these *syn* conformations occur essentially only when steric hindrance prevents the formation of the *anti* ones, or by virtue of intermolecular features (*44*).

However, evidence from studies in solution throw some doubt on this simple conception. Thus, while nuclear magnetic resonance (NMR) studies (proton shifts and spin–spin couplings) seem to indicate, in the majority of cases examined (uridine, pseudouridine, 2′-deoxyuridine, 4-thiouridine, 6-azauridine, dihydrouridine, cytidine, 6-azacytidine, adenosine, guanosine), the predominance of the *anti* form in aqueous solution (*45–55*), they do not exclude a contribution of the *syn* form. They also confirm the general predominance in solution of the *syn* conformation in nucleosides having the above-mentioned bulky substituents suitably placed. The results of optical rotatory dispersion and circular dichroism are frequently inconclusive. In an excellent survey of these studies Ts'O (*55*) stated: "Generally, the results support the notion that pyrimidine nucleosides have the *anti* conformation and have a relatively large rotational barrier between the *anti* and *syn* conformations. The results in the purine nucleosides are not conclusive" (see also refs. *56–58*).

On the other hand, recent studies on the nuclear Overhauser effect suggest in a number of cases a possible predominance or at least a relatively large proportion of the *syn* form in aqueous Me_2SO. Quantitative results have been presented in this field by Hart and Davis (*41, 42*) for purine nucleosides: while guanosine and xanthosine exist, respectively, 46% and 60% in the *syn* form, adenosine seems to exist 34% in the *anti* form and 66% in a form at the borderline between the *syn* and *anti*; inosine seems to exist in two forms, both *anti*. Moreover, the 2′,3′-isopropylidene derivatives (designated hereafter by the prefix i-) of adenosine, inosine and guanosine are said to exist, respectively, 84%, 80% and 76% in the *syn* conformation. The results are less quantitative for the pyrimidine nucleosides, although here again an appreciable proportion of the *syn* conformer is proposed for cytidine, uridine and thymidine and a possible predominance of the *syn* conformers for their isopropylidene derivatives (*40, 59–61*). A significant proportion of the *syn* form is also proposed on the basis of similar studies for i-adenosine 5′-acetate (*62*). However, the 5-halogenouridines, by the same technique, exist preferentially and largely in the *anti* form (*63*). Finally, a significant contribution if not the predominance of *syn* conformations in benzene or dioxane solutions seems also to be indicated by the measurements of dipole moments of isopropylidene derivatives of adenosine, inosine and uridine, but not of guanosine, although the interpretation of these results is subject to caution (*64*).

This situation raises explicitly the question of the *intrinsic* relative stabilities of the *anti* and *syn* forms of the nucleosides and their derivatives. It is obvious that the equilibrium between these forms may be and is influenced by environmental factors. However, understanding of the role of these factors can be obtained best if we have sufficient knowledge on *which is fundamentally the most stable form of the nucleoside*, the form it would adopt in the free state, and may perhaps be expected to adopt in inert solvents.

This problem can be dealt with by constructing conformational energy maps presenting the energy as a function of the rotation of the base with respect to the sugar around the glycosyl linkage. However, examination of the structural formulas and models clearly indicates that this problem is a complex one and that in its treatment particular attention must be paid to the possibilities of intramolecular interactions, in particular hydrogen bonding, involving on the one hand the base (more specifically N3 of the purines and O2 of the pyrimidines) and on the other the O5'—HO5' and O2'—HO2' bonds of the sugar. It is easily seen (Fig. 4) that intramolecular hydrogen bonding between the base and the O5'—HO5' bond of the sugar will favor the *syn* conformation while the existence of such bonding with the O2'—HO2' bond of the sugar will favor the *anti* conformation of the nucleosides. The reality of such intramolecular hydrogen bonding has been indicated in a number of cases: by X-ray crystallography for the interaction involving O5'—HO5' bond and the purine ring (see refs. 65, 66) and, in solution studies, utilizing infrared and other techniques for interactions involving both the O5'—HO5' and O2'—HO2' bonds in both purine and pyrimidine nucleosides (55, 67–69). However, the reality of hydrogen bonding involving the O2'—HO2' bond of the sugar, and the efficiency of its interaction with the bases, has been questioned by Sundaralingam (35), who indicates that it can occur only in a C2'-endo—C3'-exo or a C2'-endo—C1'-exo conformation of the sugar, and by Chan (70, see also 65, 71). Recently, Ts'o et al. (72) showed that the influence of the 2'-hydroxyl group of the ribose on the conformation of adenine dinucleoside monophosphates is exerted through the steric hindrance of this group, not through its hydrogen-bonding properties.

Unfortunately, these interactions have generally been omitted from the theoretical studies of the problem, including our own early PCILO calculations (66, 73, 74). In these early studies, the essential aim was to compare the results of the computations with the available X-ray crystallographic data. Now, in the great majority of these data on purine and pyrimidine nucleosides, there is no intramolecular hydrogen bond, the O5'—HO5' bond being rather used for the formation of an intramolecular hydrogen bond with a neighboring nucleoside or with a water molecule. For this practical reason, we carried out our early calculations by precluding in general the intramolecular hydrogen bond by appropriately preorienting the exocyclic CH_2OH group. Insofar as the nucleoside is concerned, the possibility of formation of such a bond depends, in the first place, on whether the orientation of C5'—O5' bond of the sugar is *gg* or *gt* or *tg*. The intramolecular hydrogen bond is possible only for the *gg* conformation and needs moreover, in that case, an appropriate orientation of the O5'—HO5' bond corresponding to $\Phi_{C5'-O5'} \approx 60°$. In general, the early PCILO calculations (*66, 73, 74*), have been carried out with the *gg* orientation of the C5'—O5' bond, which is the one most frequently observed in crystal structures (*24, 25*), but with a value of $\Phi_{C5'-O5'} \approx 180°$, which precludes an intramolecular hydrogen bond but makes an intermolecular hydrogen bond possible. It is only in the case of C2'-endo purines, in which an intramolecular hydrogen bond between the O5'—HO5' bond of the sugar and N3 of the purine base is observed in crystal structures, that the effects of this possibility on the conformational energy maps were examined (*66, 74*). The results of the calculations show that the *anti* conformations are the most stable ones in the cases in which such an intramolecular hydrogen bond was *a priori* forbidden ($\Phi_{C5'-O5'} \approx 180°$), but that the *syn* conformation is the most stable one in the case when such a bond is permitted ($\Phi_{C5'-O5'} \approx 60°$), in agreement with the available X-ray crystallographic data.

In view of a more appropriate investigation of the *intrinsic* conformational preferences of the nucleosides, a complete PCILO study has been recently carried out (*75*) exploring the influence on their conformation of the possible interactions between the base and the O5'—HO5' and O2'—HO2' bonds of the sugar. This complete study comprised three stages of computations:

1. Construction of conformational energy maps as a function of χ_{CN} and $\Phi_{C5'-O5'}$ with no interaction allowed between the base and the O2'—HO2' bond of the sugar. These computations are of direct significance for the 2'-deoxynucleosides and 2',3'-isopropylidene derivatives.

2. Construction of conformational energy maps as a function of χ_{CN} and $\Phi_{C2'-O2'}$ with no interaction allowed between the base and the O5'—HO5' bond of the sugar. These computations are of significance for cases in which the O5'—HO5' bond is unavailable for an intramolecular hydrogen bond, a situation that occurs frequently in crystals.

3. Construction of conformational energy maps as a function of χ_{CN} with $\Phi_{C5'-O5'}$ and $\Phi_{C2'-O2'}$ fixed in their preferred values. These maps thus take into account the simultaneous effect of the two possible intramolecular interactions and represent the situation in the individual free molecule. The results of the computations carried out at each stage by adopting the input geometrical data from the available X-ray crystal structures represent the most complete theoretical investigation performed so far on the subject. They are therefore presented prior to the description of

other, generally much more fragmentary and restricted, studies carried out by other procedures.

1. EFFECT OF THE ORIENTATION OF THE O5'—HO5' BOND

A series of PCILO conformational energy maps for purine and pyrimidine nucleosides having C2'-endo and C3'-endo sugar puckers as a function of χ_{CN} and $\Phi_{C5'-O5'}$ has been constructed (75) with the orientation of C5'—O5' bond fixed in the gg conformation. The preselected values of both $\Phi_{C3'-O3'}$ and $\Phi_{C2'-O2'}$ are equal to 180°. The principal results from this study are as follows.

Among the five C2'-endo nucleosides (76–80), four (deoxyguanosine, inosine molecule-2, adenosine and uridine) present a global energy minimum at $\chi_{CN} = 240°$ and $\Phi_{C5'-O5'} = 60°$ (75). A representative map corresponding to C2'-endo inosine molecule-2 is shown in Fig. 6. For C2'-endo cytidine, the global minimum occurs at $\chi_{CN} = 210°$ and $\Phi_{C5'-O5'} = 90°$ (75). In all these cases, the possibility of an intramolecular hydrogen bonding between the exocyclic CH$_2$OH group and the base results therefore in a preferential stabilization of the *syn* conformation over the *anti* one. A local minimum occurs in all cases at $60° < \chi_{CN} < 90°$ and $\Phi_{C5'-O5'} \approx 180°$, corresponding to an *anti* conformation. It may be remarked here that out of the five C2'-endo nucleosides, deoxyguanosine is the only one that exists in the *syn* form in the crystal (in complex with 5'-bromodeoxycytidine) (76).

The situation is more diverse with the C3'-endo nucleosides. Adenosine and cytidine have a global energy minimum at $\chi_{CN} = 180°$ and $\Phi_{C5'-O5'} \approx 60°$. The conformational energy map for cytidine is presented in Fig. 7. Thus the preferred conformation of χ_{CN} in these two nucleosides is *syn*. Guanosine (Fig. 8) has its global minimum at $\chi_{CN} = 90°$ associated with $\Phi_{C5'-O5'} = 150°$, at the borderline between the *syn* and *anti* conformations. Uridine (Fig. 9) has its global minimum at $\chi_{CN} = 0°$ and $\Phi_{C5'-O5'} = 180°$, thus presenting a preferred *anti* conformation. Moreover, cytidine (Fig. 7) has a local minimum at $\chi_{CN} = 0°$ and $\Phi_{C5'-O5'} = 180°$, only about 0.5 kcal/mol above the global minimum. On the other hand, guanosine (Fig. 8) and uridine (Fig. 9) possess local energy minima corresponding to *syn* forms at $\chi_{CN} \approx 180°$ and 210°, respectively, only 2 kcal/mol above the global minimum. Because of the particular case of uridine, which alone shows a definite preference for the *anti* conformation, computations have been repeated for this molecule with a different set of input data. The results of Fig. 9 were obtained with geometrical input data coming from the crystal structure of 5-bromouridine in complex with adenosine (81). When the calculations are repeated with such data coming from β-adenyl(2'-5')uridine, A(2'-5')U (78), the results are

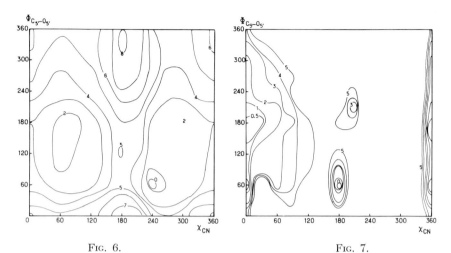

Fig. 6.   Fig. 7.

Fig. 6. PCILO conformational energy map for C2'-endo inosine (molecule-2) as a function of χ_{CN} and $\Phi_{C5'-O5'}$ with $\Phi_{C2'-O2'} = \Phi_{C3'-O3'} = 180°$. Isoenergy curves in kcal/mol with the global minimum taken as energy zero.

Fig. 7. PCILO conformational energy map for C3'-endo cytidine as a function of χ_{CN} and $\Phi_{C5'-O5'}$ with $\Phi_{C2'-O2'} = \Phi_{C3'-O3} = 180°$. Isoenergy curves in kcal/mol with the global minimum taken as energy zero.

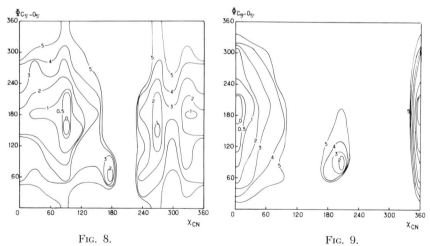

Fig. 8.   Fig. 9.

Fig. 8. PCILO conformational energy map for C3'-endo guanosine as a function of χ_{CN} and $\Phi_{C5'-O5'}$ with $\Phi_{C2'-O2'} = \Phi_{C3'-O3'} = 180°$. Isoenergy curves in kcal/mol with the global minimum taken as energy zero.

Fig. 9. PCILO conformational energy map for C3'-endo uridine as a function of χ_{CN} and $\Phi_{C5'-O5'}$ with $\Phi_{C2'-O2'} = \Phi_{C3'-O3'} = 180°$. Isoenergy curves in kcal/mol with the global minimum taken as energy zero.

practically unchanged, indicating again the global minimum in the same *anti* conformation with the same local minimum in the *syn* conformation at 2 kcal/mol higher as those in Fig. 9.

Altogether, the intramolecular interactions between the exocyclic CH_2OH group and the base seem to be more important for nucleosides having the C2'-endo sugar pucker than for those having C3'-endo pucker, and this results in a preferred *syn* conformation for the former.

2. EFFECT OF THE ORIENTATION OF THE O2'—HO2' BON

PCILO conformational energy maps have been constructed as a function of χ_{CN} and $\Phi_{C_2'-O_2'}$ for the purine and pyrimidine nucleosides discussed in the preceding section with the exception of deoxyguanosine (75). In order to preclude the interaction between the exocyclic CH_2OH group and the base, $\Phi_{C_4'-C_5'}$ and $\Phi_{C_5'-O_5'}$ were fixed at 60° and 180°, respectively. As in the preceding section, $\Phi_{C_3'-O_3'}$ is also kept fixed at 180°.

The results indicate that in all cases, with the exception of C_2'-endo uridine (Fig. 10), the global energy minimum corresponds to an *anti* conformation that is generally centered around $\chi_{CN} = 0°$. It is striking to observe that it is only in C2'-endo purines that this global minimum occurs at $\Phi_{C_2'-O_2'} = 60°$, corresponding to an interaction of H-bond type between the O2'—HO2' bond and N3 of the base. The map for C2'-endo adenosine representative for this case is shown in Fig. 11. For C2'-endo cytidine, the minimum occurs for $\chi_{CN} = 0°$ and $\Phi_{C_2'-O_2'} = 30°$. In all other cases it occurs for $\Phi_{C_2'-O_2'} = 180°$ which corresponds to the O2'—HO2' bond pointing away from the base. Two maps representative of the C3'-endo nucleosides (guanosine and cytidine) are shown in Figs. 12 and 13, respectively.

All these results indicate that in the great majority of nucleosides, in the absence of an interaction between the O5'—HO5' bond and the base, the preferred conformation is *anti* irrespective of the C2'-endo or C3'-endo sugar pucker and of the "interaction" of the O2'—HO2' bond with the base.

The only exceptional case is that of C2'-endo uridine (Fig. 10), where the global minimum occurs for $\chi_{CN} = 240°$ and $\Phi_{C_2'-O_2'} = 180°$, corresponding to a *syn* form. There is, however, a local minimum only about 0.5 kcal/mol higher, at $\chi_{CN} = 60°$ associated with the same value of $\Phi_{C_2'-O_2'} = 180°$, corresponding to the *anti* form. A second map constructed for this molecule with $\Phi_{C_3'-O_3'} = 0°$ is shown in Fig. 14. The choice of $\Phi_{C_3'-O_3'} = 0°$ stems from experimental observation of Pitha (69). The results show an interchange in the relative stabilities of the *anti* and *syn* forms: a global minimum occurs now at $\chi_{CN} = 240°$. The associated value

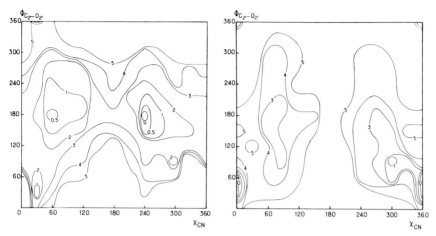

Fig. 10.   Fig. 11.

Fig. 10. PCILO conformational energy map for C2'-endo uridine as a function of χ_{CN} and $\Phi_{C2'-O2'}$ with $\Phi_{C5'-O5'} = \Phi_{C3'-O3'} = 180°$. Isoenergy curves in kcal/mol with the global minimum taken as energy zero.

Fig. 11. PCILO conformational energy map for C2'-endo adenosine as a function of χ_{CN} and $\Phi_{C2'-O2'}$ with $\Phi_{C5'-O5'} = \Phi_{C3'-O3'} = 180°$. Isoenergy curves in kcal/mol with the global minimum taken as energy zero.

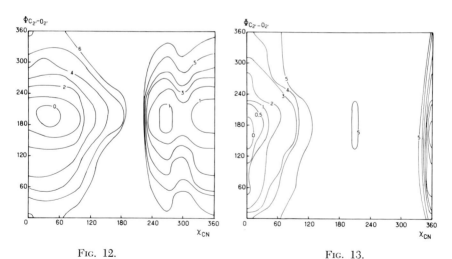

Fig. 12.   Fig. 13.

Fig. 12. PCILO conformational energy map for C3'-endo guanosine as a function of χ_{CN} and $\Phi_{C2'-O2'}$ with $\Phi_{C5'-O'} = \Phi_{C3'-O3'} = 180°$. Isoenergy curves in kcal/mol with the global minimum taken as energy zero.

Fig. 13. PCILO conformational energy map for C3'-endo cytidine as a function of χ_{CN} and $\Phi_{C2'-O2'}$ with $\Phi_{C5'-O5'} = \Phi_{C3'-O3'} = 180°$. Isoenergy curves in kcal/mol with the global minimum taken as energy zero.

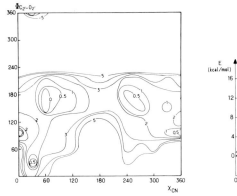

Fig. 14.

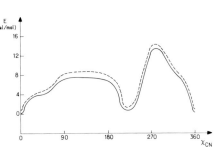

Fig. 15.

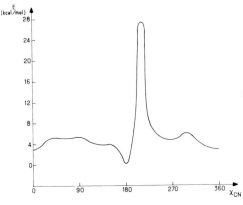

Fig. 16.

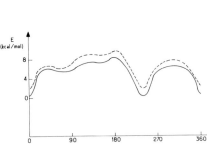

Fig. 17.

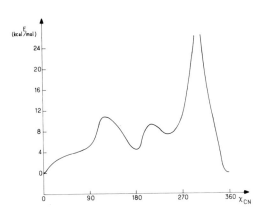

Fig. 18.

of $\Phi_{C2'-O2'} = 150°$ for these minima indicates that there is no interaction between the O2'—HO2' and O2 of the base. In sum, it is obvious that this nucleoside does not show a particular preference for either *anti* or *syn* conformation as a function of $\Phi_{C2'-O2'}$.

2. EFFECT OF SIMULTANEOUS INTERACTIONS OF THE O5'—HO5' AND O2'—HO2' BONDS WITH THE BASE

The relevant conformational energy maps have been constructed as a function of χ_{CN} with preselected values of $\Phi_{C5'-O5'}$ and $\Phi_{C2'-O2'}$ such that there is a maximum chance of interactions between the O5'—HO5' and O2'—HO2' bonds with the base. The most appropriate preselected values are $\Phi_{C4'-C5'} = 40°–60°$ (*gg* confromation), $\Phi_{C5'-O5'} = 60°$ and $\Phi_{C2'-O2'} = 60°$, with the exception of C2'-endo cytidine (Fig. 15) and C3'-endo adenosine (Fig. 16), for which $\Phi_{C5'-O5'} = 90°$, and of C2'-endo adenosine (Fig. 17) and C2'-endo cytidine (Fig. 15), for which $\Phi_{C2'-O2'} = 30°$. The computations carried out for values of 180° and 0° for $\Phi_{C3'-O3'}$ give practically identical results (Figs. 15 and 17) as far as the relative stability of the *syn* and *anti* forms is concerned; in all other computations, $\Phi_{C3'-O3'} = 0°$ has been adopted. Altogether eight maps have been constructed (75). Four of them (Figs. 15–18) are reproduced here. The principal results from these maps are as follows.

Purine nucleosides manifest equivalent energy minima for the *syn* and *anti* conformations. The only exception is C3'-endo adenosine (Fig. 16), whose global minimum occurs at $\chi_{CN} = 180°$ in the *syn* conformation, with a local minimum at $\chi_{CN} = 0°$ in the *anti* conformation at about 3 kcal/mol higher. Altogether, free purine nucleosides do not show a strong preference for either *syn* or *anti* conformation and thus are expected to be quite sensitive to the influence of external factors.

FIG. 14. PCILO conformational energy map for C2'-endo uridine as a function of χ_{CN} and $\Phi_{C2'-O2'}$ with $\Phi_{C5'-O5'} = 180°$ and $\Phi_{C3'-O3'} = 0°$. Isoenergy curves in kcal/mol with the global minimum taken as energy zero.

FIG. 15. PCILO conformational energy map for C2'-endo cytidine as a function of χ_{CN} with the global minimum taken as energy zero, with $\Phi_{C5'-O5'} = 90°$ and $\Phi_{C2'-O2'} = 30°$. Full curve: results for $\Phi_{C3'-O3'} = 180°$ and dashed curve: results for $\Phi_{C3'-O3'} = 0°$.

FIG. 16. PCILO conformational energy map for C3'-endo adenosine as a function of χ_{CN} with the global minimum taken as energy zero, with $\Phi_{C5'-O5'} = 90°$, $\Phi_{C3'-O3'} = 0°$ and $\Phi_{C2'-O2'} = 60°$.

FIG. 17. PCILO conformational energy map for C2'-endo adenosine as a function of χ_{CN} with the global minimum taken as energy zero, with $\Phi_{C5'-O5'} = 60°$ and $\Phi_{C2'-O2'} = 30°$. Solid curve: results for $\Phi_{C3'-O3'} = 180°$; dashed curve: results for $\Phi_{C3'-O3'} = 0°$.

FIG. 18. PCILO conformational energy map for C3'-endo uridine as a function of χ_{CN} with the global minimum taken as energy zero, with $\Phi_{C5'-O5'} = 60°$, $\Phi_{C3'-O3'} = 0°$ and $\Phi_{C2'-O2'} = 60°$.

The pyridine nucleosides, on the contrary, have distinct preferences: the C3'-endo nucleosides of cytidine and uridine show a definite preference for the *anti* form; C2'-endo cytidine (Fig. 15) shows a slight preference for the *anti* form; and C2'-uridine shows a distinct preference for the *syn* conformation having $\chi_{CN} = 240°$ (75).

We can now make a closer comparison of the results of each type of computation with the experimental data, which come mainly from X-ray crystallography, nuclear Overhauser effect (NOE), and nuclear magnetic resonance (NMR) studies of nucleosides in solution. Table V presents a summary of the results obtained at each stage of the calculations and offers a schematic comparison with the available experimental data.

The most significant information about the role of the torsion about the C5'—O5' bond comes from studies on compounds in which the interaction of the O2'—HO2' bond with the base is prevented. This is the situation in the 2',3'-isopropylidene derivatives of nucleosides and 2'-deoxynucleosides. An appreciable amount of data on the first type of compounds come from NOE studies (41, 42, 59–63), which indicate that all the compounds studied [i-Ino, i-Ado, i-Guo and i-Urd (i = isopropylidene)] have a strong preference for the *syn* conformation and that all the nucleosides have a C3'-endo pucker in aqueous solution. The calculations reported above also predict a general preference for the *syn* conformation with the exception of C3'-endo uridine, where the *anti* conformation is expected to be the most stable. Measurements of dipole moments of this type of molecule including i-Urd also seem to favor the existence of *syn* conformation (64). The NMR studies on 2'-deoxyuridine by Hruska (52) show that this compound is predominantly in the *anti* conformation. No information, however, is given about the nature of sugar pucker for this compound.

Compounds representative of the sole influence of the interaction of the O2'—HO2' bond with the base are more difficult to find. This group, however, may be represented to some extent by the 3':5'-cyclic nucleotides. The conformational properties of such compounds have been studied by Schweizer and Robins (45) and Smith and co-workers (82–84), who found that the ribose moiety in these compounds exists predominantly in the C3'-endo conformation. Although Schweizer and Robins (45) assigned a conformation of the base with respect to the sugar, Smith and co-workers did not commit themselves in this respect. The results obtained with the appropriate computations indicate the predominance of *anti* conformers for C3'-endo nucleosides. It is further noted that while the *anti* conformer is predicted theoretically to be more stable by about 5 kcal/mol for pyrimidine nucleosides, the differ-

TABLE V

THE syn–anti EQUILIBRIUM IN NUCLEOSIDES

Interaction of the base with	Theory		Experimental	
	Type of nucleosides	Major components	Type of compounds	Supposed conformation	Results

Interaction of the base with	Type of nucleosides	Major components	Type of compounds	Supposed conformation	Results
O5'—HO5'	2'-endo Ino,Ado,Urd,Cyd	*syn*	i-Nucleosides[a]	3'-endo	Ino,Ado,Guo,Urd: *syn*
	3'-endo Ado,Cyd	*syn*	2'-Deoxy nucleosides	3'-endo ⇌ 2'-endo	Urd: *anti*
	3'-endo Guo	*syn* ≃ *anti*	Some crystals	2'-endo	Puo: *syn*
	3'-endo Urd	*anti*	2',3'-Cyclic nucleotides	O_E, O^E	Cyd: *syn*; Urd: *anti*
O2'—HO2'	2'-endo Ino,Ado,Cyd	*anti*			
	2'-endo Urd	*syn* ≃ *anti*	3',5'-Cyclic nucleotides	3'-endo ($^3T_4$)	cUMP,cIMP,cAMP: *anti*
	3'-endo Ado,Guo,Urd,Cyd	*anti*			cGMP: *syn*
O5'—HO5' and O2'—HO2'	2'-endo Ado,Ino	*syn* ≃ *anti*	Free nucleosides	Pyd: 3'-endo ⇌ 2'-endo	Pyd,Halo-Urd: *anti*
	2'-endo Urd	*syn*			Urd: *anti* ⇌ *syn*
	2'-endo Cyd	*anti*			
	3'-endo Ado,Guo	*syn* ≃ *anti*		Puo: 2'-endo predominates	Puo: *anti* ≈ *syn*
	3'-endo Cyd,Urd	*anti*			
No interaction	3'-endo Puo,Pyd	*anti*	Majority of crystals		*anti*
	2'-endo Puo,Cyd	*anti*			
	2'-endo Urd	*syn* ≃ *anti*			
	3'-exo, 2'-exo Puo,Pyd	*anti*			

[a] i = isopropylidene.

ence in stability is much smaller (0.5–1 kcal/mol) for purine nucleosides. The experimental results of Schweizer and Robins (45) indicate a *syn* conformation. On the whole, these results are in general agreement with the predictions. Recently, detailed PCILO computations have been carried out for 3':5'-cyclic nucleotides (86).

Concerning free nucleosides in which the combined effects of the interaction of the base with the O5'—HO5' and O2'—HO2' bonds are operative, the available experimental data again agree to a large extent with our theoretical predictions. In discussing the experimental data in solution, there is one underlying difficulty, which concerns the evaluation of the predominant types of sugar puckering. There is a general belief (45, 46, 48, 49) that free pyrimidine nucleosides exist in a rapid equilibrium between the C2'-endo and C3'-endo sugar puckers. For purine nucleosides, a certain predominance of the C2'-endo pucker is admitted (42, 45). Then, the large number of NMR studies by Hruska, Smith, Schweizer and their colleagues, on different pyrimidine nucleosides devoid of bulky substituents at C6 in aqueous solution (uridine, 5-methyluridine, 5-bromouridine, 5-methylcytidine, 6-azauridine, α- and β-pseudouridines, dihydrouridine, cytidine, 5-methylcytidine, etc.) invariably point to the predominance of the *anti* form. It is only when bulky substituents are present at C6 of uracil or cytosine that the molecules are found in the *syn* form. Similarly, NOE measurements on 5-halogenouridines indicate the predominance of the *anti* conformer (63), although a more equivalent mixture of *syn* and *anti* is estimated for uridine (40). When these results are compared with the PCILO theoretical computations, they tend to indicate a certain preference for the C3'-endo puckers of sugars. On the other hand, in conformity with the more ambiguous nature of the theoretical predictions, the experimental situation is more ambiguous for free purine nucleosides, both types of results indicating a less clear-cut preference for the *syn* or *anti* conformer. Thus, while the NMR studies of Schweizer and Robins (45) on adenosine and guanosine suggest a preferred *anti* conformation for both, NOE studies of Hart and Davis (42) indicate a preferred *anti* conformation for inosine, a nearly equilibrium mixture of *syn* and *anti* conformers for guanosine, a predominance of a borderline ($\chi_{CN} = 270°$) conformation for adenosine and a predominance of *syn* conformer of xanthosine. Miles *et al.* (57) estimated, on the basis of circular dichroism (CD) studies, that guanosine exists preferentially in the *anti* conformation in water but preferentially in the *syn* conformation in alcoholic solvents or at low pH (see also 87). The vagueness of these data indicates the difficulty in making firm deductions from these experimental techniques and probably also the fluidity of the situation itself. As the equilibrium between the conformations is obviously a delicate one for purine nucleosides, they may thus possibly be easily influenced by external factors. This situation is in complete agreement with the theoretical evaluation, which indicates, especially for the C2'-endo conformers of these nucleosides, a near equivalence of the *anti* and *syn* conformers from the energetical point of view.

Finally, we arrive at the case of the majority of crystal structures of nucleosides where the interactions of the base with the O5'—HO5' and O2'—HO2' bonds is absent. Extensive PCILO computations (66, 73, 74) have been carried out with these assumptions for both pyrimidine and purine nucleosides exhibiting all the four forms of sugar, i.e., C3'-endo, C2'-endo, C3'-exo and C2'-exo puckers. The results predict (Table V) the predominance of the *anti* conformation in all the cases with the exception of C2'-endo uridine, where both *syn* and *anti* becomes equally probable. The predominance of the *anti* conformation is thus in excellent agreement with crystal structure data. The values of χ_{CN} in a number of nucleosides and nucleotides have been compiled in Table VI (see Section V, A).

Altogether, the agreement between theory and experiment for each of the situations mentioned above may be estimated to be quite satisfactory, taking into account the ambiguities of the experimental data from solution studies. In general, in the absence of interacton between the base and the exocyclic 5'—CH_2OH group of the ribose, the nucleosides tend to adopt preferentially the *anti* conformation for χ_{CN}. The existence of such interaction tends to increase the probability of the *syn* conformer. The O2'—HO2' bond of the ribose does not generally have a tendency to hydrogen bond with the base. Free pyrimidine nucleosides generally have a tendency to exist predominantly in the *anti* conformation while the free purine nucleosides have an equal tendency to adopt a *syn* or an *anti* conformation.

We can now compare the results obtained above by the PCILO method with those coming from other theoretical computations. Both classical empirical approaches (27, 88–92) and other quantum-mechanical methods (32, 93–95) have been employed, but none of these studies have taken into account explicitly the interactions of the O5'—HO5' and O2'—HO2' bonds with the base. Haschemeyer and Rich (88), using the criteria of minimum contact distances between nonbonded atoms, have concluded that both the *anti* and *syn* conformers are allowed in purine nucleosides having the C2'-endo pucker of the ribose, but that for purine nucleosides having a C3'-endo or C3'-exo pucker, the allowed *syn* region is small and hence the *anti* conformation is preferred. For pyrimidine nucleosides having a C2'-endo pucker, the *anti* as well as the *syn* conformation is allowed in their computations, while for the C3′-endo pucker, the *anti* conformation is strongly preferred. Similar computations on C3′-exo thymidine (*91*) showed that its allowed conformations were similar to those for a C2′-endo pyrimidine. Lakshminarayanan and Sasisekharan (*27*), also using the same method with slightly different criteria of allowed interatomic contacts, concluded that both *anti* and *syn* conformations are possible for purine nucleosides having C2′-endo, C3′-endo and C3′-exo puckers of the ribose, whereas for C2′-exo pucker, only the *anti* conformation is allowed. In pyrimidine nucleosides having the four above-mentioned puckers of the ribose, only the *anti* conformation is allowed. Later, using potential energy functions, Lakshminarayanan and Sasisekharan (*90*) essentially confirmed the same results, with few changes. Their results (*90*) showed that both *anti* and *syn* conformations are allowed for purine nucleosides having the four sugar puckers, whereas in pyrimidine nucleosides, only the *anti* conformation is allowed with all the sugar puckers except for the C2′-endo pucker when both *anti* and *syn* conformations are permitted. Tinoco *et al.* (*89*), using the criteria of potential energy computations, concluded that both *anti* and *syn* conformations are possible for purine nucleosides having the C2′-endo pucker of the sugar, whereas for pyrimidine nucleosides with the same sugar pucker, the *anti* conformation is preferred. Finally, Wilson and Rahman (*92*), using only nonbonded interaction energy for the computations, arrived at the same general conclusions as those of Haschemeyer and Rich (*88*) and Lakshminarayanan and Sasisekharan (*27, 90*). Their results show that both *anti* and *syn* conformations are favorable for purine nucleosides having the C2′-endo, C3′-endo and C3′-exo puckers of the sugar, and that the same situation is true for C2′-endo and C3′-exo pyrimidine nucleosides. For C3′-endo pyrimidine nucleoside, the *syn* conformation is not as favorable as the *anti* one. As pointed out earlier, none of these above-mentioned studies (*27, 88–92*) in the classical empirical approach, takes into account the possibility of an intramolecular hydrogen bond between the sugar and the base. Their results should thus be compared to those of the PCILO computations, which preclude the formation of such a bond. Under these circumstances, there seems to be a large agreement between the general conclusions drawn from these different studies (*27, 88–92*).

Along the quantum-mechanical approaches, EHT (*93, 95*), CNDO (*32*) and INDO (*32, 94*) methods have been utilized for the prediction of *syn* and *anti* conformations in nucleosides. Early EHT calculations carried out by Jordan and Pullman (*93*) predicted that the *anti* conformation is preferred for cytidine, uridine and adenosine, whereas for guanosine it is the *syn* one. They (*93*) concluded that the pyrimidine nucleosides distinctly prefer an *anti* conformation whereas the purine nucleosides should be able to adopt either the *syn* or the *anti* conformation, the energy difference between the two being of the order of 1 kcal/mol or less. Recently, Jordan (95), utilizing again the same EHT method, has studied the *syn-anti* interconversion in *syn* purine nucleosides; the results obtained seem to be in general agreement with the PCILO predictions (66), although some differences appear upon closer examination. Kang (32) utilized both INDO and CNDO methods to predict the conformational preferences in pyrimidine nucleosides. His INDO calculations for cytidine and thymidine having the C3'-endo puckers of the sugar indicate that the *anti* conformation is preferred by about 1–2 kcal/mol; CNDO results on 4-thiouridine (32) show that the *syn* conformation is about 0.4 kcal/mol more stable than the *anti* one. This author has also studied the effect of the sugar pucker on χ_{CN} in deoxycytidine, but in all his computations the carbon C5' atom of the sugar was replaced by a hydrogen atom. The same replacement was considered in a recent study (94) on the influence of hydrogen bonding between the 2'-hydroxyl group and the base on the preferred values of χ_{CN}. This replacement is unjustified, and the results obtained from such calculations have little, if any significance because of the obvious importance of the interaction of 5'—CH_2OH group with the base. Altogether, the different quantum-mechanical computations (32, 93–95) carried out outside the PCILO scheme provide some useful informations, but are fragmentary and quite incomplete.

Very recently, Lennard–Jones potential calculations have been performed by Jordan (96) on the barrier of rotation around the glycosyl bond in purine nucleosides and nucleotides. Mention must also be made of a NOE study on β-pseudouridine by Nanada *et al.* (97), which indicates that both *syn* and *anti* conformations exist in rapid equilibrium with roughly equal populations of each conformer. These workers (97) have also carried out CNDO computations on the conformation about the glycosyl bond, and they suggested the stabilization of the *syn* conformation by the formation of an intramolecular hydrogen bond between the C4—O4 and O5'—HO5' bonds.

B. The Rigidity and Flexibility of 5'-Nucleotides

The computations described in the preceding section were earlier considered to be representative of both nucleosides and nucleotides. More recently, it has appeared essential to investigate more explicitly the conformational properties of nucleotides. The interest in such exploration stems from a postulate by Sundaralingam (25, 98) that nucleotides are conformationally more "rigid" than are nucleosides. In particular, following this author, in contrast to nucleosides, which may exist in the *syn* or *anti* conformation about the glycosyl bond, nucleotides exhibit only *anti* conformation. Also, whereas three possible staggered conformations, *gg*, *gt* and *tg* about the exocyclic C4′—C5′ bond are observed in nucleosides, the 5′-nucleotides exhibit only the *gg* conformer.

Sundaralingam's concept is based entirely on the survey of X-ray crystallographic results and is apparently valid for the situation in crystals although a recent X-ray crystallographic study, on 6-azauridine 5′-

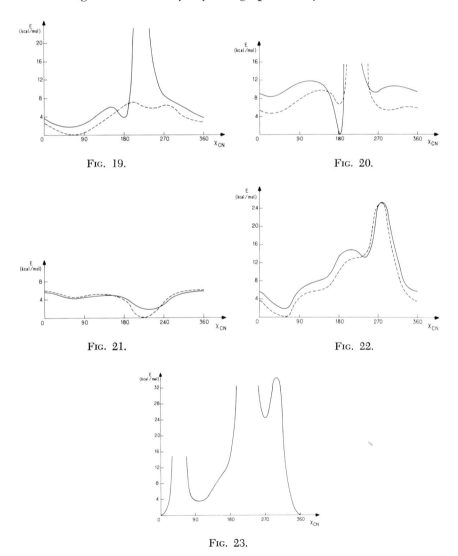

Fig. 19.

Fig. 20.

Fig. 21.

Fig. 22.

Fig. 23.

phosphoric acid (99) shows that this compound exists in *gt* conformation. Such is also the case for 5'-dGMP (100). Anyway, the question may be raised whether this rigidity of the crystal conformations of nucleotides represents their genuine, intrinsic property, which would then imply a fundamental difference with respect to nucleosides, or whether it is brought about by a particular influence of the packing forces.

With this problem in view, PCILO computations have been carried out on a number of purine and pyrimidine 5'-nucleotides (101, 102). The geometrical input data for 5'-UMP and 5'-IMP were taken directly from their crystal structures (43, 103). For 5'-AMP (104), the geometry of the ribose was assumed to be the same as that of 5'-IMP (103). For 5'-CMP and 5'-GMP, the input data come also from their crystal structures (105, 106). The computations have been carried by taking simultaneously into account, in each case, four degrees of freedom, the rotation χ_{CN} about the glycosyl bond and the rotations $\Phi_{C5'-O5'}$, $\Phi_{O5'-P}$ and Φ_{P-OIII} of the phosphate group, the rotation $\Phi_{C4'-C5'}$ being fixed in the privileged *gg* conformation (39). Thus, this study centered essentially about the rigidity or flexibility of the conformations about the glycosyl bond.

The results of the computations for 5'-AMP, -GMP, and -IMP are shown in Figs. 19, 20 and 21, respectively. The full curves correspond to the possibility of formation of an intramolecular hydrogen bond between O5'—HO5' of the phosphate and N3 of the base whereas the dashed curves correspond to the cases where this possibility is precluded. 5'-AMP clearly shows a strong preference for the *anti* conformation while 5'-GMP and 5'-IMP show a preference for the *syn* conformation. In the

FIG. 19. PCILO conformational energy map for C2'-endo adenosine 5'-phosphate as a function of χ_{CN} with $\Phi_{C4'-O5'} = 60°$ and $\Phi_{C2'-O2'} = 180°$. Solid curve: results for $\Phi_{P-OIII} = \Phi_{O5'-P} = 300°$ and $\Phi_{C5'-O5'} = 150°$; dashed curve: results for $\Phi_{P-OIII} = 90°$, $\Phi_{O5'-P} = 300°$ and $\Phi_{C5'-O5'} = 180°$. From Berthod and Pullman (102).

FIG. 20. PCILO conformational energy map for C3'-endo guanosine 5'-phosphate as a function of χ_{CN} with $\Phi_{C4'-C5'} = 60°$ and $\Phi_{C2'-O2'} = 180°$. Solid curve: results for $\Phi_{P-OIII} = 40°$, $\Phi_{O5'-P} = 200°$ and $\Phi_{C5'-O5'} = 180°$; dashed curve: results for $\Phi_{P-OIII} = \Phi_{O5'-P} = 300°$ and $\Phi_{C5'-O5'} = 240°$. From Berthod and Pullman (101).

FIG. 21. PCILO conformational energy map for C2'-endo inosine 5'-phosphate as a function of χ_{CN} with $\Phi_{C4'-C5'} = 60°$ and $\Phi_{C2'-C2'} = 180°$. Solid curve: results for $\Phi_{P-OIII} = 330°$, $\Phi_{O5'-P} = 180°$ and $\Phi_{C5'-O5'} = 200°$; dashed curve: results for $\Phi_{P-OIII} = \Phi_{O5'-P} = \Phi_{C5'-O5'} = 180°$. From Berthod and Pullman (101).

FIG. 22. PCILO conformational energy map for C2'-endo uridine 5'-phosphate as a function of χ_{CN} with $\Phi_{C4'-C5'} = 60°$ and $\Phi_{C2'-O2'} = 180°$. Solid curve: results for $\Phi_{P-OIII} = 60°$, $\Phi_{O5'-P} = \Phi_{C5'-O5'} = 180°$; dashed curve: results for $\Phi_{P-OIII} = 90°$, $\Phi_{O5'-P} = 270°$ and $\Phi_{C5'-O5'} = 180°$.

FIG. 23. PCILO conformational energy map for C3'-exo deoxycytidine 5'-phosphate as a function of χ_{CN} with $\Phi_{C4'-C5'} = 60°$, $\Phi_{P-OIII} = 270°$, $\Phi_{O5'-P} = 300°$ and $\Phi_{C5'-O5'} = 160°$.

case of 5′-GMP, the global minimum occurs at $\chi_{CN} = 180°$ whereas in the case of 5′-IMP it is around $\chi_{CN} = 210°$. In addition, 2′-GMP also shows an intrinsic tendency for a *syn* conformation (*101*).

Figure 22 reproduces the representative conformational energy curves for 5′-UMP, and it can be seen that they indicate a clear-cut preference for the *anti* conformation centered around $\chi_{CN} = 60°$. The results for 5′-dCMP are presented in Fig. 23. The search for an intramolecular hydrogen bond was unsuccessful. There is again an evident preference for the *anti* conformation centered about $\chi_{CN} = 0°$. There is, however, a secondary minimum, about 3 kcal/mol above the global one at the onset of the *syn* region.

Altogether, PCILO results indicate that 5′-GMP and 5′-IMP, show a preference for the *syn* conformation, whereas 5′-AMP, -UMP and -CMP show a preference for the *anti* conformation. These last three may thus be considered as conformationally more rigid than the corresponding nucleosides, which manifest a more diversified behavior (*75*). Sundaralingam's proposal (*25, 98*) is thus essentially valid for these last nucleotides.

The available NMR results on 5′-nucleotides in solution (*47, 53, 107–109*) confirm most of the theoretical results presented above, although some ambiguities remain. Thus, for example, Schweizer *et al.* (*107*) found all the 5′-nucleotides, including 5′-IMP and 5′-GMP, to be in the *anti* conformation. However, recent studies by nuclear Overhauser effect indicate that guanosine 5′-, 3′- and 2′-monophosphates exist in solution predominantly (50–80%) in the *syn* conformation (*110*). A *syn* conformation is also proposed on the basis of CD studies for guanosine in guanylyl(3′-5′)uridine, G-U (*111*). 5′-AMP has also been studied by nuclear Overhauser effect recently (*112*), and this compound exists in *anti* conformation in agreement with the computations.

Empirical computations have been carried out on the same problem by Yathindra and Sundaralingam with the use of potential functions (*113–115*). They indicate in agreement with the PCILO calculations that the 5′-phosphates of adenosine, cytidine, uridine and ribothymidine should exist preferentially in the *anti* conformation whereas that of guanosine should exist in the *syn* conformation. Olson (*116*), also utilizing potential energy functions, found also that 5′-AMP should exist predominantly in the *anti* conformation whereas 5′-GMP should be in the *syn* conformation.

Vasilescu *et al.* (*117*) have utilized that EHT method to study 5′-AMP. Their results, in agreement with PCILO and empirical potential energy calculations, show that the *anti* conformation about the glycosyl bond is preferred. [The preferred conformation about the exocyclic

C4′—C5′ bond is, however, *gt*, which is about 0.8 kcal/mol more stable than the *gg* conformation associated with the same χ_{CN} value. This discrepancy is evidently due to the imprecision of EHT methodology (see Section V, A)].[4]

V. Conformation about the Exocyclic C4′—C5′ Bond

Three staggered conformations about the exocyclic C4′—C5′ bond have been frequently observed in β-nucleosides and β-nucleotides and they are the *gg*, *gt* and *tg* conformations corresponding, respectively, to the values of $\Phi_{C_{4'}-C_{5'}}$ = 60°, 180° and 300° (Fig. 5). Both classical and quantum-mechanical approaches (*27, 32, 39, 92, 119–122*) have been employed to study the preferred conformation about this bond.

Along the classical, "empirical" line of approach, Lakshminarayanan and Sasisekharan used first the "hard-sphere" model (*27*) and later partitioned potential energy functions (*120*) to describe the torsions around the C4′—C5′ and C5′—O5′ bonds in sugar-phosphate monomer units having the four commonly occurring sugar puckers, C3′-endo, C2′-endo, C3′-exo and C2′-exo. However, the models used in these computations (*27, 120*) do not correspond to real nucleosides because no bases were included; also, there is a phosphorus atom attached to O5′ instead of a hydrogen atom. Wilson and Rahman (*92*) utilized nonbonded potential energy functions alone to study the torsion about the C4′—C5′ bond in the C3′-endo, C2′-endo and C3′-exo β-nucleosides. They did not consider the torsions about the C5′—O5′ bond, nor was the hydrogen attached to O5′ taken into consideration. Along the quantum-mechanical approach, Govil and Saran (*121*) utilized the EHT method to evaluate the two torsions in a ribose-phosphate unit with the sugar fixed in the C3′-endo configuration. Later, Saran and Govil (*122*) also computed, within the same method, a conformational energy map for these rotations in a dinucleotide unit. However, the bases were not taken into account. Similarly, Kang (*32*), using INDO computations for the torsion about the C4′—C5′ bond in nucleosides, also neglected the bases and substituted them by hydrogens. His computations do not include the torsion about the C5′—O5′ bond either.

In view of the frequently fragmentary and incomplete nature of the above-mentioned theoretical studies, systematic PCILO investigations

---

[4] In the crystal structure, both 5′-IMP (*118*) and 5′-dGMP (*100*) are in the *anti* conformation. From theoretical results, this situation must be ascribed to the action of the crystal packing forces. Whatever it may be, Sundaralingam's rule on the rigidity of nucleotides seems verified with no exception *for the solid state* as concerns the χ_{CN} torsion, and verified also as concerns $\Phi_{C_{4'}-C_{5'}}$ in the nucleotides of the usual nucleic acid bases.

have been carried out on the complete series of β- and α-nucleosides (39, 119) including all their constituents with different sugar puckers and also taking into account the simultaneous rotations about the C4'—C5' and C5'—O5' bonds. Similar PCILO computations have also been carried out (39) on 5'-β-nucleotides and β-dihydrouridine.

The results of all these theoretical studies may be compared with the ample experimental evidence coming mainly from the X-ray crystal structures of nucleosides and nucleotides and from recent NMR studies on the structure of nucleosides in solution (46, 123–125).

A. β-Nucleosides

1. C3'-ENDO PYRIMIDINE NUCLEOSIDES

The results of PCILO computations (39) on C3'-endo uridine are shown in Fig. 24, and this map may be considered as representative of the whole group of C3'-endo pyrimidine nucleosides. It shows a global minimum at $\Phi_{C4'-C5'} = 60°$ in the gg conformation associated with $\Phi_{C5'-O5'} = 180°$. The gt ($\Phi_{C4'-C5'} = 180°$) and tg ($\Phi_{C4'-C5'} = 300°$) conformations associated with $\Phi_{C5'-O5'} = 180°$ are about 3 and 2 kcal/mol above the global minimum, respectively. The calculations thus indicate a strong preference for the gg conformer in the C3'-endo pyrimidine nucleosides.

This theoretical estimation is fully confirmed by the available X-ray data on β-nucleosides. As indicated in Table VI, nearly all the nucleosides (thirteen out of sixteen) of this category that have been experimentally studied (31, 81, 126–134) are in the gg conformation. This includes the model compound 5-bromouridine (81) for which $\Phi_{C4'-C5'} = 36.2°$. The only exceptions are 6-azauridine (135, 136) and 3',5'-diacetyl-2'-deoxy-2'-fluorouridine (137). The crystalline asymmetric unit of 6-azauridine contains two molecules, both in gt conformation. The PCILO map constructed for 6-azauridine molecule-1 (Fig. 25) shows two global minima both occurring for the gg conformation, one associated with $\Phi_{C5'-O5'} = 180°$ (the same as in Fig. 24) and the other associated with 60°–90°. In contrast to Fig. 24, the gt and tg conformations in Fig. 25 occur only about 1 kcal/mol above the global minimum. The crystallographic conformation for 6-azauridine molecule-1 ($\Phi_{C4'-C5'} = 189.9°$ and $\Phi_{C5'-O5'} = 86.3°$) lies thus near the local minimum in the gt conformation.[5] Among the C3'-endo pyrimidine nucleotides, all except one are in the gg conformation (78, 98, 139–142), the exception being 6-azauridine 5'-phosphate trihydrate (99).

[5] Very recently, 6-azacytidine has been shown to be in the gg conformation about the C4'—C5' bond (138).

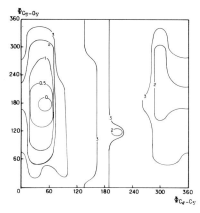
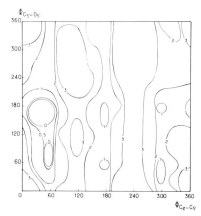

FIG. 24. FIG. 25.

FIG. 24. PCILO conformational energy map for C3'-endo uridine as a function of $\Phi_{C4'-C5'}$ and $\Phi_{C5'-O5'}$. Isoenergy curves in kcal/mol with the global minimum taken as energy zero. From Jaran et al. (39).

FIG. 25. PCILO conformational energy map for C3'-endo 6-azauridine (molecule-1) as a function of $\Phi_{C4'-C5'}$ and $\Phi_{C5'-O5'}$. Isoenergy curve in kcal/mol with the global minimum taken as energy zero.

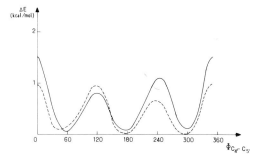

FIG. 26. Results of empirical computations of Wilson and Rahman (92) on the variation of nonbonded energy with $\Phi_{C4'-C5'}$ for C3'-endo nucleosides. Solid curve: results for 5-methyluridine; dashed curve: results for inosine monoclinic.

The results of NMR studies (46, 125, 196) also indicate that the gg conformers are the most populated ones for C3'-endo pyrimidine nucleosides in aqueous solution.

Two other theoretical computations have been presented for C3'-endo pyrimidine nucleosides. Wilson and Rahman (92), using only nonbonded interactions, have obtained the potential energy curve for 5-methyluridine (128) as a function of $\Phi_{C4'-C5'}$. Their results, shown in Fig. 26, indicate that the three gg, gt and tg conformations are energetically

TABLE VI

CONFORMATION OF THE EXOCYCLIC 5'-CH$_2$OH GROUP AND THE VALUES OF χ_{CN}
IN PURINE AND PYRIMIDINE β-NUCLEOSIDES AND β-NUCLEOTIDES

No.	Compound	$\Phi'_{C4'-C5'}$	$\Phi_{C4'-C5'}$	gg, gt or tg	χ_{CN}	anti or syn	Reference
	C3'-endo Pyrimidine Nucleosides						
1	dCyd · HCl	288.3	46.2	gg	−0.2	anti	126
2	Cyd	289.6	47.1	gg	18.4	anti	127
3	5-BrUrd · Ado	285.4	36.2	gg	20.0	anti	81
4	4-SUrd · H$_2$O	308.2	65.7	gg	263.4	syn	31
5	5MeUrd · (½ H$_2$O)	292.2	49.5	gg	29.4	anti	128
6	2,4-S$_2$Urd · H$_2$O	285.0	41.3	gg	18.1	anti	129
7	2,4-S$_2$Urd · H$_2$O	282.5	40.9	gg	19.5	anti	130
8	5-IUrd (1)	297.5	62.3	gg	11.8	anti	131
9	2-SCyd · (2 H$_2$O)	296.2	54.6	gg	20.3	anti	132
10	βUrd (1)	287.4	45.0	gg	16.8	anti	133
11	βUrd (2)	281.4	39.1	gg	23.8	anti	133
12	4-SUrd disulfide (1)	296.8	57.4	gg	16.6	anti	134
13	4-SUrd disulfide (2)	293.4	46.3	gg	18.1	anti	134
14	6-AzaUrd (1)	73.6	189.9	gt	81.3	anti	135, 136
15	6-AzaUrd (2)	68.9	186.7	gt	76.5	anti	135, 136
16	3',5'Ac$_2$-2'FUrd	166.8	281.7	tg	251.7	syn	137
	C3'-endo Pyrimidine Nucleotides						
17	5'-dTMP(Ca) · (6 H$_2$O)	297.3	57.3	gg	43.5	anti	139
18	pU moiety of A(2'-5')U	298.3	56.9	gg	5.0	anti	78
19	Up moiety of U-A · (½ H$_2$O) (1)	297.3	54.0	gg	20.0	anti	98
20	Up moiety of U-A · (½ H$_2$O) (2)	292.5	49.9	gg	10.3	anti	98
21	Up moiety of U-A · (½ H$_2$O) (1)	294.0	53.4	gg	12.1	anti	140
22	Up moiety of U-A · (½ H$_2$O) (2)	297.5	58.3	gg	19.5	anti	140
23	pC moiety of G-C	—	51.0	gg	25.0	anti	141
24	pU moiety of A-U (1)	—	57.0	gg	29.0	anti	141, 142
25	pU moiety of A-U (2)	—	57.0	gg	30.0	anti	141, 142
26	6-Aza5'UMP · (3 H$_2$O)	71.0	188.0	gt	~84.5	anti	99
	C3'-endo Purine Nucleosides						
27	Ado · 5-BrUrd	286.2	40.0	gg	12.4	anti	81
28	Puromycin · (2 HCl) · (5 H$_2$O)	295.1	54.0	gg	19.3	anti	143, 144
29	(3'-Phosphonomethyl)Ado · EtOH	291.5	49.8	gg	28.1	anti	145
30	Ado	60.1	176.9	gt	9.9	anti	146
31	Ino (monoclinic)	74.7	191.0	gt	12.4	anti	147
32	Ado moiety of B$_{12}$ coenzyme	77.2	194.3	gt	74.7	anti	148
33	dGuo · actinomycin (1)	75.1	184.8	gt	88.5	anti	149
	C3'-endo Purine Nucleotides						
34	3'-AMP · (2 H$_2$O)	56.7	171.7	gt	3.8	anti	150
35	5'-AMP · (H$_2$O)	282.0	40.0	gg	25.6	anti	104

TABLE VI (Continued)

No.	Compound	$\Phi'_{C4'-C5'}$	$\Phi_{C4'-C5'}$	gg, gt or tg	χ_{CN}	anti or syn	Reference
36	5'-GMP · (3 H$_2$O)	288.2	45.8	gg	12.4	anti	106
37	ATP(Na$_2$) · (3 H$_2$O) (1)	312.0	67.0	gg	69.4	anti	151
38	5'-IMP · H$_2$O	293.6	52.3	gg	20.2	anti	118
39	pA moiety of (U-A) · (½ H$_2$O) (1)	292.2	52.3	gg	48.7	anti	98
40	pA moiety of (U-A) · (½ H$_2$O) (2)	297.7	55.8	gg	35.5	anti	98
41	pA moiety of (U-A) · (½ H$_2$O) (1)	300.0	54.8	gg	37.0	anti	140
42	pA moiety of (U-A) · (½ H$_2$O) (2)	295.2	53.6	gg	44.1	anti	140
43	Gp moiety of G-C	—	—	gg	13.0	anti	141
44	Ap moiety of A-U (1)	—	—	gg	7.0	anti	141, 142
45	Ap moiety of A-U (2)	—	—	gg	2.0	anti	141, 142
46	A^{1}p moiety of A^1-A^2-A^3	296.7	52.6	gg	8.0	anti	152
47	pA^{2}p moiety of A^1-A^2-A^3	297.6	56.0	gg	28.0	anti	152
48	pA^3 moiety of A^1-A^2-A^3	302.6	61.3	gg	26.7	anti	152
	C2'-endo Pyrimidine Nucleosides						
49	5-BrdCyd · dGuo	308.6	60.8	gg	58.9	anti	76
		73.1	185.3	gt			
50	5-BrUrd	294.0	52.0	gg	52.5	anti	79
51	5-BrUrd · Me$_2$SO	294.0	55.0	gg	62.2	anti	153
52	5-FdUrd	173.0	292.0	tg	59.0	anti	154
53	5-BrdUrd	50.2	166.7	gt	47.2	anti	79
54	5-ClUrd	293.7	52.3	gg	51.4	anti	155
55	5-IUrd (2)	182.9	294.6	tg	55.8	anti	131
56	6-MeUrd (1)	291.0	51.4	gg	250.9	syn	156, 157
57	6-MeUrd (2)	61.9	180.0	gt	252.7	syn	156, 157
58	dUrd (1)	173.5	290.8	tg	26.3	anti	158, 159
59	dUrd (2)	167.6	285.6	tg	28.4	anti	158, 159
60	5-IdUrd	292.3	51.4	gg	63.3	anti	160
61	3'Ac,4S-dThd	291.7	51.4	gg	54.0	anti	161, 162
62	H$_2$Urd · (½ H$_2$O) (1)	51.3	169.1	gt	65.5	anti	123, 124
63	H$_2$Urd · (½ H$_2$O) (2)	176.7	294.8	tg	57.1	anti	123, 124
		282.8	40.7	gg			
64	H$_2$Urd · (½ H$_2$O) (1)	52.0	169.6	gt	65.3	anti	163
65	H$_2$Urd · (½ H$_2$O) (2)	186.3	306.3	tg	57.8	anti	163
		162.0	282.0	tg			
		287.7	42.7	gg			
66	2'-ClUrd	290.1	50.0	gg	61.7	anti	164
67	3-DeazaUrd	296.1	55.1	gg	52.3	anti	165, 166
68	5-CldUrd	50.5	168.1	gt	41.9	anti	167

(continued)

TABLE VI (*Continued*)

No.	Compound	$\Phi'_{C4'-C5'}$	$\Phi_{C4'-C5'}$	gg, gt or tg	χ_{CN}	anti or syn	Reference
69	5-HO-Urd	294.3	45.5	gg	39.3	anti	168
70	dThd(5'-COOH) (1)	159.1	277.3	tg	42.4	anti	169
71	dThd(5'-COOH) (2)	151.6	267.0	tg	53.0	anti	169
	C2'-endo Pyrimidine Nucleotides						
72	3'-CMP (monoclinic)	286.3	45.5	gg	39.3	anti	170
73	3'-CMP (orthorhombic)	285.3	43.8	gg	41.8	anti	80
74	2'-CMP · (6 H₂O) (1)	294.0	55.0	gg	45.0	anti	171
75	2'-CMP · (6 H₂O) (2)	293.0	52.0	gg	51.0	anti	171
76	3'-UMP(Na₂) · (4 H₂O)	282.6	42.7	gg	45.1	anti	172
77	5'-UMP(Ba) · (7 H₂O)	292.6	55.2	gg	43.2	anti	43
78	Urd-3'-[OP(S, OH)OMe] (1)	288.7	40.5	gg	38.2	anti	173, 174
79	Urd-3'-[OP(S, OH)OMe] (2)	280.8	40.0	gg	43.5	anti	173, 174
80	pdT moiety of pdT-dT	282.0	41.0	gg	34.0	anti	175
81	pdTp moiety of pdT-dT	292.0	46.0	gg	27.0	anti	175
	C2'-endo Purine Nucleosides						
82	Ino · (2 H₂O) (1)	304.7	64.0	gg	120.1	syn	77
83	Ino · (2 H₂O) (2)	286.6	47.0	gg	49.1	anti	77
84	Ino (orthorhombic) (1)	297.3	56.7	gg	~238.0	syn	176
85	Ino (orthorhombic) (2)	302.4	64.2	gg	~234.5	syn	176
86	Guo · (2 H₂O) (1)	309.1	67.9	gg	122.5	syn	77
87	dGuo · 5-BrdCyd	289.4	42.8	gg	211.3	syn	76
88	dGuo · Actinomycin (2)	65.9	184.4	gt	93.0	~anti	149
89	6-SGuo	64.2	184.8	gt	65.7	anti	177
90	2-Me₂Guo	64.9	182.5	gt	256.1	syn	178
91	Ado · HCl	300.7	61.0	gg	43.1	anti	179
92	3'-AcAdo	299.4	58.0	gg	227.0	syn	65
93	8-BrAdo	286.1	45.4	gg	240.6	syn	180
94	8-BrGuo	295.5	54.3	gg	229.9	syn	180
95	8-BrGuo	294.0	54.0	gg	231.0	syn	181
96	6-SIno (1)	297.5	55.4	gg	225.0	syn	182
97	6-SIno (2)	297.0	57.0	gg	216.3	syn	182
98	Formycin · HBr · H₂O	46.6	162.4	gt	210.7	syn	183, 184
99	Formycin · H₂O	57.6	175.8	gt	109.8	syn	185
100	Tubercidin	61.9	180.0	gt	71.8	anti	186
101	Tubercidin	62.0	181.7	gt	73.0	anti	187
102	(5'-MeNH₂⁺I⁻)Ado · H₂O	43.0	158.3	gt	208.8	syn	188, 189
103	5'-BrAdo · riboflavin	176.4	283.5	tg	254.3	syn	190
	C2'-endo Purine Nucleotides						
104	5'-IMP(Ba) (1)	302.0	49.0	gg	46.0	anti	191
105	5'-IMP(Ba) (2)	303.0	51.0	gg	34.0	anti	191

TABLE VI (Continued)

No.	Compound	$\Phi'_{C4'-C5'}$	$\Phi_{C4'-C5'}$	gg, gt or tg	χ_{CN}	anti or syn	Reference
106	5'-IMP(Na₂)	299.5	56.0	gg	43.0	anti	191
107	5'-IMP(Na) · (8 H₂O)	297.2	59.4	gg	40.9	anti	103
108	ADP(Rb) · (3 H₂O)	297.8	57.4	gg	38.0	anti	192
109	ATP(Na₂) · (3 H₂O) (2)	302.2	48.6	gg	38.8	anti	151
110	A2'p moiety of A(2'-5')U · (4 H₂O)	287.2	45.3	gg	54.6	anti	78
111	5'-dAMP · (6 H₂O)	287.8	46.8	gg	63.5	anti	193
	C3'-exo Compounds						
112	dThd	56.1	172.8	gt	39.1	anti	194
113	5'-dCMP · H₂O	297.0	56.9	gg	−6.0	anti	105
114	dAdo · H₂O	68.1	186.9	gt	10.9	anti	195

equivalent. This result is much less satisfactory than PCILO results (Fig. 24), which distinguish between the three conformers, predicting gg to be the most stable one in agreement with the X-ray data mentioned above.

As said earlier, Kang (32) used the INDO method to compute the conformation about the C4'—C5' bond in C3'-endo pyrimidine nucleosides, but the pyrimidine bases were omitted in his computations. It is only the geometry of the sugars that was taken from crystallographic data on the nucleosides or nucleotides: 4-thiouridine (31), cytidine (127) and calcium thymidine 5'-phosphate (139). His results showed that the three gg, gt and tg conformations are equally stable for 4-thiouridine, the gg conformation more stable by about 2 kcal/mol than the gt and tg ones in cytidine, and the gt and tg more stable by about 2.5 kcal/mol than the gg in calcium thymidine 5'-phosphate. These results are, thus, only moderately successful.

2. C3'-ENDO PURINE NUCLEOSIDES

The model compound chosen for the PCILO computations on C3'-endo purine nucleosides (39) was monoclinic inosine (147). The results (Fig. 27), indicate that the global minimum occurs for the gg conformation ($\Phi_{C4'-C5'} = 60°$) associated with $\Phi_{C5'-O5'} = 180°$. The gt conformation is 1 kcal/mol above the global minimum with $\Phi_{C5'-O5'} = 180°$, but there is a local minimum for the gt conformation only about 0.5 kcal/mol above the global one associated with $\Phi_{C5'-O5'} = 270°–310°$. This local minimum corresponds closely to the observed conformation of the model compound (*147*): $\Phi_{C4'-C5'} = 191°$ and $\Phi_{C5'-O5'} = 269°$. There is also a large area included within the 1 kcal/mol isoenergy curve for both the *gg* and *gt* regions. This close similarity of the *gg* and *gt* conformations represents the most striking difference between the theoretical results of Figs. 24 and 27 and thus between the predicted conformational properties of C3'-endo purine and pyrimidine nucleosides, giving evidence for the effect of the base. The *tg* conformation in Fig. 27 remains at the same level of energy at about 2 kcal/mol as in Fig. 24.

The above deductions from PCILO computations are again corroborated by the available experimental data. Out of seven C3'-endo purine nucleosides studied by X-ray crystallography, three nucleosides (*81, 143–145*) are in the *gg* conformation and the remaining four nucleosides (*146–149*) are in the *gt* conformation. No molecules of this category have so far been discovered in the *tg* conformation. Among the nucleotides of this category, all (*98, 104, 106, 118, 140–142, 151, 152*) are in the *gg* conformation with the exception of adenosine 3'-phosphate (*150*), which exists in the *gt* conformation.

The only other theoretical calculation made for the C3'-endo purine nucleosides is that of Wilson and Rahman (*92*), carried out, in fact, for the same inosine molecule (*147*) as the PCILO one. Their results (Fig. 26) show that all three *gg*, *gt* and *tg* conformations about the exocyclic C4'—C5' bond are equally stable and thus again overlook one of the essential features of the real situation.

3. COMPARISON WITH OTHER COMPUTATIONS ON C3'-ENDO RIBOSO

A few computations have been carried out for C3'-endo ribose and ribose-phosphate units (*27, 120–122*). Although these molecules do not reflect the situation in nucleosides, it is nevertheless useful to look briefly at the results obtained by such calculations.

The hard-sphere model calculations (*27*) for C3'-endo ribose indicated that three regions of $\Phi_{C4'-C5'}$ may be sterically allowed, namely, 40°–70°, 170°–220° and 260°–320°, corresponding to the *gg*, *gt* and *tg* conformations, respectively. The predicted value of $\Phi_{C5'-O5'}$ was in the region 150°–210°. When nonbonded interactions alone were first taken into consideration in more refined partitioned potential functions calculations (*120*), the three minima remained practically unchanged with equal stability for the three conformations. This situation persisted when electrostatic interactions were included in the computation (*120*), the *gg* being perhaps 0.1–0.2 kcal/mol more stable than the *gt* and *tg* conformations. A conformational energy map was also made (*120*) for torsions about the C4'—C5' and C5'—O5' bonds for a ribose with a phosphorus

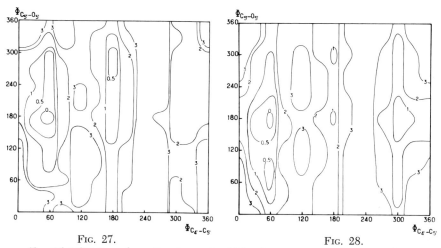

Fig. 27.   Fig. 28.

Fig. 27. PCILO conformational map for C3'-endo inosine monoclinic as a function of $\Phi_{C4'-C5'}$ and $\Phi_{C5'-O5'}$. Isoenergy curves in kcal/mol with the global minimum taken as energy zero. From Jaran et al. (39).

Fig. 28. PCILO conformational energy map for C2'-endo deoxyuridine as a function of $\Phi_{C4'-C5'}$ and $\Phi_{C5'-O5'}$. Isoenergy curves in kcal/mol with the global minimum taken as energy zero. From Jaran et al. (39).

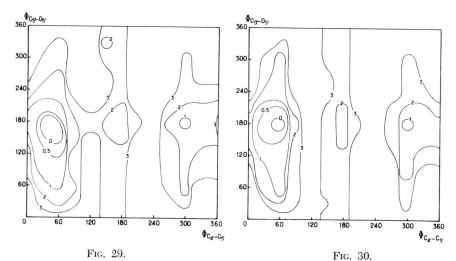

Fig. 29.   Fig. 30.

Fig. 29. PCILO conformational energy map for C2'-endo inosine molecule-1 as a function of $\Phi_{C4'-C5'}$ and $\Phi_{C5'-O5'}$. Isoenergy curves in kcal/mol with the global minimum taken as energy zero. From Jaran et al. (39).

Fig. 30. PCILO conformational energy map for C2'-endo inosine molecule-2 as a function of $\Phi_{C4'-C5'}$ and $\Phi_{C5'-O5'}$. Isoenergy curves in kcal/mol with the global minimum taken as energy zero. From Jaran et al. (39).

attached to O5; the most stable conformation predicted was for $\Phi_{C4'-C5'}$ = 300° and $\Phi_{C5'-O5'}$ = 180°. This result is contrary to experimental observation and has been discussed (197).

Finally, mention must be made of the EHT computations (121, 122) on the torsions around the C4′—C5′ and C5′—O5′ bonds in C3′-endo ribose: the three regions corresponding to the gg, gt and tg conformations were predicted to be equally stable. Similarly, the conformational energy map (122) made for the torsion angles $\Phi_{C4'-C5'}$ and $\Phi_{C5'-O5'}$ for a dinucleotide did not distinguish a global minimum associated with $\Phi_{C4'-C5'}$ = 60° and $\Phi_{C5'-O5'}$ = 180°, but predicted the three regions gg, gt and tg associated with $\Phi_{C5'-O5'}$ = 180° to be equally stable. The EHT results are thus obviously less precise than the PCILO ones and do not represent progress with respect to the empirical ones.

4. C2′-ENDO PYRIMIDINE NUCLEOSIDES

The model compound chosen for PCILO computations on C2′-endo pyrimidine nucleosides was deoxyuridine with its geometrical input data coming from the crystal structure of 5-bromo-2′-deoxyuridine (79). The results (Fig. 28) show that the global minimum occurs, as for the C3′-endo pyrimidine nucleosides, at $\Phi_{C4'-C5'}$ = 60° (gg conformation) and $\Phi_{C5'-O5'}$ = 180°. There is a large area included within the 1 kcal/mol isoenergy curve above the global minimum for that conformation, similar to that in Fig. 24, but with $\Phi_{C5'-O5'}$ varying from 20° to 300°. A local gg minimum also appears at about 0.5 kcal/mol above the global one, but associated with $\Phi_{C5'-O5'}$ = 60°–100°. The most striking feature of Fig. 28, in particular in comparison with Fig. 24, is that the local energy minima corresponding to the gt ($\Phi_{C4'-C5'}$ = 180°) and tg ($\Phi_{C4'-C5'}$ = 300°) conformations associated with $\Phi_{C5'-O5'}$ = 180° are now only 1 kcal/mol above the global minimum. The transition from a C3′-endo ribose to C2′-endo ribose in pyrimidine nucleosides should thus be accompanied by a relative destabilization of the gg conformation with respect to the gt and tg ones.

The available crystallographic data confirm the theoretical PCILO results most strikingly. It is easily seen from Table VI that the three gg, gt and tg conformations about the exocyclic C4′—C5′ bond are experimentally observed in C2′-endo pyrimidine nucleosides. Out of 23 C2′-endo pyrimidine nucleosides, 12 (76, 79, 123, 124, 153, 155–157, 160–166, 168) exist in the gg conformation, 6 (76, 79, 123, 124, 155, 157, 163, 167) in the gt conformation and 8 (123, 124, 131, 154, 158, 159, 163, 169) in the tg conformation. The proportion of the gg form thus changes from about 81% for the C3′-endo pyrimidine nucleosides to about 52% for the C2′-endo pyrimidine nucleosides. The model compound 5-

bromo-2′-deoxyuridine (79) is *gt* with $\Phi_{C4'-C5'} = 166.7°$. It is also interesting to note that for two compounds, the complex of 5-bromo-2′-deoxycytidine with 2′-deoxyguanosine (76) and 6-methyluridine (156, 157), both the *gg* and *gt* conformations, and for one compound, dihydrouridine hemihydrate (123, 124, 163), the three *gg*, *gt* and *tg* conformations have been observed crystallographically. The ten nucleotides of this category (43, 80, 170–175) studied by X-ray crystallography are all in the *gg* conformation (see Table VI).

It is especially interesting to note that very extensive NMR studies on nucleosides in solution involving the determination of $J_{1'2'}$, $J_{3'4'}$, $J_{4'5'}$ and $J_{4'5''}$ vicinal H–H coupling constants carried out by Hruska and his colleagues (46, 125) lead also to the general result in excellent agreement with the main conclusion, namely, that the *gg* conformer is destabilized by the C2′-endo conformation of the sugar in pyrimidine nucleosides with respect to the related C3′-endo nucleosides. No other theoretical computations have been carried out explicitly for the C2′-endo pyrimidine nucleosides.

5. C2′-ENDO PURINE NUCLEOSIDES

Because of the existence of this type of nucleoside in the crystals in the *syn* (65, 76, 77, 176, 178, 180–185, 188–190) and *anti* (77, 149, 177, 179, 186, 187) conformations about the glycosyl linkage, two PCILO maps, representing the two possibilities, have been constructed (39). The model compounds originate from the study on inosine dihydrate (77) in which molecule-1 occurs in the *syn* conformation ($\chi_{CN} = 120.1°$) while molecule-2 is in the *anti* conformation ($\chi_{CN} = 49.1°$). The two maps shown in Figs. 29 and 30 are very similar, and both indicate a global minimum at $\Phi_{C4'-C5'} = 60°$ and $\Phi_{C5'-O5'} = 180°$, and thus the *gg* conformation. The zero kcal/mol isoenergy curve is particularly wide for the *syn* conformer. The *gt* and *tg* conformations are, on both maps, 2 and 1 kcal/mol, respectively, above the global minimum. The particularly interesting feature of these maps is the stabilization of the *gg* conformation with respect to the *gt* and *tg* ones, as compared with the situation in the C2′-endo pyrimidine and C3′-endo purine nucleosides. The C2′-endo purine nucleosides, on the basis of Figs. 29 and 30, are predicted to behave similarly to the C3′-endo pyrimidine nucleosides while the C3′-endo purine nucleosides behave similarly to the C2′-endo pyrimidine nucleosides.

The observed crystal data are again in full agreement with the theoretical deductions. Among the *syn* nucleosides, 11 (65, 76, 77, 176, 180–182) are in the *gg* conformation, including the model compound inosine dihydrate molecule-1 ($\Phi_{C4'-C5'} = 64°$ and $\Phi_{C5'-O5'} = 73°$). Four compounds (*178, 183–185, 188, 189*) are in the *gt* conformation. Two of them, formycin hydrobromide monohydrate (*183, 184*) and formycin monohydrate (*185*), are, in fact, analogs of adenosine but with a "C-C" glycosyl linkage; the third, 5′-methylammonium-5′-deoxyadenosine iodide monohydrate (*188, 189*), has a nitrogen atom (N5′) instead of O5′. Finally, the last compound having a *syn* conformation, a complex of 5′-bromo-5′-deoxyadenosine and riboflavin (*190*), is in the *tg* conformation. However, this compound is again exceptional, having a bromine at position O5′. The fact that the majority of the *syn* nucleosides have the *gg* conformation arises from to the possibility of efficient intramolecular hydrogen-bond formation between the exocyclic 5′—CH$_2$OH group and the N3 of the purine bases in such an arrangement. Among the *anti* nucleosides, two (*77, 179*), including the model compound inosine dihydrate molecule-2 are in the *gg* conformation, having $\Phi_{C4'-C5'} = 47°$ and $\Phi_{C5'-O5'} = 102°$. Three *anti* nucleosides (*149, 177, 186, 187*) are in the *gt* conformation; of these, one, tubercidin (*186, 187*), is again an analog of adenosine. All the eight nucleotides (*78, 103, 151, 191–193*) of C2′-endo purines are, however, in the *gg* and *anti* conformations.

The only other theoretical computations done for C2′-endo purine nucleosides are those of Wilson and Rahman (*92*), who used only the nonbonded interaction. The computations were carried out on the same inosine dihydrate molecules-1 and -2 as were the PCILO ones. Their results (Fig. 31) indicate that the three *gg*, *gt* and *tg* minima are of practically equal depths. The analysis of the PCILO and experimental results shows unequivocally the deficiency of this empirical approach.

6. COMPARISON WITH OTHER COMPUTATIONS ON C2′-ENDO RIBOSE

The other existing computations on the general problem of C2′-endo ribose without specification of the attached base were carried out by the hard-sphere model (*27*) or the partitioned potential energy functions (*120*). The hard-sphere computations predicted three regions, 60°–70°, 170°–210° and 270°–310° for $\Phi_{C4'-C5}$, and the 150°–210° region for $\Phi_{C5'-O5'}$, to be sterically allowed. These computations thus allow the three *gg*, *gt and tg* confirmations. When the partitioned potential functions in the approximation of nonbonded interactions alone are taken into consideration (*120*), three minima of almost equal depths, similar to the computations of ref. 92, are predicted to be stable. When electrostatic interactions are included in the calculations, the *tg* conformation is predicted to be more stable than the *gg* and *gt* conformations by about 0.8 kcal/mol. This result, contrary to the PCILO results and to the available experimental data, is thus quite unsatisfactory.

7. 5,6-DIHYDROURIDINE

β-Dihydrouridine (*123, 124*) has a saturated base. A PCILO conformational energy map of β-5,6-dihydrouridine has been constructed for this compound (*39*) in order to verify if there is a change in the predicted stabilities of the conformations about the exocyclic C4'—C5' bond with respect to the map of 2'-endo deoxyuridine (Fig. 28) due to the saturation of the C5—C6 bond. Crystallographically, molecule-1 of β-dihydrouridine occurs in the *gt* conformation while molecule-2 occurs both in the *tg* and *gg* conformation in the ratio of 7:1 (*123, 124*). The results of PCILO computations on molecule 2 are presented in Fig. 32 and they are practically identical to those of Fig. 28. In particular, the *gg* conformation is still more stable than the *gt* and *tg* ones by about 1 kcal/mol. Although a full conformational energy map for molecule-1 was not made, a probe into the stabilities of the three *gg*, *gt* and *tg* conformers revealed the same results as those presented in Fig. 32. The same dihydrouridine compound has been also studied by another group of workers (*163*) by X-ray crystallography and results very similar to those of refs. *123, 124* were observed.

In the presence of these results, the question may be raised as to whether this discrepancy is a genuine one or whether the PCILO results are fundamentally correct, insofar as an unperturbed dihydrouridine molecule is concerned, the discrepancy being due to the influence of the environmental crystal packing forces. It seems that this second possibility may represent the situation. The NMR studies on dihydrouridine in aqueous solution (*49*), thus in conditions more easily comparable to the calculated ones, indicate that the compound is present as a mixture of the three conformations, but that the *gg* conformation about the C(4')—C(5') bond is preferred in the molecule over the *gt* and *tg* ones.

8. C3'-EXO PYRIMIDINE NUCLEOSIDES

For C3'-exo pyrimidine nucleosides, the PCILO computations have been carried out on the only crystallographically studied compound: thymidine (*194*) having $\chi_{CN} = 39.1°$. The results presented in Fig. 33, show that the global minimum occurs for $\Phi_{C4'-C5'} = 60°$ (*gg* conformation) and $\Phi_{C5'-O5'} = 180°$. The *gt* and *tg* conformations are both 1 kcal/mol higher than the global minimum. The observed crystallographic conformation is *gt* with $\Phi_{C4'-C5'} = 172.8°$ and $\Phi_{C5'-O5'} = 108.0°$, which means that the crystallographic conformation lies within the 2 kcal/mol isoenergy curve. The only other compound whose crystal structure is known is a 5'-nucleotide, deoxycytidine 5'-phosphate (*105*), and this compound

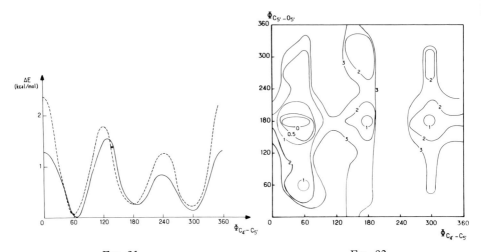

Fig. 31.   Fig. 32.

Fig. 31. Results of empirical computations of Wilson and Rahman (92) on the variation of nonbonded energy with $\Phi_{C4'-C5'}$ for C2'-endo nucleosides. Solid curve: results for inosine molecule-1; dashed curve: results for inosine molecule-2.

Fig. 32. PCILO conformational energy map for C2'-endo dihydrouridine molecule-2 as a function of $\Phi_{C4'-C5'}$ and $\Phi_{C5'-O5'}$. Isoenergy curves in kcal/mol with the global minimum taken as energy zero.

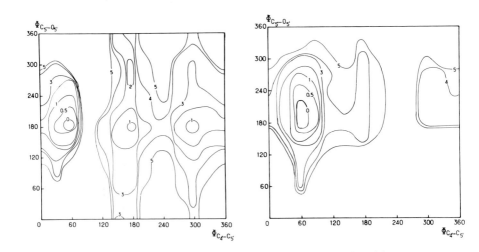

Fig. 33.   Fig. 34.

Fig. 33. PCILO conformational energy map for C3'-exo thymidine as a function of $\Phi_{C4'-C5'}$ and $\Phi_{C5'-O5'}$. Isoenergy curves in kcal/mol with the global minimum taken as energy zero.

Fig. 34. PCILO conformational energy map for C3'-exo deoxyadenosine as a function of $\Phi_{C4'-C5'}$ and $\Phi_{C5'-O5'}$. Isoenergy curves in kcal/mol with the global minimum taken as energy zero.

occurs in the *gg* conformation with $\Phi_{C4'-C5'} = 56.9°$. Because of the limited number of experimental data, no definite judgment can be made on the validity of PCILO predictions on the basis of crystallographic results. In this case again, NMR results on thymidine in solution indicate that this compound exists preferentially in the *gg* conformation (F. E. Hruska, personal communication). Therefore, it seems reasonable that the departure from this conformation in the crystal must be due to the influence of packing forces.

The only other theoretical computations made for thymidine are those of Wilson and Rahman (92); their results show that both *gt* and *tg* conformations are equally more stable than the *gg* one by about 0.8 kcal/mol. Although this result is apparently in better agreement with the crystallographic conformation than are the PCILO results, it is, nevertheless, for reasons stated above, fundamentaly less satisfactory.

9. C3'-EXO PURINE NUCLEOSIDES

The only compound of this category studied crystallographically is deoxyadenosine (195), and the PCILO results on it are presented in Fig. 34. The map shows a clear global minimum at $\Phi_{C4'-C5'} = 60°$ (*gg*) and $\Phi_{C5'-O5'} = 180°$. The *gt* and *tg* conformations are situated at plateaus about 4 kcal/mol above the global minimum. The crystallographic conformation is *gt* with $\Phi_{C4'-C5'} = 186.9°$ and $\Phi_{C5'-O5'} = 193.2°$, which means that the experimental conformation is 4 kcal/mol above the global minimum. In this case again, however, NMR studies in solution on the same deoxyadenosine molecule indicate a preference for the *gg* conformer (46, 125), which is estimated to be present in 62%. Therefore, the agreement between the theoretical evaluation and the observation in solution confirms the fundamental correctness of the theory and the presence of the *gt* conformer in the crystal may be considered as being due to the effect of environmental forces.

The only theoretical computation on C3'-exo purine nucleoside is that of Wilson and Rahman (92), whose results for deoxyadenosine are similar to those they found for thymidine, but the computations carried out including only nonbonded interactions are again questionable.

10. COMPARISON WITH OTHER CALCULATIONS ON C3'-EXO RIBOSE

Theoretical computations on C3'-exo ribose, without base, were carried out by the hard-sphere (27) and partitioned potential functions (120) procedures. Hard-sphere computations predicted two regions of stability for $\Phi_{C4'-C5'} = 170°-220°$ and $260°-310°$, thus corresponding to the *gt* and *tg* conformations. The *gg* conformation was completely disallowed. When potential functions representing nonbonded interactions alone are considered (120), the results show that the minima associated with the *gt* and *tg* conformations are of equal depth and about 0.6 kcal/mol deeper than the minimum associated with the *gg* conformation ($\Phi_{C4'-C5'} = 80°$). However, when electrostatic interactions are also included (120) in the computation, the minimum associated with the *tg* conformation becomes the deepest with a difference of 1 kcal/mol with respect to the *gg* conformation. The *gt* conformation is intermediate, about 0.6 kcal/mol higher than the global minimum for the *tg* conformation.

This summary shows that these empirical results have neither the advantage of the PCILO computations of agreeing with the experimental situation in solution nor the benefit of accounting for the situation in the solid state, as no compound of this category has as yet been found in the *tg* conformation in the crystals. An argument has been put forward (120) that intermolecular hydrogen bonds stabilize the *gt* conformation over the *tg* one.

B. α-Nucleosides

Only a few α-nucleosides (148, 198–204) have been studied by X-ray crystallography, and a conformational analysis of these compounds in crystals has been reported by Sundaralingam (205). As concerns the torsion angle χ_{CN} in these nucleosides, the only theoretical calculations available are the PCILO ones (73, 74), which show that these molecules prefer the *anti* conformation, in agreement with experiment. As concerns the conformation about the exocyclic C4'—C5' bond, the only theoretical computations available in the literature are again the PCILO ones (119) performed on C2'-exo α-pseudouridine (200) and C3'-exo α-D-2'-amino-2'-deoxyadenosine (201).

Figure 35 shows the PCILO map for α-pseudouridine, which has a C2'-exo pucker for the sugar ring in its crystal structure (200). The global minimum occurs for the *gg* conformation ($\Phi_{C4'-C5'} = 60°$), associated with $\Phi_{C5'-O5'} = 300°$; the *gt* and *tg* conformations associated with $\Phi_{C5'-O5'} = 300°$ are both 1 kcal/mol above the *gg* conformation. The crystallographic conformation for this compound (200) is *gt* with $\Phi_{C4'-C5'} = 189.7$ and $\Phi_{C5'-O5'} = 296.9°$, which will lie, when plotted on the map of Fig. 35, within the 1 kcal/mol isoenergy curve associated with this secondary energy minimum. However, NMR studies in solution (48, 51, 125, 196, 206) indicate that both α- and β-pseudouridines exist predominantly in the *gg* conformation, although *gt* and *tg* conformers are appreciably populated. These solutions studies thus corroborate the PCILO results, and the experimental crystallographic *gt* conformation may be considered as due to crystal packing forces. Three other

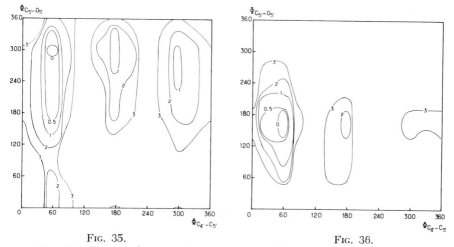

Fig. 35.   Fig. 36.

Fig. 35. PCILO conformational energy map for C2'exo α-pseudouridine as a function of $\Phi_{C4'-C5'}$ and $\Phi_{C5'-O5'}$. Isoenergy curves in kcal/mol with the global minimum taken as energy zero.

Fig. 36. PCILO conformational energy map for C3'-exo α-D-2'-amino-2'-deoxyedenosine as a function of $\Phi_{C4'-C5'}$ and $\Phi_{C5'-O5'}$. Isoenergy curves in kcal/mol with the global minimum taken as energy zero.

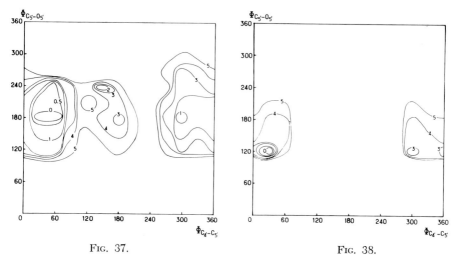

Fig. 37.   Fig. 38.

Fig. 37. PCILO conformational energy map for C2'-endo inosine 5'-phosphate as a function of $\Phi_{C4'-C5'}$ and $\Phi_{C5'-O5'}$. Isoenergy curves in kcal/mol with the global minimum taken as energy zero. Combination 7 of Table VII. From Berthod and Pullman (102).

Fig. 38. PCILO conformational energy map for C2'-endo inosine 5'-phosphate as a function of $\Phi_{C4'-C5'}$ and $\Phi_{C5'-O5'}$. Isoenergy curves in kcal/mol with the global minimum taken as energy zero. Combination 1 of Table VII. Berthod and Pullman (102).

compounds in the C2′-exo α-nucleosides category have been observed crystallographically in the gg conformation. These are vitamin B_{12} wet (*198*), vitamin B_{12} air-dry (*199*) and vitamin B_{12} 5′-phosphate (*204*) with $\Phi_{C4'-C5'}$ equal to 74°, 66° and 53°, respectively.

The results of PCILO computations on C3′-exo α-D-2′-amino-2′-deoxyadenosine monohydrate (*201*), in Fig. 36, indicate that the global minimum for the gg conformation occurs at $\Phi_{C4'-C5'} = 60°$, associated with $\Phi_{O5'-O5'} = 180°$. The gt and tg conformations are, respectively, 2 and 3 kcal/mol higher than the gg conformation. The crystallographic conformation for this compound (*201*) is gt with $\Phi_{C4'-C5'} = 171.2°$ and $\Phi_{O5'-O5'} = 101.3°$; it thus falls within the 3 kcal/mol isoenergy curve. Unfortunately, no NMR studies in solution are available for this compound. However, the situation is very similar to that for the C3′-exo β-deoxyadenosine, for which PCILO computations predict the gg conformation as the most stable one (*39*) while the crystallographic conformation is gt (theoretically, 4 kcal/mol above the gg conformation) but for which NMR results in solution (*46, 125*) confirm the predominance of the gg conformer. The gg conformation is thus intrinsically the most stable one for α-D-2′-amino-2′-deoxyadenosine, and the crystallographic gt conformation must be attributed to environmental factors. There is another C3′-exo α-nucleoside whose crystal structure is known: bis[1-(2-deoxy-α-D-ribofuranosyl)-5-uracilyl] disulfide[6] (*202*), which has two molecules in its crystalline asymmetric unit with molecule-1 in tg ($\Phi_{C4'-C5'} = 290°$) and molecule-2 in the gg ($\Phi_{C4'-C5'} = 68°$) conformations.

Finally, it should be mentioned that there are two other α-nucleosides for which no theoretical computations have been performed. These compounds are vitamin B_{12} coenzyme (*148*), which exists in the gg conformation ($\Phi_{C4'-C5'} = 53°$) and 5-[1-(2-deoxy-α-D-ribofuranosyl)]uracilyl methyl sulfide (*203*), which has two molecules in the crystalline asymmetric unit; molecule-1 is in the gg ($\Phi_{C4'-C5'} = 48°$) and molecule-2 is in the gt ($\Phi_{C4'-C5'} = 170°$) conformation. While the sugar pucker of vitamin B_{12} coenzyme (*148*) is C3′-endo, the two molecules of the other compound (*203*) have C4′-endo sugar puckers.

C. 5′-β-Nucleotides

Table VI lists all the 5′-β-nucleotides studied by X-ray crystallography. With the exceptions of 6-azauridine 5′-phosphate (*99*) and deoxyguanosine 5′-phosphate (*100*) (not listed in Table VI) which exist in the gt conformation, all the other compounds (*43, 78, 98, 103–106, 118, 139–142, 151, 152, 175, 191–193*) exist in the gg conformation about the exocyclic C4′—C5′ bond. As far as the glycosyl torsion angle χ_{CN} is

[6] More correctly named 5,5′-dithiobis[1-(2-deoxy-α-D-ribofuranosyl)uracil]—Ed.

concerned, all are in the *anti* conformation. These experimental results led to the concept of the "rigidity" of 5'-β-nucleotides (25, 98) with respect to the two torsion angles concerned. The rigidity or flexibility of the torsion angle χ_{CN} has been discussed in Section IV, and we analyze here only the results of the theoretical computations performed on the conformation about the exocyclic C4'—C5' bond in 5'-β-nucleotides.

Both classical (113–115) and quantum mechanical (39, 117) methods have been employed to predict the preferred conformation about this bond. Along the classical approach, Yathindra and Sundaralingam (113–115) have utilized the partitioned potential energy functions. Their results on adenosine 5'-phosphate (113) show that the *gg* conformation is the preferred one over the *gt* and *tg* conformations. For guanosine 5'-phosphate, their results (114) predict a predominance of the *gg* conformation followed by the *gt* conformation. Yathindra and Sundaralingam (115) have also studied three pyrimidine 5'-nucleotides, namely, rCMP, rUMP and rTMP, by empirical partitioned potential energy functions; their results indicate that, in all of them, the *gg* conformation about the exocyclic C4'—C5' bond is preferred over the *gt* and *tg* conformations. Thus the results of the partitioned potential functions confirm the type of rigidity of 5'-nucleotides postulated by Sundaralingam (25, 98).

Along the quantum-mechanical approach, the PCILO and EHT methods (39, 117) have been employed to study the preferred conformation about the exocyclic C4'—C5' bond in 5'-nucleotides, but the PCILO computations are the most detailed ones. The model compound in PCILO computations (39) was C2'-endo inosine 5'-phosphate. Three values of torsion angles $\Phi_{\text{O5'-P}}$ and $\Phi_{\text{P-OIII}}$ (see Fig. 3) corresponding to the classical staggered conformations (60°, 180° and 300°) are possible; in order to preselect these two torsion angles, a PCILO probe was carried out by adopting the existing crystal data for $\Phi_{\text{O5'-P}}$ and $\Phi_{\text{P-OIII}}$, but trying also other possible values (39). The results of such a probe are presented in Table VII, and it can be seen that the most stable combination is No. 7 with $\Phi_{\text{O5'-P}} = 291°$, $\Phi_{\text{P-OIII}} = 74.3°$ and $\Phi_{\text{C5'-O5'}} = 180°$ followed by combination No. 9. The next more stable combinations are Nos. 1 and 3, only about 0.9 kcal/mol less stable than combination No. 7. Two conformational energy maps were constructed corresponding to the combinations Nos. 7 and 1.

Figure 37 shows the PCILO map constructed with the combination No. 7 and it can be seen that the global minimum occurs at the *gg* conformation with $\Phi_{\text{C4'-C5'}} = 60°$ and $\Phi_{\text{C5'-O5'}} = 180°$. The *gt* and *tg* conformations, associated with $\Phi_{\text{C5'-O5'}} = 180°$ are, respectively, 3 and 1 kcal/mol above the global minimum. Figure 38, constructed with com-

TABLE VII

Results of the Probe for the Preselection of Torsional Angles for the Study of C2'-Endo Inosine 5'-Phosphate[a]

No.	Preselected torsions	Energy (kcal/mol) (with respect to the individual global minimum)			Energy (kcal/mol) (with respect to the most stable arrangement)
		gg ($\Phi_{C4'-C5'}$ = 60°)	gt ($\Phi_{C4'-C5'}$ = 180°)	tg ($\Phi_{C4'-C5'}$ = 300°)	
1	$\Phi_{O5'-P}$ = 69.2°, Φ_{P-OIII} = 74.3°, $\Phi_{C5'-O5'}$ = 180°	0.0	2.1	0.3	0.9
2	$\Phi_{O5'-P}$ = 69.2°, Φ_{P-OIII} = 180°, $\Phi_{C5'-O5'}$ = 180°	0.0	2.3	0.4	3.3
3	$\Phi_{O5'-P}$ = 69.2, Φ_{P-OIII} = 300°, $\Phi_{C5'-O5'}$ = 180°	0.0	2.3	0.3	0.9
4	$\Phi_{O5'-P}$ = 176°, Φ_{P-OIII} = 74.3°, $\Phi_{C5'-O5'}$ = 180°	0.0	2.5	1.0	1.2
5	$\Phi_{O5'-P}$ = 176°, Φ_{P-OIII} = 180°, $\Phi_{C5'-O5'}$ = 180°	0.0	2.5	0.9	3.9
6	$\Phi_{O5'-P}$ = 176°, Φ_{P-OIII} = 300°, $\Phi_{C5'-O5'}$ = 180°	0.0	2.4	0.9	1.5
7	$\Phi_{O5'-P}$ = 291°, Φ_{P-OIII} = 74.3°, $\Phi_{C5'-O5'}$ = 180°	0.0	3.4	1.4	0.0
8	$\Phi_{O5'-P}$ = 291°, Φ_{P-OIII} = 180°, $\Phi_{C5'-O5'}$ = 180°	0.0	3.3	1.4	2.6
9	$\Phi_{O5'-P}$ = 291°, Φ_{P-OIII} = 300°, $\Phi_{C5'-O5'}$ = 180°	0.0	3.4	1.6	0.5

[a] From Saran et al. (39).

bination No. 1 (Table VII) shows considerable modifications. In Fig. 38, the global minimum occurs for the gg conformation at $\Phi_{C4'-C5'}$ = 30° associated with $\Phi_{C5'-O5'}$ = 120°; thus, there is a displacement of the global minimum, that of Fig. 37. This displacement is due to the strong interaction between the hydrogen atom attached to C8 of inosine and one of the negatively charged oxygen atoms of the phosphate group. In this case, the gg conformation is strongly preferred over the gt and tg conformations. Altogether, the PCILO results (39) thus predict also the predominance of the gg conformation over the gt and tg ones in 5'-nucleotides in agreement with a large number of X-ray crystallographic data as compiled in Table VI.

Another quantum-mechanical computation has been carried out by Vasilescu et al. (117), who have utilized the EHT method to study the conformation of adenosine 5'-phosphate. Their results indicate that the *gt* conformation is preferred over the *gg* one by about 0.8 kcal/mol. This result is in contrast with a large body of experimental X-ray crystallographic results and also with the other theoretical computations (*39, 113–115*), which clearly show that the *gg* conformation is preferred over the *gt* and *tg* conformations. This discrepancy of the EHT method is obviously due to the imprecision of the methodology.

Finally, mention must be made of the NMR study on 5′-nucleotides in solution by Wood *et al.* (*207*), who investigated the preferred conformations about the C4′—C5′ and C5′—O5′ in these nucleotides. Their results indicate that the *gg* conformer is the most populated in all 5′-nucleotides with the exception of 6-azauridine 5′-phosphate, which has only 20% of the *gg* conformer. The conformation about the C5′—O5′ bond is *trans* ($\Phi_{C5'-O5'} \approx 180°$) in all nucleotides. These recent solution results (*207*) lend thus further support to the concept of rigidity in 5′-nucleotides as far as conformations about C4′—C5′ and C5′—O5′ bonds are concerned.

VI. The Backbone Structure of Di- and Polynucleotides

The conformational study of the backbone structure of dinucleotides is an essential step toward understanding the conformation of polynucleotides and nucleic acids. It concerns the knowledge of conformational preferences about the five torsion angles of the backbone, namely: $\Phi_{C4'-C5'}$, $\Phi_{C5'-O5'}$, $\Phi_{O5'-P}$, $\Phi_{P-O3'}$, and $\Phi_{O3'-C3'}$ (see Fig. 1). This topic has been the center of many theoretical studies in recent years by both classical potential energy functions (*29, 116, 120, 208–211*) and quantum-mechanical methods (*121, 122, 197, 212–219*). In the quantum mechanical procedures, four different molecular orbitals methods have been employed: EHT (*121, 122, 215*), PCILO (*197, 212–214*), CNDO (*216, 217*) and *ab initio* (*213, 218, 219*). Altogether, four different models—ribose phosphate (*120, 121*), dimethyl monophosphate (*208, 213, 218*), disugar monophosphate (*122, 197, 212, 214–217, 219*) and dinucleoside monophosphate (*29, 116, 209–211*)—have been utilized. The results, in most cases, have been presented in the form of conformational energy maps as a function of two consecutive torsion angles, while the remaining three are kept fixed in some preselected values. We summarize the results obtained by different procedures and compare them with the experimental results coming mainly from the X-ray crystallographic studies. In addition, this section also deals with the results of PCILO studies (*213*) on the fundamental role of the geometry of the phosphate group in determining the preferred conformation around two P—O ester bonds in dinucleoside monophosphates and polynucleotides.

A. Conformational Energy Maps about Two Consecutive Torsion Angles

Four types of conformational energy maps have been constructed by the PCILO procedure corresponding to the four possible combinations of two consecutive torsion angles of the backbone ($\Phi_{P-O3'}-\Phi_{O5'-P}$), ($\Phi_{O3'-C3'}-\Phi_{P-O3'}$), ($\Phi_{O5'-P}-\Phi_{C5'-O5'}$) and ($\Phi_{C5'-O5'}-\Phi_{C4'-C5'}$). For the construction of these maps, preselected values based both on Sundaralingam's compilation (24) and upon indications of various preliminary computations have been adopted for the torsion angles not involved in the particular map under construction. These preselected values, generally, are 300° (preferred), 60° and 180° for $\Phi_{O5'-P}$ and $\Phi_{P-O3'}$; 60° (preferred), 180° and 300° for $\Phi_{C4'-C5'}$; 180° for $\Phi_{C5'-O5'}$ and 240° for $\Phi_{O3'-C3'}$. The up to date X-ray crystallographic data on nucleotides, di-, and trinucleoside phosphates and polynucleotides have been compiled in Table VIII. For some polynucleotides and models of nucleic acids, more than one set of data of the backbone torsion angles have been indicated, which correspond to subsequent refinements of the different models proposed for them. For 3'- and 5'-nucleotides, the terminal torsion angles have been calculated by including the terminal hydrogen atoms. For A(2'-5')U, the torsion angles $\Phi_{P-O3'}$ and $\Phi_{O3'-C3'}$ correspond, respectively, to $\Phi_{P-O2'}$ and $\Phi_{O2'-C2'}$ because of the (2'-5')-linkage instead of the usual (3'-5')-linkage. On the PCILO conformational energy maps, the experimentally observed X-ray crystallographic conformations have been plotted with the filled circles (●) referring to the compounds whose remaining torsion angles are close to the preselected ones, and the open circles (○) to the compounds in which these angles are farther away from the preselected ones or are partially or even completely unknown. Only C3'-endo–C3'-endo combination of the sugar puckers were considered in the PCILO computations (197, 212) (whereas the empirical computations of Olson and Flory (209), described in the next section, consider both C3'-endo–C3'-endo and C2'-endo—C2'-endo combinations for the sugar puckers).

1. ($\Phi_{P-O3'}-\Phi_{O5'-P}$) CONFORMATIONAL ENERGY MAP

Figure 39 shows the PCILO map constructed for the disugar monophosphate model (197, 212) with $\Phi_{O3'-C3'} = 240°$, $\Phi_{C5'-O5'} = 180°$ and $\Phi_{C4'-C5'} = 60°$, and with the C3'-endo conformation for the sugars. The map shows a global minimum centered around $\Phi_{P-O3'} = \Phi_{O5'-P} = 300°$ with a large area of conformational stability included within 1 or 2 kcal/mol isoenergy curves.

The experimental conformations plotted on this map indicate a num-

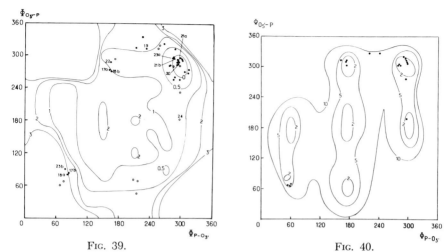

FIG. 39. FIG. 40.

FIG. 39. PCILO ($\Phi_{P-O3'}-\Phi_{O5'-P}$) conformational energy map with $\Phi_{O3'-C3'} = 240°$, $\Phi_{C5'-O5'} = 180°$ and $\Phi_{C4'-C5'} = 60°$. Isoenergy curves in kcal/mol with the global minimum taken as energy zero. From Pullman et al. (197).

FIG. 40. EHT ($\Phi_{P-O3'}-\Phi_{O5'-P}$) conformational energy map of Saran and Govil (122) constructed with $\Phi_{O3'-C3'} = 240°$, $\Phi_{C5'-O5'} = \Phi_{C4'-C5'} = 180°$.

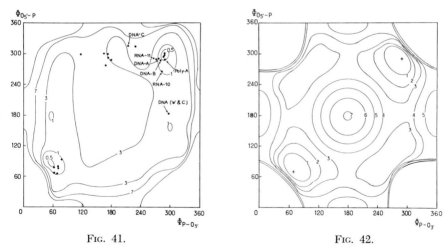

FIG. 41. FIG. 42.

FIG. 41. CNDO ($\Phi_{P-O3'}-\Phi_{O5'-P}$) conformational energy map of Tewari et al. (216) constructed with $\Phi_{O3'-C3'} = 240°$, $\Phi_{C5'-O5'} = 180°$ and $\Phi_{C4'-C5'} = 60°$.

FIG. 42. Ab initio ($\Phi_{P-O3'}-\Phi_{O5'-P}$) conformational energy map of Newton (218) constructed on the model of dimethyl phosphate. + signs indicate the global minima.

TABLE VIII
BACKBONE TORSION ANGLES IN NUCLEOTIDES, DI- AND TRINUCLEOSIDE PHOSPHATES AND POLYNUCLEOTIDES

No.	Compound	$\Phi_{C3'-C4'}$	$\Phi_{C4'-C5'}$	$\Phi_{C5'-O5'}$	$\Phi_{O5'-P}$	$\Phi_{P-O3'}$	$\Phi_{O3'-O3'}$	Reference No.
			3'-Nucleotides					
1	3'-AMP · (2 H$_2$O)	83.3	171.7		283.3	295.4	237.3	150
2	3'-CMP (monoclinic)	150.2	45.5		61.0	68.0	248.9	170
3	3'-CMP (orthorhombic)	152.0	43.8		284.8	170.7	268.5	80
4	3'-UMP(Na$_2$) · (4 H$_2$O)	144.2	42.7		—	—	254.2	172
5	(3'-Phosphonomethyl)Ado	81.5	49.9		233.4	297.4	195.0	145
6a	Urd-3'[OP(S,OH)OMe] (1)	159.0	40.5		46.5	213.6	269.1	173, 174
6b	Urd-3'[OP(S,OH)OMe] (2)	156.4	40.0		69.0	215.1	274.3	173, 174
			5'-Nucleotides					
7	5'-dTMP(Ca) · (6 H$_2$O)	82.0	57.3	204.3	—	—	—	139
8	5'-AMP · H$_2$O	75.5	40.0	177.2	291.0	—	—	104
9	5'-GMP · (3 H$_2$O)		45.8	194.2	279.7	—	—	106
10a	ATP(Na$_2$) · (3 H$_2$O) (1)	86.1	67.0	224.0	64.2	162.8	—	151
10b	ATP(Na$_2$) · (3 H$_2$O) (2)	151.0	48.6	224.0	308.5	290.6	—	151
11	5'-IMP · H$_2$O	84.9	52.3	155.7	—	—	—	118
12	5'-UMP(Ba) · (7 H$_2$O)	145.0	55.2	176.0	181.4	—	—	43
13	5'-IMP(Na) · (8 H$_2$O)	147.9	59.4	171.6	—	—	—	103
14	ADP(Rb) · (3 H$_2$O)	151.0	57.4	149.0	72.0	207.0	—	192
15	5'-dAMP · (6 H$_2$O)	132.5	46.8	196.7	302.2	—	—	193
16	5'-dCMP · H$_2$O	153.3	56.9	166.1	69.1	74.3	—	105
			Dinucleoside Monophosphates					
17a	(U-A) · (½ H$_2$O) (1)	93.5(A) 80.5(U)	52.3	191.9	272.2	162.7	223.5	98
17b	(U-A) · (½ H$_2$O) (2)	82.7(A) 86.8(U)	55.8	202.3	83.8	83.9	202.0	98
18a	(U-A) · (½ H$_2$O) (1)	85.0(A) 86.3(U)	54.8	203.0	81.9	81.1	205.8	140
18b	(U-A) · (½ H$_2$O) (2)	93.4(A) 76.9(U)	53.6	191.7	270.7	163.6	224.4	140
19	A(2'-5')U · (4 H$_2$O)	83.8(U) 144.0(A)	56.9	169.9	313.4	232.5	243.7	78

CONFORMATION OF NUCLEIC ACIDS 279

20	G-C		51.0	186.0	284.0	291.0	209.0	141
21a	A-U (1)		57.0	177.0	289.0	294.0	213.0	141, 142
21b	A-U (2)		57.0	169.0	296.0	285.0	220.0	141, 142
		Dinucleotides						
22	dT-dT moiety of pdT-dT	158.0(dT²)	41.0	187.0	288.0	163.0	252.0	175
		155.0(dT¹)						
		Trinucleoside Diphosphates						
23a	A¹-A² moiety of A¹-A²-A³	81.0(A²)	56.0	162.2	297.4	283.1	222.7	152
		82.3(A¹)						
23b	A²-A³ moiety of A¹-A²-A³	78.9(A³)	61.3	188.3	92.7	77.2	209.2	152
		81.0(A²)						
		Nucleic Acids and Polynucleotides						
24	DNA-10 (Watson and Crick)	100.0	155.0	179.0	182.0	296.0	186.0	220
25a	B-DNA-10	130.0	58.0	212.0	281.0	282.0	147.0	221
25b	B-DNA-10	145.8	34.2	194.1	318.4	257.4	170.3	222
25c	B-DNA-10	156.4	36.4	214.0	313.2	264.9	155.0	223
25d	B-DNA-10	146.2	26.8	159.9	334.7	224.1	194.3	224
26	C-DNA-9.3	141.0	48.0	143.0	315.0	212.0	211.0	225
27a	A-DNA-11	76.1	67.0	167.0	283.0	279.0	221.0	226
27b	A-DNA-11	76.1	44.8	188.9	296.2	297.8	200.2	222
27c	A-DNA-11	82.6	45.5	207.9	275.4	313.6	177.7	223
28a	RNA-10(α- and β-A-RNA)	80.0	88.0	188.0	257.0	285.0	203.0	227
28b	α-A-RNA-10 (reovirus)	79.6	76.4	198.3	257.0	295.7	193.9	222
28c	β-A-RNA-10 (reovirus)	78.6	83.8	179.7	261.0	287.1	212.5	222
29a	RNA-11 (α- and β-A-RNA)	75.0	69.0	165.0	282.0	277.0	223.0	228
29b	α-A-RNA-11 (reovirus)	92.6	49.5	186.5	294.0	293.0	201.8	222
29c	β-A-RNA-11 (reovirus)	88.6	48.9	185.9	294.2	293.1	202.2	222
29d	α-A-RNA-11 (reovirus)	83.2	53.5	204.6	271.7	307.9	181.0	229
29e	β-A-RNA-11 (reovirus)	83.2	51.6	208.9	268.0	314.0	173.7	229
30a	Poly(rA · rU); (β-A-RNA-11)	78.8	57.7	185.4	288.3	292.1	204.9	26
30b	Poly(rA · rU); (β-A-RNA-11)	83.5	47.4	180.1	297.9	286.4	208.3	229, 230
31a	Poly(rI · rC); (A′-RNA-12)	79.0	50.2	195.3	291.0	301.9	192.5	26
31b	Poly(rI · rC); (A′-RNA-12)	82.3	43.9	192.9	294.8	300.9	192.4	229, 230
32	Poly(rI · dC)-12	77.7	56.3	183.5	288.6	295.3	193.9	231
33	Poly(rAH⁺ · rAH⁺)-8	82.0	69.0	168.0	285.0	293.0	216.0	232
34a	Poly(s²U) α-chain	79.0	40.8	155.6	322.5	266.0	221.3	233
34b	Poly(s²U) β-chain	72.0	52.8	163.5	313.4	287.0	218.9	233

ber of representative points clustering around the global minimum. The points represent essentially the different proposed models of nucleic acids and polynucleotides and also include two dinucleoside monophosphates, G-C and A-U molecules-1 and -2 (Nos. 20, 21a, 21b of Table VIII) and one unit of a trinucleoside diphosphate (152): A^1-A^2 (compound No. 23a). The DNA-10 (Watson and Crick) represented by compound No. 24 of Table VIII lies away from this region but within the 1 kcal/mol isoenergy curve. Compound No. 19, a (2'-5')-linked dinucleoside phosphate lies near to the region of the global minimum. Two dinucleoside phosphates, Nos. 17a, 18b and 22, lie near $\Phi_{P-O3'} = 180°$ and $\Phi_{O5'-P} = 300°$, within the 1 kcal/mol isoenergy curve. There are a couple of representative points (compounds Nos. 17b, 18a and 23b) lying near $\Phi_{P-O3'} = \Phi_{O5'-P} = 90°$, which on the map of Fig. 39 are at relatively higher energy levels. Another PCILO map, constructed (197, 212) with $\Phi_{C4'-C5'} = 180°$ while the other torsion angle remain the same as in Fig. 39, shows practically two equivalent global minima centered around $(\Phi_{P-O3'}, \Phi_{O5'-P}) = (300°, 300°)$ and $(90°, 90°)$. Although, in a strict sense, Nos. 17b, 18a and 23b do not correspond to the preselection for this map, they would lie on it in the region of the global minimum centered around $(\Phi_{P-O3'}, \Phi_{O5'-P}) = (90°, 90°)$. The reason for the high value of the energy in the region around $(90°, 90°)$ in Fig. 39 is the replacement of the 5'-CH_2OH group by a methyl group, done for computational convenience (197). When, however, the 5'-CH_2OH group and a more realistic value of 200° for $\Phi_{O3'-C3'}$ are considered, the region $\Phi_{P-O3'} = \Phi_{O5'-P} = 90°$ becomes conformationally allowed (214) and the above-mentioned Nos. 17b, 18a and 23b lie in the allowed regions.

The EHT computations (122, 215) predict (Fig. 40) the existence of seven equivalent regions of energy minima located around $(\Phi_{P-O3'}, \Phi_{O5'-P}) = (60°, 60°)$, $(60°, 180°)$, $(180°, 60°)$, $(180°, 180°)$, $(180°, 300°)$, $(300°, 180°)$ and $(300°, 300°)$. This means that although EHT indicates minima in the stable regions, it is completely unselective with respect to these minima. Further, it places at the same level regions of high energy and, in fact, unoccupied by the experimental conformations. Obviously, the EHT procedure is less precise than the PCILO one. It may be added that the EHT map was constructed with $\Phi_{C4'-C5'} = 180°$, but there is little probability that the calculations with $\Phi_{C4'-C5'} = 60°$ would improve the results significantly. Recently CNDO computations (216, 217) have been carried out on the same model as the EHT ones, and the results (Fig. 41) predict two global minima of equal stability centered around $(\Phi_{P-O3'}, \Phi_{O5'-P}) = (290°, 290°)$ and $(70°, 70°)$. These CNDO results are thus quite comparable to the PCILO ones (197, 212, 214). The *ab initio* computations (218) carried out on dimethyl monophosphate indicate, as shown in Fig. 42, two global minima of equal stability around $(\Phi_{P-O3'}, \Phi_{O5'-P}) = (285°, 285°)$ and $(75°, 75°)$, similar to the CNDO computations (216, 217). In these calculations, the region around (180°, 180°) is about 7 kcal/mol higher than the global minima. Thus, all the three more precise molecular orbital methods, namely, PCILO (197, 212, 214), CNDO (216, 217) and the *ab initio* (218), predict high energy for the region around $(\Phi_{P-O3'}, \Phi_{O5'-P}) = (180°, 180°)$, and this fact is corroborated by the observed conformations in this region.

In the classical potential energy function procedures (208), the $(\Phi_{P-O3'}-\Phi_{O5'-P})$ map was constructed for dimethyl phosphate as the reference compound. The results are very similar to those of EHT method and indicate seven energy minima located in the same regions as those of EHT. Following the details of the computational procedure, some or other regions represent global minima. Thus, when only nonbonded and torsional interactions are considered, the global minima are located at $(\Phi_{P-O3'}, \Phi_{O5'-P}) = (60°, 60°)$ and $(300°, 300°)$; when electrostatic interactions are included in the computations with the dielectric constant $\epsilon = 4$, the global minimum shifts to $(180°, 180°)$; when $\epsilon = 10$, however, the global minima move again to $(60°, 60°)$ and $(300°, 300°)$. These results suffer thus from the existence of too great a number of local minima, some of which lie in regions of apparently little significance, and from the difficulty in choosing unambiguously the preferred ones.

Very recently, Olson and Flory (29, 209, 210), using classical potential energy functions on a model of a dinucleoside monophosphate, presented conformational energy maps for torsions about the adjacent bonds of the chain backbone. Aimed at refining the empirical computations of Sasisekharan and Lakshminarayanan (208) through the evaluation of the statistical weights of the low-energy domains of the conformational energy maps, they lead to what can be only considered as very unsatisfactory results. Trying to go beyond a number of equivalent energy minima as in Sasisekharan and Lakshminarayanan's computations (208), the calculations of Olson and Flory (29, 209, 210) select the most probable one. This leads, in general, to a pronounced disagreement with the available experimental information. Their $(\Phi_{P-O3'}-\Phi_{O5'-P})$ conformational energy map constructed with C3'-endo–C3'-endo sugar pucker combination is presented in Fig. 43. This map shows the global minimum located at $(\Phi_{P-O3'}, \Phi_{O5'-P}) = (180°, 180°)$ for which no conformations have yet been observed experimentally. As mentioned above, all refined molecular orbital methods (PCILO, CNDO and *ab initio*) predict high energies in this region. Olson and Flory attribute the disagreement between their findings and the experimental data to the neglect of "specific" base–base stacking interactions. This, however, cannot be the predominant reason, as may be seen from the fact that the PCILO (197, 212, 214) and CNDO (216, 217) calculations, which also do not include stacking of the bases (being carried out on a model without bases), predict conformational energy minima that are in general agreement with the experimental data, including those involving polynucleotides. Also, the *ab initio* computations (218), which were carried out on the even much simpler dimethyl phosphate model, predict a high energy in the global minimum region of Olson and Flory's map (Fig. 43). The situation does not improve much when C2'-endo pucker is considered for both sugars in their computations: the global minimum now shifts to ($\Phi_{P-O3'}$, $\Phi_{O5'-P}$) = (210°, 60°), for which no models of polynucleotides and nucleic acids have been experimentally observed.

2. ($\Phi_{O3'-C3'}-\Phi_{P-O3'}$) CONFORMATIONAL ENERGY MAP

The PCILO conformational energy map (197, 212), constructed with $\Phi_{O5'-P} = 300°$, $\Phi_{C5'-O5'} = 180°$ and $\Phi_{C4'-C5'} = 60°$, is presented in Fig. 44. This map shows that the global minimum occurs at ($\Phi_{O3'-C3'}$, $\Phi_{P-O3'}$) = (180°, 270°) with the associated zone of low energy (<0.5 kcal/mol) extending from $\Phi_{O3'-C3'} = 180°$ to 240°. There are again large surfaces enclosed within the 1 and 2 kcal/mol isoenergy curves. The experimental conformations, mainly the different models of polynucleotides and nucleic acids, cluster around ($\Phi_{O3'-C3'}$, $\Phi_{P-O3'}$) = (210°, 290°) and are thus very close to the global minimum. B-DNA-10 (Table VIII, No. 25a) lies near the 3 kcal/mol isoenergy curve. The subsequent crystallographic refinements for this compound (Nos. 25b, 25c and 25d) place it well within the allowed conformational region with No. 25b, very close to the global minimum. Three other compounds: A-DNA-11 (No. 27c), α-A-RNA-11 (reovirus) and β-A-RNA-11 (reovirus) (Nos. 29d and 29e) lie outside the conformationally allowed regions, although earlier results for these compounds (Nos. 27b, 29b and 29c, respectively) fall well within them. The compounds Nos. 17a, 18b, 19, 21a and 21b, whose other torsion angles are close to the preselected ones, lie within the allowed energy regions. Compound No. 23a, which forms one unit of A^1-A^2-A^3, (152) lies near the global minimum while pdT^1-dT^2 (175) lies within 2 kcal/mol isoenergy curve.

Another PCILO map constructed with $\Phi_{O5'-P} = 60°$, $\Phi_{C5'-O5'} = 180°$ and $\Phi_{C4'-C5'} = 60°$ (197) is presented in Fig. 45. The global minimum, now shifts to ($\Phi_{O3'-C3'}$, $\Phi_{P-O3'}$) = (180°–210°, 90°). Three representative points, corresponding to Nos. 17b, 18a and 23b, lie very close to the global minimum, in excellent agreement with theory.

The EHT results (122, 215), which agree broadly with the PCILO results (197, 212), indicate two equivalent global minima located at

CONFORMATION OF NUCLEIC ACIDS 283

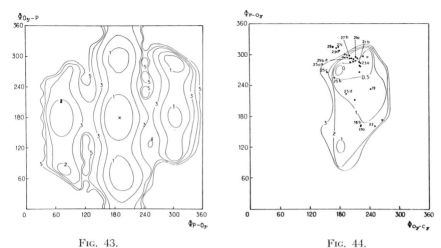

Fig. 43.   Fig. 44.

Fig. 43. Empirical ($\Phi_{P-O3'}-\Phi_{O5'-P}$) conformational energy map of Olson and Flory (209) constructed with $\Phi_{O3'-C3'} = 215°$, $\Phi_{C5'-O5'} = \Phi_{C4'-C5'} = 180°$. X indicates the global minimum.

Fig. 44. PCILO ($\Phi_{O3'-C3'}-\Phi_{P-O3'}$) conformational energy map with $\Phi_{O5'-P} = 300°$, $\Phi_{C5'-O5'} = 180°$ and $\Phi_{C4'-C5'} = 60°$. Isoenergy curves in kcal/mol with the global minimum taken as energy zero. From Pullman et al. (197).

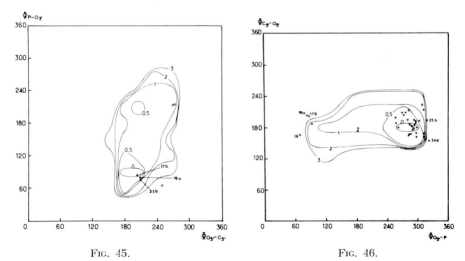

Fig. 45.   Fig. 46.

Fig. 45. PCILO ($\Phi_{O3'-C3'}-\Phi_{P-O3'}$) conformational energy map with $\Phi_{O5'-P} = 60°$, $\Phi_{C5'-O5'} = 180°$ and $\Phi_{C4'-C5'} = 60°$. Isoenergy curves in kcal/mol with the global minimum taken as energy zero. From Pullman et al. (197).

Fig. 46. PCILO ($\Phi_{O5'-P}-\Phi_{C5'-O5'}$) conformational energy map with $\Phi_{O3'-C3'} = 240°$, $\Phi_{P-O3'} = 300°$ and $\Phi_{C4'-C5'} = 60°$. Isoenergy curves in kcal/mol with the global minimum taken as energy zero. From Pullman et al. (197).

($\Phi_{O3'-C3'}$, $\Phi_{P-O3'}$) = (240°, 300°) and (240°, 180°). No conformational energy map involving the two torsion angles $\Phi_{O3'-C3'}$ and $\Phi_{P-O3'}$ has been constructed by the CNDO method, but a one-dimensional energy curve as a function of $\Phi_{O3'-C3'}$ with preselected values of $\Phi_{O5'-P} = \Phi_{P-O3'} = 290°$, $\Phi_{C5'-O5'} = 180°$ and $\Phi_{C4'-C5'} = 60°$ has been presented (216). The CNDO results indicate a very flat minimum around $\Phi_{O3'-C3'} = 260°$, and the region of stability extends from 200° to 300°. The three-bond NMR coupling constants in poly(U) have been measured (234, 235), and a value of $\Phi_{O3'-C3'} = 240°$, deduced from the coupling constants through the Karplus equation (216, 234), is in excellent accord with the computations.

In the classical potential energy functions computations, the results of Lakshminarayanan and Sasisekharan (120) are less satisfactory; they suggest that the potential energy is independent of $\Phi_{P-O3'}$ in the ($\Phi_{O3'-C3'}$, $\Phi_{P-O3'}$) variation and that the global minimum occurs at $\Phi_{O3'-C3'} = 180°$, with a second local minimum about 1.9 kcal/mol above, at $\Phi_{O3'-C3'} = 300°$, and a barrier of 3 kcal/mol between the two at 240°. These results do not account satisfactorily for the observed situation. Olson and Flory's results (209) show two stable regions centered around ($\Phi_{O3'-C3'}$, $\Phi_{P-O3'}$) = (180°, 180°) and (180°, 300°) for the C3'-endo–C3'-endo sugar pucker combination. For the C2'-endo–C2'-endo sugar pucker combination, the stable region extends from $\Phi_{P-O3'} = 150°$ to 270° associated with $\Phi_{O3'-C3'} = 210°$. These results are more satisfactory than those of Lakshminarayanan and Sasisekharan (120).

3. ($\Phi_{O5'P}$–$\Phi_{C5'-O5'}$) Conformational Energy Map

Figure 46 shows the PCILO map (197, 212) constructed with $\Phi_{O3'-C3'} = 240°$, $\Phi_{P-O3'} = 300°$ and $\Phi_{C4'-C5'} = 60°$. The global minimum occurs at $\Phi_{O5'-P} = 270°–300°$ and $\Phi_{C5'-O5'} = 180°$. There is a large 0.5 kcal/mol isoenergy curve surrounding it. All the representative compounds with other torsion angles close to the preselected ones cluster around the global minimum, except two (Nos. 25b and 34a), which lie outside the low energy contours but in any case quite close to the 3 kcal/mol isoenergy curve. Another PCILO map constructed with the same preselected values as Fig. 46, but with $\Phi_{P-O3'} = 180°$ (Fig. 47), shows two equivalent global minima located at ($\Phi_{O5'-P}$, $\Phi_{C5'-O5'}$) = (270°–300°, 180°) and (60°, 180°) and again the representative compounds lie in the regions of stability. Three (Nos. 16, 17b and 18a), whose other torsion angles are not close to the preselected ones and which lie outside the energy contours of map in Fig. 46, lie within the regions of stability in Fig. 47.

The EHT computations (122, 215), with $\Phi_{O3'C3'} = 240°$, $\Phi_{P-O3'} =$

300° and $\Phi_{C4'-C5'} = 180°$, predict three equivalent global minima centered around $(\Phi_{O5'-P}, \Phi_{C5'-O5'}) = (180°, 120°), (180°, 180°)$ and $(300°, 180°)$. Differing from the PCILO results, they do not select the $(300°, 180°)$ minimum as the deepest one and place it thus at the same level as the $(180°, 120°)$, which does not occur in PCILO computations (*197*). The experimental data are in favor of the PCILO results. Although no two-dimensional conformational energy may has been constructed by the CNDO method, a one-dimensional energy curve as a function of $\Phi_{C5'-O5'}$ has been presented (*217*). The results show that the stability region for $\Phi_{C5'-O5'}$ extends from 160° to 260°, with a sharp global minimum at 260°. This particular CNDO result is quite unsatisfactory because most of the experimental conformations are observed to cluster around 180°.

The classical potential energy computations of Lakshminarayanan and Sasisekharan (*120*) indicate nine regions of predicted stability centered around $(\Phi_{O5'-P}, \Phi_{C5'-O5'}) = (60°, 60°), (60°, 180°), (60°, 300°), (180°, 60°), (180°, 180°), (180°, 300°), (300°, 60°), (300°, 180°)$ and $(300°, 300°)$, and they place the global minimum in the $(180°, 180°)$ region. These results thus do not reflect the experimental situation. Olson and Flory (*209*), also using potential energy functions, predict nine regions of stability (Fig. 48) for C3'-endo–C3'-endo combination of sugar puckers. It is seen from Fig. 48 that the global minimum occurs at $(\Phi_{O5'-P}, \Phi_{C5'-O5'}) = (180°, 180°)$, similar to the results of Lakshminarayanan and Sasisekharan (*120*), and that these recent empirical computations do not constitute an improvement over the older ones. When the C2'-endo–C2'-endo combination for the sugars is considered, the regions of stability are reduced from nine to three centered around $(\Phi_{O5'-P}-\Phi_{C5'-O5'}) = (60°, 180°), (180°, 180°)$ and $(300°, 180°)$, the global minimum again occurring at $(180°, 180°)$. The classical potential energy computations (*120, 209*) are thus completely at variance with the crystallographic situation.

4. $(\Phi_{C5'-O5'}-\Phi_{C4'-C5'})$ Conformational Energy Map

The results of PCILO computations (*197, 212*) carried out with $\Phi_{O3'-C3'} = 240°, \Phi_{P-O3'} = \Phi_{O5'-P} = 300°$ are presented in Fig. 49. The map shows a global minimum occurring at $(\Phi_{C5'-O5'}, \Phi_{C4'-C5'}) = (180°, 60°)$ and the surrounding 3 kcal/mol isoenergy curve extends to $\Phi_{C4'-C5'} = 300°$. There is a local minimum at $(180°, 180°)$ about 2 kcal/mol above the global one. All the representative experimental compounds are in the vicinity of the global minimum or at least in the zone of stability. DNA-10 (Watson and Crick compound No. 24) lies outside the stability zone and near the local minimum at $(180°, 180°)$. Another PCILO map—

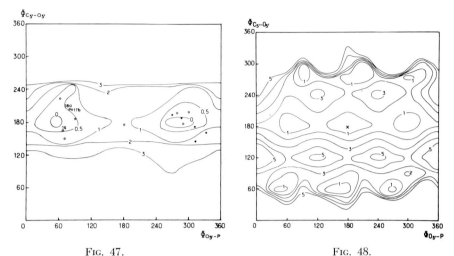

Fig. 47.   Fig. 48.

Fig. 47. PCILO ($\Phi_{O5'-PC5'}-\Phi_{C5'-O5'}$) conformational energy map with $\Phi_{O3'-C3'} = 240°$, $\Phi_{P-O3'} = 180°$ and $\Phi_{C4'-C5'} = 60°$. Isoenergy curves in kcal/mol with the global minimum taken as energy zero. From Pullman et al. (197).

Fig. 48. Empirical computational ($\Phi_{O5'-P}-\Phi_{C5'-O5'}$) conformational energy map of Olson and Flory (209) constructed with $\Phi_{O3'-C3'} = 215°$, $\Phi_{P-O3'} = \Phi_{C4'-C5'} = 180°$. X indicates the global minimum.

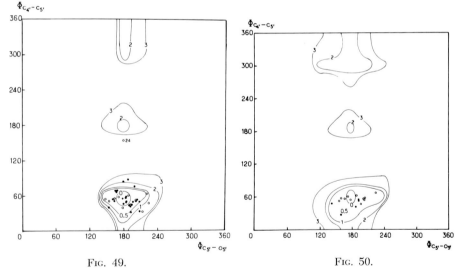

Fig. 49.   Fig. 50.

Fig. 49. PCILO ($\Phi_{C5'-O5'}-\Phi_{C4'-C5'}$) conformational energy map with $\Phi_{O3'-C3'} = 240°$, $\Phi_{P-O3'} = \Phi_{O5'-P} = 300°$. Isoenergy curves in kcal/mol with the global minimum taken as energy zero. From Pullman et al. (197).

Fig. 50. PCILO ($\Phi_{C5'-O5'}-\Phi_{C4'-C5'}$) conformational energy map with $\Phi_{O3'-C3'} = 240°$, $\Phi_{P-O3'} = 180°$ and $\Phi_{O5'-P} = 300°$. Isoenergy curves in kcal/mol with the global minimum taken as energy zero. From Pullman et al. (197).

constructed (197) with $\Phi_{P-O3'}$ changed to 180° while the other two preselected torsion angles remain the same as in Fig. 49—is presented in Fig. 50. The two maps (Figs. 49 and 50) are quite similar and all the representative experimental conformations lie near the global minimum.

The EHT results (122, 215) shown in Fig. 51 predict three equivalent regions of energy minima centered around $(\Phi_{C5'-O5'}, \Phi_{C4'-C5'}) = (180°, 60°), (180°, 180°)$ and $(180°, 300°)$. However, these minima, in particular the one at $(180°, 60°)$, are very narrow and one has to consider high energy contours ($\sim$10 kcal/mol) in order to encircle the experimental points. Contrary to PCILO results and to experimental evidence, the narrowness of the minimum at $(180°, 60°)$ does not suggest this minimum as the most important one. Although the CNDO conformational energy map (Fig. 52) shows results (216) similar to the PCILO ones, the energy levels are much higher. Also, the map shows a deep global minimum located at $(260°, 50°)$ in contrast to a flat one in PCILO computations centered at $(180°, 60°)$. As we have seen earlier, the experimental crystallographic conformations confirm the correctness of PCILO results. A further confirmation comes from NMR studies of poly(U) (216, 234) in aqueous solution, which indicate that these conformational angles in poly(U) are close to $(180°, 60°)$.

The potential energy computations of Lakshminarayanan and Sasisekharan (120) indicate four principal minima, three of which are located at $\Phi_{C5'-O5'} = 180°$ with $\Phi_{C4'-C5'} = 60°, 180°$ and $300°$ and the fourth at $(\Phi_{C5'-O5'}, \Phi_{C4'-C5'}) = (300°, 300°)$. Two secondary minima appear at $(90°, 60°)$ and $(90°, 180°)$ and the global minimum is placed at $(180°, 300°)$, thus quite different from PCILO and CNDO results and further in a region devoid of experimental conformations. Olson and Flory's map (209) for C3'-endo–C3'-endo combination of sugar puckers (Fig. 53) finds that the global minimum occurs around $(\Phi_{C5'-O5'}, \Phi_{C4'-C5'}) = (280°, 180°)$, for which not a single experimental conformation has yet been observed. The authors (209) again attributed this discrepancy to the absence of specific base–base interaction. That this, however, is not the reason is clear from the results of PCILO (197, 212) and CNDO (216) computations carried out on a model in which the bases have not been included. When the C2'-endo–C2'-endo combination for the sugars was utilized by the same authors, the global minimum occurred around $(180°, 60°)$, and only this result is in agreement with the experimental conformations. Among the possible reasons for the failure of Olson and Flory's computations (209) is that these authors have made a particularly unfortunate selection of preferred torsion angles associated individually with the different bonds, and have used them as fixed values of the torsion angles not involved in the particular map under construc-

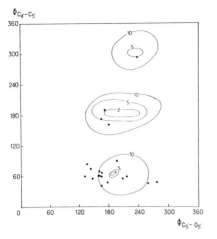
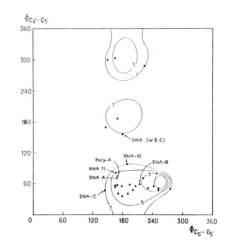

Fig. 51. Fig. 52.

Fig. 51. EHT ($\Phi_{C5'-O5'}-\Phi_{C4'-C5'}$) conformational energy map of Saran and Govil (122) constructed with $\Phi_{O3'-C3'} = 240°$, $\Phi_{P-O3'} = \Phi_{O5'-P} = 300°$.

Fig. 52. CNDO ($\Phi_{C5'-O5'}-\Phi_{C4'-C5'}$) conformational energy map of Tewari et al. (216) constructed with $\Phi_{O3'-C3'} = 240°$, $\Phi_{P-O3'} = \Phi_{O5'-P} = 290°$.

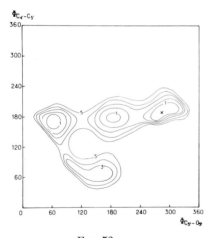
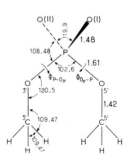

Fig. 53. Fig. 54.

Fig. 53. Empirical computational ($\Phi_{C5'-O5'}-\Phi_{C4'-C5'}$) conformational energy map of Olson and Flory (209) constructed with $\Phi_{O3'-C3'} = 215°$, $\Phi_{P-O3'} = \Phi_{O5'-P} = 180°$. X indicates the global minimum.

Fig. 54. Torsion angles and geometry of dimethyl monophosphate. From Newton (218).

tion. Thus, they use preferentially the values of 180° for $\Phi_{P-O3'}$, $\Phi_{O5'-P}$ and $\Phi_{C4'-C5'}$, while the preferred values of these angles are 300°, 300° and 60°, respectively.

Recently, Clementi and Popkie (219) performed an *ab initio* computation on the conformation of a disugar monophosphate as a function of the torsion angle $\Phi_{C4'-C5'}$. Unfortunately, they used the same relatively improbable preselected values as Olson and Flory (209) for the remaining four torsion angles of the backbone. Moreover, they also attached a hydrogen atom to one of the oxygens of the phosphate group. In these conditions, it is certainly not astonishing that their results are not satisfactory. They indicated a global minimum for $\Phi_{C4'-C5'} = 300°$ followed closely by a local one at $\Phi_{C4'-C5'} = 60°$ (about 0.13 kcal/mol higher) separated by a barrier of 3.26 kcal/mol. As discussed previously, the value of 60° is the most frequently observed crystallographically and is also predicted as the most stable one by PCILO computation. Very recently Lie and Clementi (236) have extended their *ab initio* computations to other torsion angles of the backbone.

Stellman *et al.* (211) utilized classical potential energy functions to determine the conformation and crystal packing scheme for guanylyl (3'-5')cytidine, (G-C). The lowest energy conformation of the isolated molecule has the torsion angles in the range of helical RNAs and the sugar pucker combination is C3'-endo–C3'-endo.

5. Relative Stabilities of Models of Nucleic Acids and Polynucleotides

The energies corresponding to the conformations of the backbone structures associated with the models proposed for different nucleic acids and polynucleotides have been calculated by the EHT (122, 216), CNDO (216) and PCILO (197) methods. Although the physical meaning of such restricted calculations, which omit the influence of bases, is limited, they nevertheless give an indication of the relative stabilities of the different backbones. Since the publication of the first PCILO results (197), a number of refinements in the models proposed for the nucleic acids has been made, and in Table VIII all the subsequent refinements are listed. To check whether these refinements improve the results, PCILO computations have been repeated for all the compounds Nos. 24–34b listed in Table VIII. In these computations the C3'-endo–C3'-endo diribose monophosphate model (214) is used. The results are presented in Table IX, together with the results of earlier PCILO computations (197) and of the EHT (216) and CNDO (216) methods. It is seen that both CNDO (216) and earlier PCILO (197) results predict

TABLE IX
COMPARISON OF COMPUTED ENERGIES FOR THE BACKBONE STRUCTURE OF
PROPOSED MODELS OF NUCLEIC ACIDS AND POLYNUCLEOTIDES

Compound Nos. in Table VIII	Energy (kcal/mol)[a]			
	EHT (216)	CNDO (216)	PCILO (197)	PCILO (present)
24	27.5	4.9	3.6	5.1
25a	159.6	24.7	3.8	2.6
25b	—	—	—	0.9
25c	—	—	—	1.9
25d	—	—	—	2.3
26	1.0	3.5	1.7	2.3
27a	0.0	0.7	0.3	1.1
27b	—	—	—	0.2
27c	—	—	—	2.3
28a	5.1	4.6	2.8	3.3
28b	—	—	—	2.4
28c	—	—	—	2.8
29a	—	—	—	1.3
29b	—	—	—	0.1
29c	—	—	—	0.0
29d	—	—	—	1.0
29e	—	—	—	1.9
30a	—	—	—	0.4
30b	—	—	—	0.0
31a	—	—	—	0.6
31b	—	—	—	0.7
32	—	—	—	0.1
33	3.3	0.0	0.0	1.1
34a	—	—	—	0.7
34b	—	—	—	0.4

[a] Energy of the most stable conformation taken as energy zero in each method.

the same order of relative stabilities for different nucleic acids. The EHT results (122) that have been corrected (216) are much different.

The present PCILO results in addition indicate the relative stability of the different crystallographic refinements for the same compound. For example, of the four refinements for B-DNA-10, corresponding to Nos. 25a–25d, the conformation of compound 25b (222) has the lowest energy. This has been further checked by computing energies for a C2'-endo–C2'-endo di(deoxyribose) monophosphate model. Of the three refinements for A-DNA-11 (Nos. 27a–27c), the conformation of compound 27b (222) has the lowest energy. This, again, has been further checked by carrying out PCILO computations on a C3'-endo–C3'-endo di(deoxyriboso) monophosphate. Further, A-DNA-11 (No. 27b) is more stable than B-DNA-10 (No. 25b). This order of stability is also given by EHT (*216*), CNDO (*216*) and PCILO (*197*) results. RNA-11 (compound No. 29a) is more stable than RNA-10 (compound No. 28a), in all the computations [the results of EHT (*122, 216*), CNDO (*216*) and PCILO (*197*) for RNA-11 are not reported in Table IX, as some of the torsion angles for this compound were incorrectly reported in ref. 25 and were subsequently taken for computing the energies]. Table IX shows that the refinements corresponding to Nos. 29b and 29c for α- and β-A-RNA-11 (reovirus) correspond more closely to the computations for the backbone stability than do the latest ones, 29d and 29e (*299*). Anyway, these refinements look better than those corresponding to the 10-fold RNA (reovirus; Nos. 28b and 28c). For poly(rA·rU) (β-A-RNA-11), the latest refinement (*229, 230*) is still better than the previous one (*26*). Finally, for poly(rI·rC) (A'-RNA-12), both refinements (*26, 229, 230*) lead to nearly similar energies of the backbone. Thus, it can be said that in general, the subsequent refinements in the crystallographic models of nucleic acids improve the resulting conformation from the viewpoint of the energy of the backbone.

B. The Geometry of the Phosphate Group: Key to the Conformation of Polynucleotides?

The preceding survey of theoretical and experimental (in particular X-ray) data indicate that one of the main flexibility axes if not the main one, of the backbone of nucleic acids, polynucleotides and their constituents resides in the torsions around the two P—O ester bonds, $\Phi_{P-O3'}$ and $\Phi_{O5'-P}$. Recent crystallographic studies on a number of dinucleoside monophosphates have confirmed this situation and suggested a series of basic patterns. The crystal structure of U-A has been obtained by two independent groups of authors: Rubin *et al.* (*98*) and Sussman *et al.* (*140*). U-A crystallizes into two independent molecules having different conformations: one of them, referred to as U-A1 by Rubin *et al.* and U-A2 by Sussman *et al.*, has an extended conformation that can be converted into a right-handed helical structure by a rotation around P—O3' bond, while the second molecule, referred to as U-A2 by Rubin *et al.* and inversely as U-A1 by Sussman *et al.*, presents a left-handed turn in its sugar-phosphate backbone. The crystal structure of A(2'-5')U (*78*), which present a 2'-5' linkage instead of a 3'-5' linkage, exhibits only one conformation in the sugar-phosphate backbone consisting of a right-handed helical form. More recently, crystal structures of the dinucleoside monophosphates. G-C (*141*) and A-U (*141, 142*) have been determined. The crystalline asymmetric unit of A-U contains two independent molecules referred to as A-U1 and A-U2. All three molecules (*141, 142*) crystallize in antiparallel right-handed forms.

The existence of these different patterns raises the question of the factors responsible for their occurrence in a given situation and, more basically perhaps, of the intrinsic conformational preferences of the phosphate ester bonds of di- and polynucleotides. The results of theoretical computations on this subject have been discussed in Section VI, A. As indicated there, *ab initio* computations on symmetrical dimethyl monophosphate (*218*) already predict the *gauche-gauche* helical conformation to be the most stable one and give in this case an equivalent importance to the two possibilities: $\Phi_{P-O3'} = \Phi_{O5'-P} = 270°–300°$ (right-handed twist) and $\Phi_{P-O3'} = \Phi_{O5'-P} = 60°–90°$ (left-handed twist). The PCILO computations (*197, 212*) performed for an unsymmetrical disugar monophosphate give a similar overall picture but distinguish between the two helical conformations.

Obviously, dimethyl phosphate already contains the essential information about the conformational abilities of dinucleoside monophosphates. The possibility appears, therefore, that the different experimental conformations observed for the different molecules of this type may be due simply to the perturbations produced by the nucleosides in the geometry of the phosphate group, in particular the induction of an asymmetry in this geometry. To some extent, the nucleosides may be considered in this approach as "substituents" replacing the methyl groups of the dimethyl monophosphate. Such a "substitution" is bound to produce modifications in the geometry of the phosphate group, and this modification will perturb the conformational energy maps of the symmetrical dimethyl monophosphate. The question then arises: What role does this geometrical perturbation play in the determination of the conformation of dinucleoside monophosphates?

In order to answer this question, PCILO conformational energy maps for dimethyl monophosphate (Fig. 54) have been constructed (*213*), using as input data the different geometries of the phosphate group that occur in the known crystal structures of U-A1 (*98*), U-A2 (*98*) and A(2'-5')U (*78*) listed in Table X. The results of the conformational energy maps lead to a striking correlation between the internal structure of the phosphate group and the conformation of the ester P—O bonds of the backbone.

In Fig. 54, the atom O(I) is defined as the first oxygen atom encountered when one goes from the O3'—P bond in a clockwise direction around the O5'—P bond, looking from O5' to P. For simplicity, the values of 90° and 270° of the torsion angles $\Phi_{P-O3'}$ and $\Phi_{O5'-P}$ are designated, respectively, by the letters g^+ and g^-. Combinations of these two letters

TABLE X

VALENCE ANGLES OF THE PHOSPHATE GROUP IN THE CRYSTAL STRUCTURES OF DINUCLEOSIDE MONOPHOSPHATES

Dinucleoside monophosphate	O(I)PO(II)	O3'PO5'	O3'PO(I)	O3'PO(II)	O5'PO(I)	O5'PO(II)	Angle of symmetry[a]	Reference No.
U-A1	118.04	100.33	110.60	108.80	110.70	106.80	109.25	*98*
U-A2	120.99	102.05	110.30	107.50	104.80	109.40	108.05	*98*
A(2'-5')U	117.68	103.00	104.73	112.60	109.79	108.17	108.79	*78*
G-C	120.60	104.30	103.90	110.70	110.80	105.60	107.70	*141*
A-U1	119.80	105.00	102.10	110.60	113.30	105.30	107.78	*141, 142*
A-U2	121.10	103.70	103.00	110.90	112.70	104.30	107.68	*141, 142*
A¹-A² unit of A¹-A²-A³	119.53	105.20	103.49	111.81	111.25	104.84	107.81	*152*
A²-A³ unit of A¹-A²-A³	121.28	103.96	108.18	104.79	106.72	110.59	107.58	*152*

[a] Symmetrical values of O3'PO(I), O3'PO(II), O5'PO(I) and O5'PO(II) valence angles corresponding to the respective values of O(I)PO(II) and O3'PO5'.

designate the values of $\Phi_{P-O3'}$ and $P_{O5'-P}$, the first referring to $\Phi_{P-O3'}$ and the second to $\Phi_{O5'-P}$; e.g., g⁻g⁻ means $\Phi_{P-O3'} = \Phi_{O5'-P} = 270°$ etc.

The results of $(\Phi_{P-O3'}-\Phi_{O5'-P})$ conformational energy maps of dimethyl monophosphate constructed (213) with the different geometries of the free phosphate oxygens, corresponding to the crystal structures of U-A1 (98), U-A2 (98) and A(2'-5')U (78) and presented in Figs. 55–57, show that the global energy minimum occurs in each case near the crystallographic conformations of the aforesaid dinucleoside monophosphates. These results indicate that the geometrical arrangement of the free phosphate oxygens is a determining factor in the selection of the most stable conformations. Calculated for dimethyl monophosphate, the conformations nevertheless correspond closely to those observed in the crystals of dinucleoside monophosphate having the same phosphate geometries.

1. CORRELATION BETWEEN THE PHOSPHATE GROUP VALENCE AND THE STABILITIES OF THE RIGHT- AND LEFT-HANDED HELICAL CONFORMATIONS

The results of the preceding section may be stated as manifestations of a broader correlation between the arrangement of the free phosphate oxygens and the energies of the right-handed ($\Phi_{P-O3'} = \Phi_{O5'-P} = 270°$) and left-handed ($\Phi_{P-O3'} = \Phi_{O5'-P} = 90°$) folded conformations of dinucleoside monophosphates (213). This proposal requires the introduction of the notion of "symmetrical angles." The symmetrical dimethyl monophosphate having the geometry specified in Fig. 54 [i.e., the O3'PO5' and O(I)PO(II) valence angles equal, respectively, to 102.6° and 119.9°] implies that the O3'PO(I) = O3'PO(II) = O5'PO(I) = O5'PO(II) = 108.48°, and this value is termed the "symmetrical angle." The correlation between the phosphate valence angles and energies of the folded forms is then established by computing the energies of the folded forms for different values of O5'PO(I) and O5'PO(II) [or, O3'PO(I) and O3'PO(II)] around the symmetrical value of 108.48°, all other geometrical parameters being kept constant as specified above. It must be realized that the O5'PO(I) and O5'PO(II) valence angles are geometrically correlated when all the other geometrical parameters are maintained fixed: once the O5'PO(I) valence angle is specified, the O5'PO(II) valence angle is determined. The results of such computations (213) presented in Table XI indicate that the right-handed folded conformation is favored over the left-handed one when the value of the O5'PO(I) angle is larger than the symmetrical value of 108.48°, and that the inverse situation holds when this valence angle has a value smaller than the symmetrical value. The correlation between the O3'PO(I) and O3'PO(II) valence

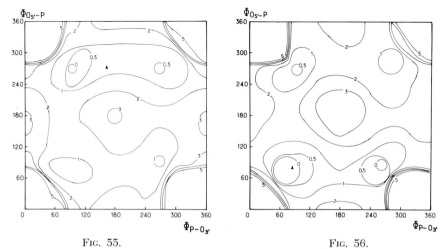

Fig. 55.   Fig. 56.

Fig. 55. PCILO ($\Phi_{P-O3'}-\Phi_{O5'-P}$) conformational energy map of dimethyl monophosphate constructed with the geometry of the phosphate group of U-A1 (98). Isoenergy curves in kcal/mol with the global minimum taken as energy zero. ▲ indicates experimental conformation of U-A1. Perahia et al. (213).

Fig. 56. PCILO ($\Phi_{P-O3'}-\Phi_{O5'-P}$) conformational energy map of dimethyl monophosphate constructed with the geometry of the phosphate group of U-A2 (98). Isoenergy curves in kcal/mol with the global minimum taken as energy zero. ▲, Experimental conformation of U-A2. From Perahia et al. (213).

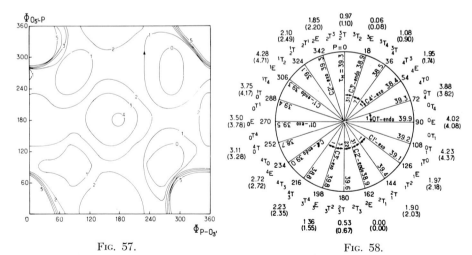

Fig. 57.   Fig. 58.

Fig. 57. PCILO ($\Phi_{P-O3'}-\Phi_{O5'-P}$) conformational energy map of dimethyl monophosphate constructed with the geometry of the phosphate group of A(2'-5')U (78). Isoenergy curves in kcal/mol with the global minimum taken as energy zero. ▲, Experimental conformation of A(2'-5')U. From Perahia et al. (213).

Fig. 58. The conformational wheel in the pseudorotational representation for D-ribose and D-deoxyribose (see text for details). From Saran et al. (37).

TABLE XI

Variation of the PCILO Energies of Dimethyl Phosphate with the O5'PO(I) and O5'PO(II) Valence Angles, the Remaining Valence Angles of the Phosphate Group Being Kept Constant[a,b]

Valence angles		Energies (kcal/mol)[c]		Favored conformations
O5'PO(I)	O5'PO(II)	g^-g^-	g^+g^+	
112.88	104.01	0.17	1.61	g^-g^-
112.00	104.91	0.00	1.14	g^-g^-
111.12	105.80	−0.11	0.75	g^-g^-
110.24	106.70	−0.15	0.43	g^-g^-
109.36	107.59	−0.11	0.18	g^-g^-
(Sym) 108.48	(Sym) 108.48	0.00	0.00	
107.59	109.36	0.18	−0.11	g^+g^+
106.70	110.24	0.43	−0.15	g^+g^+
105.80	111.12	0.75	−0.11	g^+g^+
104.91	112.00	1.14	0.00	g^+g^+
104.01	112.88	1.61	0.17	g^+g^+

[a] From Saran et al. (119).
[b] The constant values considered are O(I)PO(II) = 119.9°, O3'PO5' = 102.6°, O3'PO(I) = O3'PO(II) = 108.48°.
[c] The energy of the symmetrical conformation is taken as energy zero.

angles and the energies of folded conformations is then easily established from Table XI: the O5'PO(I) and O5'PO(II) valence angles, must be substituted, respectively, by O3'PO(II) and O3'PO(I), with the fixed symmetrical value of 108.48° given to O5'PO(I) and O5'PO(II) valence angles. The general rule that follows from these results is that *a right-handed folded conformation (g⁻g⁻) may be expected to be the most stable one of a dinucleoside monophosphate when O5'PO(I) and O3'PO(II) angles are greater than the associated symmetrical angle, and a left-handed folded conformation (g⁺g⁺) to be the most stable one when O5'PO(I) and O3'PO(II) angles are smaller than the associated symmetrical angle.* A more indefinite conformation should occur when one of the two angles O5'PO(I) or O3'PO(II) is greater and the other one is smaller than the associated symmetrical angle. The above correlation, which has been established for a fixed given geometry, holds for any specified geometry.

2. Confirmation of the Rule from Observed Crystallographic Conformations

The predictions of the rule are in excellent agreement with available experimental results on dinucleoside monophosphates listed in Table

X. The rule predicts a g^+g^+ conformation for U-A2, which has O5'PO(I) = 104.8° and O3'PO(II) = 107.5°, both of which are smaller than the symmetrical value (which is in this case 108.05°); a g^-g^- conformation for A(2'-5')U, which has O5'PO(I) = 109.79° and O3'PO(II) = 112.60°, both of which are greater than the symmetrical value (which is in this case 108.79°); a nonfolded conformation for U-A1, whose O5'PO(I) = 110.70° is greater and whose O3'PO(II) = 108.80°, is smaller than the symmetrical value (which is in this case 109.25°). Further confirmation comes from recent studies on G-C (*141*), A-U1 (*141, 142*) and A-U2 (*141, 142*). The rule predicts the g^-g^- conformations for the three compounds: in all of them the O5'PO(I) angles are equal, respectively, to 110.8°, 113.3° and 112.7° and the O3'PO(II) angles are equal, respectively, to 110.7°, 110.6° and 110.9° and are greater than the symmetrical angles equal, respectively, to 107.7°, 107.78° and 107.68°. The three molecules in fact occur in the crystals in a right-handed folded conformation. Very recently, a trinucleoside diphosphate, A¹-A²-A³ has been studied by X-ray crystallography (*152*). The rule predicts a g^-g^- conformation for the A¹-A² unit and a g^+g^+ conformation for the A²-A³ unit (see Table X). The crystallographic conformations for the two units indeed show that the former is g^-g^- ($\Phi_{P-O3'}$ = 283.1° and $\Phi_{O5'-P}$ = 297.4°) and the latter g^+g^+ ($\Phi_{P-O3'}$ = 77.2° and $\Phi_{O5'-P}$ = 92.7°), in excellent agreement with the predictions.

In addition to the above-mentioned data on dinucleoside monophosphates, the analysis of the phosphate group geometry in the crystal structures of diethyl- and dimethylammonium phosphates (*237–243*) confirms further the predictions of the rule (*213, 244*). Table XII, which lists such an analysis, contains the observed crystallographic conformations around the P—O ester bonds and the conformations predicted by the rule. It is observed that the agreement between the experimental conformations and the predicted ones is excellent. In this case, it is the surrounding metals or organic structures that bring about the distortion of the phosphate group geometry.

3. FURTHER *ab initio* COMPUTATIONS ON DIMETHYL MONOPHOSPHATE

Because of the obvious general importance of the conclusions reached in the preceding section, its essential results have been verified by further *ab initio* computations (*213*). These computations were carried out by the STO 3G procedure (*215–218*) employing the Gaussian 70 program (*249*). The high computational time restricted the examination to only a limited number of conformations, mainly the g^-g^- and g^+g^+ ones for the three dimethyl monophosphates whose phosphate geometries correspond to those of U-A1, U-A2 and A(2'-5')U. The results are given

TABLE XII

VALENCE ANGLES OF THE PHOSPHATE GROUP IN THE CRYSTAL STRUCTURES OF DIETHYL PHOSPHATES WITH OBSERVED AND PREDICTED CONFORMATIONS FOR THE $\Phi_{O5'-P}$ AND $\Phi_{P-O3'}$ TORSION ANGLES

Diethyl phosphate complexed with	Crystallographic conformation	O(I)P O(II)	O3'PO5'	O3'PO(I)	O3'PO(II)	O5'PO(I)	O5'PO(II)	Angle of symmetry[a]	Predicted conformation by the present criteria	Reference No.
Barium (+)	$\Phi_{O5'-P} = 68.20$ g^+g^+ $\Phi_{P-O3'} = 71.60$	121.60	103.50	111.90	104.60	103.40	110.50	107.58	g^+g^+	237
Barium (−)	$\Phi_{O5'-P} = 291.80$ g^-g^- $\Phi_{P-O3'} = 288.40$	121.60	103.50	104.60	111.90	110.50	103.40	107.58	g^-g^-	237
Putrescinium	$\Phi_{O5'-P} = 77.10$ g^+g^+ $\Phi_{P-O3'} = 71.80$	119.40	105.90	111.30	104.30	104.30	111.0	107.70	g^+g^+	238
Propylguanidium	$\Phi_{O5'-P} = 297.13$ g^-g^- $\Phi_{P-O3'} = 293.03$	116.80	107.30	103.50	112.90	111.00	105.10	108.09	g^-g^-	239
Arginine	$\Phi_{O5'-P} = 201.65$ g^-g^- $\Phi_{P-O3'} = 294.02$	118.00	105.00	101.00	115.00	113.00	105.00	108.27	g^-g^-	240
Silver	$\Phi_{O5'-P} = 68.00$ g^+g^+ $\Phi_{P-O3'} = 125.00$	117.20	102.40	112.31	105.49	107.30	111.25	109.07	g^+g^+	241
Magnesium	$\Phi_{O5'-P} = 77.60$ g^+g^+ $\Phi_{P-O3'} = 87.40$	117.60	108.20	110.70	105.09	105.00	110.07	107.70	g^+g^+	242
Dimethyl-ammonium phosphate (+)	$\Phi_{O5'-P} = 57.45$ g^+g^+ $\Phi_{P-O3'} = 62.34$	117.53	105.19	111.47	105.36	104.19	112.57	108.76	g^+g^+	243
Dimethyl-ammonium phosphate (−)	$\Phi_{O3'-P} = 302.55$ g^-g^- $\Phi_{P-O3'} = 297.66$	117.53	105.19	105.36	111.47	112.57	104.19	108.76	g^-g^-	243

[a] Symmetrical values of O3'PO(I), O3'PO(II), O5'PO(I) and O5'PO(II) valence angles corresponding to the respective values of O(I)PO(II) and O3'PO5'.

in Table XIII along with the PCILO results (213). It can be observed that the *ab initio* computations agree completely with PCILO results as concerns the prediction of the stable conformation. The relative values of the energies of the stable regions seem to be somewhat exaggerated in the case of A(2′-5′)U and less so in the case of U-A2 by the PCILO results; in the case of U-A1, the results of the two methods are almost identical.

Since the PCILO ($\Phi_{P-O3'}-\Phi_{O5'-P}$) conformational energy maps on dimethyl monophosphates (213) show a relatively stable area in the g^+g^- region and also since the crystallographic conformation of U-A1 is close to this region, *ab initio* computations have been carried out to investigate this region better (213). For this, the symmetrical geometry of dimethyl monophosphate as indicated in Fig. 54 has been chosen. Further, since the two methyl groups are near each other in the g^+g^- conformation, the effect of the rotation of the C3′ methyl group upon the stability of this region has been studied and compared with similar computations for the g^-g^- conformation. The orientation of the C5′ methyl group has been taken as staggered in all cases, while three positions have been considered for the C3′ methyl: staggered, eclipsed, and intermediate. The results, presented in Table XIV, indicate that the differences in energies due to the different orientations of the C3′ methyl group in the g^-g^- conformation are within 0.8 kcal/mol, while in the case of the g^+g^- conformation, they spread up to 2.3 kcal/mol. This situation is due to the fact that the two methyl groups are far apart in the g^-g^- region but close to each other in the g^+g^- region. The lowest energy in the g^+g^- region is now only about 1 kcal/mol above the global one and consists of an eclipsed conformation for the C3′ methyl and a staggered one for the C5′ methyl group. This region, which is 3

TABLE XIII
COMPARISON OF PCILO AND *Ab Initio* ENERGIES FOR DIMETHYL PHOSPHATE IN THE FOLDED REGIONS[a]

Phosphate geometry from	*Ab initio* energies (kcal/mol)[b]		PCILO energies (kcal/mol)[a]	
	g^-g^-	g^+g^+	g^-g^-	g^+g^+
U-A1	0	0.56	0	0.47
U-A2	0	−1.01	0	−1.30
A(2′-5′)U	0	0.60	0	1.04

[a] From Saran et al. (119).
[b] The energy of g^-g^- has been taken equal to energy zero.

TABLE XIV

Ab Initio Energies for g^-g^- and g^+g^- Conformations of Dimethyl Phosphate for Different Conformations of the C3' Methyl Group[a]

Conformation of the C3' methyl group	Ab initio energies (kcal/mol)[a]	
	g^-g^-	g^+g^-
Staggered	0.74	1.98
Eclipsed	0.23	1.02
30° between eclipsed and staggered	0	1.25

[a] From Saran et al. (119).
[b] The energy of the most stable conformation has been taken as energy zero.

kcal/mol above the global minimum in the *ab initio* computations of Newton (218), can in fact go down as low as 1 kcal/mol depending upon the conformation of the methyl group, and does not seem thus to be energetically improbable. In fact, the U-A1 molecule, which has a hydrogen atom attached to the C3' atom very close to an eclipsed position, lies close to this region.

As far as the g^-g^- region is concerned, the preferred conformation corresponds to a methyl group intermediate between the staggered and eclipsed positions (see Table XIV). The majority of dinucleoside monophosphates crystallographically studied have, in fact, the hydrogen atom attached to the C3' atom in a manner similar to this conformation (see ref. *141*, Table 3).

The general rule established above for dimethyl monophosphate has been further tested by carrying out systematic PCILO computations on the conformations of (3'-5')- and (2'-5')-linked diribose as well as di(deoxynucleoside) monophosphates (214) with various sugar puckers. The results indicate the validity of the rule in majority of the cases (214). Extensive experimental studies on dinucleoside monophosphates in solution having different phosphodiester linkages are available in the literature (72, 250–262).

Recently, the solution conformation of m⁶A-U has been analyzed by Altona et al. (263) using PMR technique. They concluded, on the basis of the proton–proton coupling constants, that the torsion angles $\Phi_{C4'-C5'}$, $\Phi_{C5'-O5'}$ and $\Phi_{O3'-O3'}$ in m⁶A-U are, respectively, 60°, 180°, 210°–220°. These results are in excellent agreement with PCILO computations (197, 212).

At this point, mention must be made of a proposal put forward by Kim et al. (264) to distinguish seven basic conformations of nucleic acid structural units corresponding to what these authors consider to be the seven most probable combinations of the $\Phi_{P-O3'}-\Phi_{O5'-P}$ torsion angles: 90°–270°, 170°–270°, 290°–280°, 110°–200°, 250°–160°, 80°–80°, 180°–80°. The selection was arrived at by a somewhat arbitrary combination of theoretical and experimental data with, moreover, an unhappy choice of the theoretical references. Although recognizing that the simplified assumption of an ethane-type torsional barrier for the phosphoester bond, which would result in 9 possible energy minima for $\Phi_{P-O3'}-\Phi_{O5'-P}$ (60°, 180°, 300° each) is inadequate in this case, they do nevertheless keep it as the background of their considerations, influenced obviously as they are by the results of the empirical computation of Olson and Flory (209). Then they eliminate some of them: the ($\Phi_{P-O3'}$, $\Phi_{O5'-P}$) = (180°, 180°) minimum, although it is the global one in Olson and Flory's computations, because it has never been observed crystallographically and because it is a high-energy zone in molecular orbital computation, and the (60°, 300°) minimum because it is a high-energy zone in Olson and Flory's computations! All the remaining combinations are maintained, although somewhat displaced with respect to the standard values of the angles in spite of the fact that some of them do not occur experimentally and do not represent energy minima in the same molecular orbital computations. This arbitrariness makes it difficult to ascertain the significance, if any, of the proposal.

Very recently, a number of X-ray studies have been reported (265–268) on the structure of polynucleotides and their complexes.

VII. Conformation of the Sugar Ring: The Pseudorotational Representation

The conformation of sugar rings in nucleosides and nucleotides as studied by X-ray crystallography has been described in Section III. Recently, the conformation of the sugar ring has been investigated within the pseudorotational concept (28, 36–38). This concept was first introduced in 1947 by Kilpatrick et al. (269) in the description of the conformation of cyclopentane. Their results showed an "indefiniteness" of the cyclopentane conformation, the angle of maximum puckering rotating around the ring without any substantial change in the potential energy. Later, in 1959, Pitzer and Donath (270), refining their earlier calculations, showed that the presence of substituents in cyclopentane induces potential energy barriers that restrict the free pseudorotation. Altona et al. (271–276) have explored the consequences of this limited pseudorotation for the conformation of several five-membered ring systems. Finally, Altona and Sundaralingam (36) have shown the usefulness of the pseudorotational concept in a slightly modified form for the description of the sugar rings in β-nucleosides and nucleotides. In their study, each conformation of the sugar ring is unequivocally determined in terms of two parameters: the phase angle of pseudorotation, P, and the degree of pucker, τ_M. Another useful utilization of the concept of pseudorotation in connection with NMR studies of the conformation of ribonucleosides and arabinonucleosides in solution is due to Hruska et al. (277) and Hall et al. (278). Recently, Altona and Sundaralingam (279) suggested an improved method for interpreting PMR coupling constant data in solution based on the pseudorotational concept and X-ray solid state data.

Both classical and quantum mechanical approaches (28, 37, 38, 280–282) have been employed to study the conformation of sugar rings in β-nucleosides. Sasisekharan (28), working within the scheme of the empirical partitioned potential energy functions, used four parameters to describe the pseudorotation in ribose and deoxyribose and then applied the pseudorotational concept to β-pyrimidine and purine nucleosides. Similarly, Lugovskoi et al. (38), also utilizing classical potential energy functions, chose two parameters to describe the pseudorotation in β-D-ribose and in a somewhat restricted fashion applied this methodology to β-pyrimidine and purine nucleosides (280–281).

In the field of the quantum-mechanical methods, the PCILO procedure has been utilized to investigate the conformation of the sugar ring within the pseudorotational representation (37). Govil and Saran (282) used the EHT and CNDO procedures to describe, with the help of two parameters, the conformations of β-D-ribose. These authors did not call, however, upon the pseudorotational concept but remained within the more common representations of the conformations. We shall, in this section, discuss the various results obtained by different methods.

A. Pseudorotational Parameters, P and τ_M

The five torsion angles (θ_0, θ_1, θ_2, θ_3 and θ_4) of cyclopentane along the pseudorotational pathway are described by

$$\theta_j = \theta_m \cos (P + j\delta) \qquad (54)$$

where $j = 0, 1, 2, 3, 4$ and $\delta = 144°$. For $j = 0$, Eq. 54 becomes:

$$\theta_0 = \theta_m \cos P \qquad (55)$$

It is evident from Eq. 55, that θ_0 passes through the values of θ_m, 0, $-\theta_m$, 0 and θ_m when P goes through a full pseudorotational cycle (0°–

360°), and so a pseudorotation over $P = 180°$ produces the mirror image of the original ring. From Eq. 54 one gets

$$\tan P = \frac{[(\theta_2 + \theta_4) - (\theta_1 + \theta_3)]}{2\theta_0(\sin 36 + \sin 72)} \tag{56}$$

and, knowing the five torsion angles about the sugar ring bonds, one can easily evaluate P from Eq. 56. For negative values of θ_0, 180° is to be added to the calculated value of P from Eq. 56. A standard conformation is chosen for $P = 0°$ which corresponds to the maximum positive value of C1—C2′—C3′—C4′ ($= \tau_2$) torsion angle. The simple relationships between θ's and τ's are: $\theta_0 = \tau_2$, $\theta_1 = \tau_3$, $\theta_2 = \tau_4$, $\theta_3 = \tau_0$, $\theta_4 = \tau_1$ and $\theta_m = \tau_m$. Figure 58 shows the continuously varying P values from 0°–360°. The symmetrical twist (T) and envelope (E) conformations occur, respectively, at even and odd multiples of 18° in P, and they are also indicated in the figure. The C3′-endo and C2′-endo conformations occur, respectively, at P equal to 18° and 162°.

B. Conformational Properties of Ribose and Deoxyribose

The results of the PCILO computations (37) on the conformational energies of the ribose and deoxyribose rings as a function of the pseudorotational parameter P are presented in Fig. 58. The outermost numbers on the conformational wheel represent the values of the energies of the ribose (in kcal/mol) with respect to the global minimum taken as energy zero. The energies of the deoxyribose are given in parentheses. The values of τ_m associated in the PCILO computations (37) with each P are also indicated in Fig. 58 along the corresponding radii of the conformational wheel. The distribution of the observed experimental compounds from X-ray crystal structure studies is shown by numbers placed along the arrows in the populated P values (36, 283).

The results of Fig. 58 show the existence of two nearly equivalent global energy minima occurring at P equal to 18° and 162° and corresponding, respectively, to the C3′-endo and C2′-endo conformations. The experimental data cluster around these selected conformations in very good agreement with the theoretical results. The areas of higher energies are unoccupied with one exception, an O1′-endo conformation found in the crystal structure of dihydrothymidine (284). The two populated conformational zones are in fact separated by energy barriers on both sides, which in the isolated furanose ring are nearly equivalent, of the order of 4 kcal/mol. This order of magnitude seems to be confirmed by experiment (26, 277).

Hruska et al. (277) have shown that the preference for the C3′-endo and C2′-endo conformations arises from the fact that these puckerings involve bond rotations that stagger all substituents on the furanose ring. They have also pointed out that interconversions within the two stable zones centered around the C3'-endo and C2'-endo conformations require a sign change for τ_0 and τ_4 but not for τ_1, τ_2 and τ_3. This situation accounts for the fact that the experimental values of τ_0 and τ_4 found in the crystals of nucleosides and nucleotides (24) display a continuous range of values centered approximately at 0° (ranging from —30° to 30°), while τ_1, τ_2 and τ_3 manifest a forbidden range around 0°. These observations substantiate the validity of the division of the conformational wheel of the sugar into two stable pseudorotational zones separated by two barrier zones.

Sasisekharan (28), using classical potential energy functions, also found that there are two equally stable conformational zones around the C3'-endo and C2'-endo conformations and that the barrier between the two minima along the pseudorotational path is about 2–3 kcal/mol. Lugovskoi and Dashevskii (38) carried out their computations in two distinct approximations. When the electrostatic interactions are completely neglected, their results show two energy minima corresponding to the C3'-endo and C2'-endo conformations, the latter being more restricted than the former. The transition from one stable conformation to the other along the pseudorotational pathway occurs through a barrier of 2.5 kcal/mol. However, when the electrostatic interactions are included, a step that in principle represents a refinement, the minimum at the C2'-endo conformation is strikingly destabilized, hardly representing even a local minimum, about 4 kcal/mol above the C3'-endo conformation, which represents now the only stable arrangement. The transition between the C3'-endo and C2'-endo conformations, which could occur both ways around the pseudorotational pathway, now becomes forbidden along the C2'-exo–C3'-exo road (barriers of about 13 kcal/mol) and remains possible only through the C4'-exo–C1'-exo road with a barrier of 5 kcal/mol. This example shows the difficulties connected with the utilization of empirical potential functions.

The only other quantum-mechanical computation on the conformation of β-D-ribose is that of Govil and Saran (282), which made use of the EHT and CNDO methods but did not involve the pseudorotational concept. Their results (282) indicate also that the C3'-endo and C2'-endo conformations represent the stable arrangements for the sugar ring.

C. Relation between P and χ_{CN}

PCILO conformational energy maps have been constructed (37) as a function of the pseudorotational parameter P and the glycosyl torsion angle χ_{CN} for β-purine and pyrimidine nucleosides.

1. β-Purine Nucleosides

Adenosine and deoxyadenosine have been chosen to represent β-purine nucleosides. The results of the calculations on β-adenosine (37) are shown in Fig. 59, from which it is seen that the global minimum occurs at $P = 162°$ (corresponding to the C2'-endo conformation) for χ_{CN} varying from 70° to 100°. There is a large area included within the 1 kcal/mol isoenergy curve associated with this global minimum. Three local minima about 1 kcal/mol higher than the global minimum occur, two of them associated with the same $P = 162°$ and the third with $P = 18°$, corresponding to the C3'-endo conformation with χ_{CN} varying from 70° to 125°.

Figure 59 contains also the representation of all the presently known experimental results from X-ray crystal studies on β-purine ribonucleosides and nucleotides, and all the representative points are concentrated around or in the vicinity of the calculated global and the lowest local energy minima. These results account very well for the fact that while both *anti* and *syn* conformations are found for the C2'-endo derivatives, only *anti* conformations are observed for the C3'-endo purine nucleosides. Moreover, Fig. 59 also indicates that the possible transition between the C3'-endo and C2'-endo conformations, if occurring, should do so through the C4'-exo–C1'-exo pathway with an energy barrier of about 5 kcal/mol. The other C2'-exo–C3'-exo pathway is practically forbidden.

Figure 60 shows the conformational energy map for β-deoxyadenosine, which is similar to the map of Fig. 59, with the global minimum at $P = 162°$ and $\chi_{CN} = 75°-95°$. The experimental conformations lie near the global and local energy minima. One compound, deoxyadenosine monohydrate (195), which has $P = 198°$ corresponding to the C3'-exo conformation, lies within the 3 kcal/mol isoenergy curves.

The results from classical potential energy functions (28), are, broadly speaking, similar to the PCILO results showing comparable energy minima and energy barriers. However, these calculations (28) put all the energy minima at the same level and thus do not distinguish between them and, in particular, do not pick up the global minimum. From a quantitative viewpoint, the classical potential energy functions procedures are thus less satisfactory.

2. β-Pyrimidine Nucleosides

For β-pyrimidine nucleosides, the PCILO computations (37) were carried out on β-uridine and deoxyuridine. Figure 61 presents the results for β-uridine, which show that a global minimum occurs for $P = 18°$, corresponding to the C3'-endo conformation with χ_{CN} in the *anti*

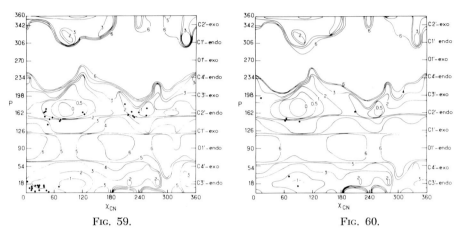

Fig. 59.   Fig. 60.

Fig. 59. PCILO conformational energy map for adenosine nucleoside as a function of P and χ_{CN}. Isoenergy curves in kcal/mol with the global minimum taken as energy zero. ●, X-ray crystallographic data. From Saran et al. (37).

Fig. 60. PCILO conformational energy map for deoxyadenosine nucleoside as a function of P and χ_{CN}. Isoenergy curves in kcal/mol with the global minimum taken as energy zero. ●, X-ray crystallographic data. From Saran et al. (37).

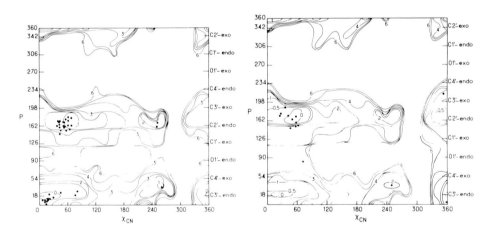

Fig. 61.   Fig. 62.

Fig. 61. PCILO conformational energy map for uridine nucleoside as a function of P and χ_{CN}. Isoenergy curves in kcal/mol with the global minimum taken as energy zero. ●, X-ray crystallographic data. From Saran et al. (37).

Fig. 62. PCILO conformational energy map for deoxyuridine nucleoside as a function of P and χ_{CN}. Isoenergy curves in kcal/mol with the global minimum taken as energy zero. ●, X-ray crystallographic data. From Saran et al. (37).

region, varying from 0° to 45°. There are two local minima about 1 kcal/mol above the global minimum, one associated with the same value of P at $\chi_{CN} = 260°$ and the other one at $P = 162°$, corresponding to the C2'-endo conformation, in the *anti* region, with $\chi_{CN} = 50°-85°$. Besides these, there are other local minima in the *syn* region associated with $P = 18°$ and $162°$ but situated 2-3 kcal/mol above the global minimum.

All the experimental conformations studied by X-ray crystallography cluster around the two fundamental *anti* energy minima at $P = 18°$ and $162°$. Three representative points are found in the *syn* region: 4-thiouridine monohydrate (*31*) with a C3'-endo conformation of the sugar, and two molecules of 6-methyluridine (*156, 157*) with a C2'-endo conformation. They occupy the secondary energy minima predicted for this region. The direction of the easiest transition between the C3'-endo and C2'-endo conformations is via the C4'-exo–C1'-exo pathway through a barrier of 6-kcal/mol, similar to the case of adenosine (see Fig. 59).

The PCILO results for β-deoxyuridine (*37*) are presented in Fig. 62, which shows two equivalent global minima associated with $P = 18°$ and $162°$, and for χ_{CN} in the *anti* region. There is a particularly large area within the 0.5 kcal/mol isoenergy curve around the global minimum at $P = 162°$. The *syn* regions associated with both $P = 18°$ and $162°$ present local energy minima at 2 kcal/mol above the global ones.

The experimentally observed crystallographic conformations cluster exclusively around the large global minimum at $P = 162°$. There is one compound, deoxycytidine hydrochloride (*126*), associated with the global minimum at $P = 18°$. Deoxycytidine 5'-phosphate monohydrate (*105*), which has a C3'-exo sugar pucker, lies close to 2 kcal/mol isoenergy, not far from the global minimum at $P = 162°$. The only known compound with an O1'-endo conformation, dihydrothymidine (*284*), lies in a relatively high-energy region. Finally, one compound, 3',5'-diacetyl-2'-deoxy-2'-fluorouridine (*137*) is found in the *syn* conformation ($\chi_{CN} = 251.7°$) associated with $P \approx 36°$. The barrier for transition from the C3'-endo to the C2'-endo conformation is 5 kcal/mol compared to 6 kcal/mol for β-uridine (Fig. 61), and this transition is possible only through the C4'-exo–C1'-exo pathway.

The computations of Sasisekharan (*28*) fail to distinguish the global energy minima among the three minima obtained, one in the *anti* region for the C3'-endo conformation and two in the *anti* and *syn* regions for the C2'-endo conformation. 4-Thiouridine monohydrate (*31*) lies in Sasisekharan's map (*28*) in a high-energy zone devoid of a local minimum. However, the transition from the C2'-endo to the C3'-endo conformations through the C2'-exo–C3'-exo pathway is forbidden (*28*), as in the PCILO calculations.

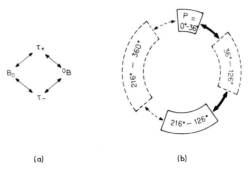

FIG. 63. Classes of states of the pseudorotational itinerary of the sugar ring in nucleosides. (a) Following Hruska et al. (277); (b) proposed by Saran et al. (37).

Altogether, the satisfactory agreement between the PCILO computations and the X-ray crystallographic conformations indicate the general significance of the global and local energy minima derived from the computations. Further, the results point to the usefulness of the pseudorotational concept in its application to the conformation of the furanose ring. It shows simultaneously the continuity of the deformability of the sugar ring along the pseudorotational pathway and the existence of two zones of stable conformations separated by large zones of relatively unstable ones. This result is particularly important for the comprehension of the conformational equilibrium of nucleosides in solution. Hruska et al. (277) deduced, from their above-mentioned abundant studies of vicinal proton–proton couplings in the sugar rings of nucleosides, that the furanose rings exist in dynamic equilibria involving well-defined states corresponding to the two stable conformational zones centered around the C3'-endo and C2'-endo conformations, which they denote by the symbols τ_+ and τ_-, respectively. These two states are separated by two barrier zones, corresponding to the two pathways C1'-exo–C4'-exo (which they denote by the symbol $^\circ B$) and C3'-exo–C2'-exo (which they denote by the symbol B_0). However, these authors gave a symmetrical representation of the overall situation as in Fig. 63a. An unsymmetrical representation, such as that given in Fig. 63b, seems to describe more adequately the real situation in nucleosides.

VIII. Related Subjects

A number of related subjects have been treated by molecular orbital procedures. The dimensions imposed to this chapter do not permit us to describe them in detail. We would like, however, to add here the relevant references to the principal such subjects so that the interested reader may consult them easily.

Detailed PCILO computations have been carried out on the conformation of 2′:3′- and 3′:5′-cyclic nucleotides (*86*). The experimental studies on these compounds have been performed by X-ray crystallography (*285–296*), by ^{13}C and ^{1}H NMR techniques (*45, 82–84, 297–299*) as well as by nuclear Overhauser effect study (*300*). Other theoretical computations on 3′:5′-cyclic nucleotides have appeared in the literature; these involve both other quantum-mechanical methods (*301, 302*) as well as the empirical classical potential energy calculations (*301, 303*). Besides the above, two other experimental studies (*304, 305*) have been performed on cyclic nucleotides. Another important class of biological molecules consists of arabinosyl nucleosides, which show strong antibacterial, antiviral and carcinostatic effects, and the discovery of uracil- and thymine-β-D-arabinofuranosides in the sponges of the species *Cryptotethya crypta* by Bergman and Feeney in 1950 (*306*) has led to many syntheses of arabinosyl nucleosides (*307, 308*). Quite a few X-ray crystallographic studies (*309–322*) as well as solution studies (*323–327*) on arabinosyl- and cyclo arabinosyl nucleosides have been reported in the literature. A complete and detailed PCILO study on these compounds has been recently presented (*328*), and there is an excellent agreement with available X-ray crystallographic data. A further confirmation of the PCILO results has been provided by the very recent X-ray crystallographic studies on 9-β-D-arabinofuranosyl adenine (*329a,b*).

PCILO computations have been carried out on the conformation of ATP (*330*), on puromycin and the terminal aminoacyladenosine moiety of tRNA (*331*), on N^6-substituted adenines with cytokinin activity such as N^6-(Δ^2-isopentenyl)adenine (*332*) and on pyridine coenzymes and their constituents (*333*). Puromycin and the terminal aminoacyladenosine of tRNA have also been studied by the empirical procedure (*334*). The iterative Extended Hückel Method was used for conformational studies on formycin (*335*).

Particular mention must be made of a large series of molecular orbital computations on the ring currents (*336, 337*) and chemical shifts (*338–340*) in the purine and pyrimidine bases of the nucleic acids, on the short- and long-range proton–proton coupling constants in purine and pyrimidine nucleosides (*341, 342*) and on P—C and P—H spin–spin coupling constants in nucleotides (*343*). These computations, particularly those of ring currents, have found applications in the elucidation of conformational problems related to nucleic acids and their components (*82, 344–346*), structure of tRNA (*347–350*), structure of oxidation–reduction coenzymes (*351–355*), actinomycin D binding to DNA and constituents (*356–360*), stereospecificity of the genetic code (*361*), etc.

Finally, it may be worth noting a number of attempts on the compilation of stacking patterns of purine and pyrimidine bases in crystals (*362–365*).

IX. Concluding Remarks

The overall survey of the results presented in this paper leads to a number of general conclusions.

Concerning the different computational techniques applied to the study of the conformational properties of the nucleic acids and their constituents, it is quite obvious that they have met with quite unequal success. The empirical methods, based on the principle of partitioned potential functions, achieve altogether a relatively poor performance and seem even, in a number of important circumstances, to be drastically unsatisfactory. Some of the poor performances, such as the inability to predict correctly the *syn–anti* equilibrium about the glycosyl link or to distinguish between the three principal conformations (*gg*, *gt* and *tg*) about the exocyclic C4′—C5′ bond, may perhaps be at least partially attributed to the use of a truncated or inappropriate models: omission of the possible interactions of the O5′—HO5′ and O2′—HO2′ bonds with the bases in the study of χ_{CN}, or the omission of the base and of the H atom attached to O5′ in the study of the *gg*, *gt* and *tg* equilibrium. Others, however, must have deeper reasons. Such is the case, e.g., with the results of the empirical computations on the preferred values of the torsion angles of the backbone. Not only is the procedure unable to select the preferred minima (e.g., the g^+g^+ and g^-g^- minima on the $\Phi_{P-O3'}-\Phi_{O5'-P}$ conformational energy map) (*208*), but, when refined (?), leads to altogether erroneous predictions (e.g., the preference for a *tt* conformation about the P—O bonds) (*209*). Together with the obvious dependence of the results on the details of the potential functions used, as evidenced, e.g., in the results on the pseudorotational cycle in ribose (*38, 280, 281*), this situation points to the limited applicability of the empirical techniques, as presently used, in this field of research.[7]

[7] In another recent empirical computation of the ($\Phi_{P-O3'}-\Phi_{O5'-P}$) conformational energy map in a dinucleoside monophosphate, Yathindra and Sundaralingam (*366*) placed the global energy minimum at ($\Phi_{P-O3'}$, $\Phi_{O5'-P}$) = (60°, 180°), a still different result, in contradiction with the refined quantum-mechanical results and without connection with experimental data. The same authors (*366*) showed that this energy minimum disappears and a global one appears at $\Phi_{P-O3'} = \Phi_{O5'-P} = 290°$ in a dinucleoside triphosphate, and they consider therefore that it is the terminal phosphates that

A similar comment may also be made about some of the quantum-mechanical procedures, in particular about the simplest of the all-valence electrons methods, the Extended Hückel Theory. Thus this theory is also unable to distinguish, in the examples to which it was applied, the appropriate conformational preferences about say, the C4′—C5′ bond in nucleosides or the P—O bonds of the backbone of dinucleoside monophosphates. Those acquainted with the quantum-mechanical methodologies will not be astonished by this situation, the unsuitability of the Extended Hückel Theory for the study of conformational problems being nowadays illustrated by many striking examples. The CNDO and INDO procedures have been only scarcely used for conformational studies in nucleic acids and their constituents. When utilized with truncated models, as in some of the studies on the conformational preferences about the exocyclic C4′—C5′ bond in nucleosides (32), these methods seem incapable of distinguishing among the three possibilities: gg, gt or tg. This, however, can be due as much to the choice of an inappropriate model as to an intrinsic deficiency of the method. When a better representation of the molecular reality is adopted, as in some studies on conformational preferences of the backbone torsion angles (216, 217), reasonable results are obtained.

On the other hand, the overall results of the PCILO computations seem to be remarkably consistent with experimental evidence. The situation is the more worth stressing as the computations have nearly always been carried out for free molecules while the experimental evidence comes from crystal or solution studies. The general agreement, frequently even in fine details, between the predictions of the PCILO method and the results of X-ray crystallographic investigations clearly indicates that the conformations predicted as the most stable ones for free molecules are generally preserved in the crystals, which means that the crystal packing forces only have a moderate influence in this case, if any, on the intramolecular interactions determining these conformations. Still more strikingly, perhaps, in the few cases in which the theoretical PCILO predictions do not agree completely with the X-ray crystallographic results (as, for example, for the preferred conformations about the C4′—C5′ exocyclic bond in dihydrouridine or α-pseudouridine), they are verified by NMR results in solution, which seems to indicate that in these

---

determine the preference of the g^-g^- conformation. At the same time, however, their map for dinucleoside triphosphates shows a local energy minimum practically degenerated with the global one at $\Phi_{P-O3'} = \Phi_{O5'-P} = 180°$. One could thus similarly consider that the terminal phosphates favor the tt conformation. This result confirms that the empirical procedures seem unable at present to produce a coherent and significant chart of data in this respect.

particular cases the crystal packing forces are responsible for the discrepancy between theory and experiment. A special mention must be made in this respect of the PCILO investigations of the problem of the *syn–anti* equilibrium about the glycosyl linkage. By "freezing" in a fixed selected orientation or, on the contrary, by allowing rotational freedom to some of the appropriate groupings on the molecular skeleton, the theory mimics to some extent the situations likely to occur in different environmental conditions and gives thereby indications of the conformational preferences likely to occur in these different situations.

Finally, it must be remembered that, by producing conformational energy maps, the computations furnish some exceptionally rich information. Thus, X-ray results generally indicate only one (sometimes, but rarely, more) frozen conformation. NMR results, while able to indicate the population of conformers in solution, generally give only an indication of some of the torsion angles studied. The theoretical conformational energy maps, on the other hand, indicate both the preferred and secondary stable conformations and even the energy barriers between them, a quantity not easily obtainable experimentally and nearly unknown for the principal torsion angles in mono- or polynucleosides or nucleotides. The computations give thus a wider picture of the overall conformational possibilities of the compounds studied than most other techniques. Because flexibility may be an intrinsic characteristic of interactions between biological substances, this may be an important advantage for their understanding.

Altogether, the combination of the theoretical and experimental investigations of the conformational properties of the nucleic acids and their constituents leads to a fair understanding of the conformational preferences and possibilities of this type of molecule with respect to the principal torsion angles. Among the particularly significant contributions of the quantum-mechanical computations are their demonstration of the relatively larger (larger than originally thought) coexistence of *syn* and *anti* forms about the torsion angle χ_{CN}, the confirmation of the concept of rigidity of nucleotides as compared to nucleosides with, however, an indication of the limits of this concept, the demonstration of the preponderant role of phosphate geometry in the conformational destiny of dinucleoside monophosphates and probably of polynucleotides in general, and a refinement of the description of the conformational properties of the sugar rings with the help of the pseudorotational concept. The study of the conformational properties of the nucleic acids and their constituents represents thus an outstanding example of the utility of a close collaboration between quantum-mechanical theory and experiment.

Acknowledgments

The authors wish to express their thanks to Dr. Alberte Pullman for her help in the presentation of the quantum-mechanical methods. The contributions of Dr. Hélène Berthod and Mr. David Parahia to the PCILO results described in this paper are also gratefully acknowledged. Thanks are also due to Dr. Girjesh Govil and Dr. Wolfram Saenger for communication of data prior to publication. Finally, the authors wish to express their sincere thanks for permission to reproduce some of the tables and figures in this article to: North-Holland Publishing Co., Amsterdam, for Figs. 19–23, published in *FEBS Letters* **30**, 231, 1973; *ibid.*, **33**, 147, 1973; Elsevier Scientific Publishing Co., Amsterdam, for Tables VII, X–XIV and Figs. 24, 27–30, 32–39, 44–47, 49, 50, 54–60, published in *Biochimica et Biophysica Acta* in 1972–74; H. R. Wilson for Figs. 26 and 31, published in *Journal of Molecular Biology* **56**, 129, 1971; G. Govil for Figs. 41 and 52, published in *Journal of Theoretical Biology* **46**, 229, 1974; The American Chemical Society and M. D. Newton for Fig. 42, published in *Journal of the American Chemical Society* **95**, 256, 1973; John Wiley and Sons, Inc., New York, and P. J. Flory for Figs. 43, 48 and 53 published in *Biopolymers* **11**, 25, 1972 and Springer-Verlag, Berlin, in Figs. 58–63 published in *Theoretica Chimica Acta,* **30**, 31, 1973.

References

1. B. Pullman and A. Pullman, *Advan. Protein Chem.* **16**, 347 (1974).
2. C. C. J. Roothaan, *Rev. Mod. Phys.* **23**, 69 (1951).
3. R. Hoffman, *J. Chem. Phys.* **39**, 1397 (1963).
4. R. Hoffman, *J. Chem. Phys.* **40**, 2745 (1964).
5. R. Pariser and R. G. Parr, *J. Chem. Phys.* **21**, 466 (1953).
6. J. A. Pople, D. P. Santry and G. A. Segal, *J. Chem. Phys.* **43**, S129 (1965).
7. J. A. Pople, D. L. Beveridge and P. A. Dobosh, *J. Chem. Phys.* **47**, 2026 (1967).
8. J. A. Pople and G. A. Segal, *J. Chem. Phys.* **43**, S136 (1965).
9. J. A. Pople and G. A. Segal, *J. Chem. Phys.* **44**, 3289 (1966).
10. S. Epstein, *Phys. Rev.* **28**, 695 (1926).
11. R. K. Nesbet, *Proc. Roy. Soc., Ser. A* **230**, 312, 322 (1955).
12. J. E. Lennard-Jones, *Proc. Roy. Soc., Ser. A* **198**, 1 (1949).
13. S. F. Boys, *Proc. Roy. Soc., Ser A* **201**, 125 (1950).
14. S. F. Boys, *Proc. Roy. Soc., Ser A* **217**, 235 (1953).
15. S. F. Boys, *Rev. Mod. Phys.* **32**, 296 (1960).
16. J. M. Foster and S. F. Boys, *Rev. Mod. Phys.* **32**, 300 (1960).
17. R. K. Nesbet, *Advan. Chem. Phys.* **9**, 321 (1965).
18. S. Diner, J. P. Malrieu and P. Claverie, *Theoret. Chim. Acta* **13**, 1 (1969).
19. S. Diner, J. P. Malrieu, F. Jordan and M. Gilbert, *Theoret. Chim. Acta* **15**, 100 (1969).
20. A. Masson, B. Levy and J. P. Malrieu, *Theoret. Chim. Acta* **18**, 193 (1970).
21. G. Del Re, *JCS*, 4031 (1958).
22. C. Giessner-Prettre and A. Pullman, *Theoret. Chim. Acta* **18**, 14 (1970).
23. In "Conformation of Biological Molecules and Polymers," 5th Jerusalem Symp. Quantum Chem. Biochem. (E. D. Bergmann and B. Pullman, eds.), p. 815. Academic Press, New York, 1973.
24. M. Sundaralingam, *Biopolymers* **7**, 821 (1969).
25. M. Sundaralingam, in "Conformation of Biological Molecules and Polymers,"

5th Jerusalem Symp. Quantum Chem. Biochem. (E. D. Bergmann and B. Pullman, eds.), p. 417. Academic Press, New York, 1973.
26. S. Arnott, *Progr. Biophys. Mol. Biol.* **21**, 265 (1970).
27. A. V. Lakshminarayanan and V. Sasisekharan, *BBA* **204**, 49 (1970).
28. V. Sasisekharan, in "Conformation of Biological Molecules and Polymers," 5th Jerusalem Symp. Quantum Chem. Biochem. (E. D. Bergmann and B. Pullman, eds.), p. 247. Academic Press, New York, 1973.
29. W. K. Olson and P. J. Flory, *Biopolymers* **11**, 1 (1972).
30. J. Donohue and K. N. Trueblood, *JMB* **2**, 363 (1960).
31. W. Saenger and K. H. Scheit, *JMB* **50**, 153 (1970).
32. S. Kang, *JMB* **58**, 297 (1971).
33. M. Sundaralingam, *JACS* **87**, 599 (1965).
34. M. Sundaralingam and L. H. Jensen, *JMB* **13**, 930 (1965).
35. M. Sundaralingam, in "The Purines: Theory and Experiment," 4th Jerusalem Symp. Quantum Chem. Biochem. (E. D. Bergmann and B. Pullman, eds.), p. 44. Academic Press, New York, 1972.
36. C. Altona and M. Sundaralingam, *JACS* **94**, 8205 (1972).
37. A. Saran, D. Perahia and B. Pullman, *Theoret. Chim. Acta* **30**, 31 (1973).
38. A. A. Lugovskoi and V. G. Dashevskii, *Mol. Biol., USSR (Engl. Ed.)* **6**, 354 (1972).
39. A. Saran, B. Pullman and D. Perahia, *BBA* **287**, 211 (1972).
40. R. E. Schirmer, J. P. Davis, J. H. Noggle and P. A. Hart, *JACS* **94**, 2561 (1972).
41. J. P. Davis and P. A. Hart, *Tetrahedron* **28**, 2883 (1972).
42. P. A. Hart and J. P. Davis, in "Conformation of Biological Molecules and Polymers," 5th Jerusalem Symp. Quantum Chem. Biochem. (E. D. Bergmann and B. Pullman, eds.), p. 297. Academic Press, New York, 1973.
43. E. Shefter and K. N. Trueblood, *Acta Crystallogr.* **18**, 1067 (1965).
44. W. Saenger, *Proc. 1st Eur. Biophys. Congr, Vienna, p.* 289 (1971).
45. M. P. Schweizer and R. K. Robins, in "Conformation of Biological Molecules and Polymers," 5th Jerusalem Symp. Quantum Chem. Biochem. (E. D. Bergmann and B. Pullman, eds.), p. 329. Academic Press, New York, 1973.
46. F. E. Hruska, in "Conformation of Biological Molecules and Polymers," 5th Jerusalem Symp. Quantum Chem. Biochem. (E. D. Bergmann and B. Pullman, eds.), p. 345. Academic Press, New York, 1973.
47. I. C. P. Smith, B. J. Blackburn and T. Yamane, *Can. J. Chem.* **47**, 513 (1969).
48. F. E. Hruska, A. A. Grey and I. C. P. Smith, *JACS* **92**, 4088 (1970).
49. R. Deslauriers, R. D. Lapper and I. C. P. Smith, *Can. J. Biochem.* **49**, 1279 (1971).
50. F. E. Hruska, K. K. Ogilvie, A. A. Smith and H. Wayborn, *Can. J. Chem.* **49**, 2449 (1971).
51. A. A. Grey, I. C. P. Smith and F. E. Hruska, *JACS* **93**, 1765 (1971).
52. F. E. Hruska, *Can. J. Chem.* **49**, 2111 (1971).
53. T. Schleich, B. J. Blackburn, R. D. Lapper and I. C. P. Smith, *Bchem.* **11**, 137 (1972).
54. M. Ikehara, S. Uesugi and K. Yoshida, *Bchem* **11**, 830 (1972).
55. P. O. P. Ts'o, in "Fine Structures of Proteins and Nucleic Acids" (G. D. Fasman and J. N. Timascheff, eds.), p. 49. Dekker, New York, 1970.
56. D. W. Miles, S. J. Hahn, R. K. Robins, M. J. Robins and H. Eyring, *J. Phys. Chem.* **72**, 1483 (1968).

57. D. W. Miles, L. B. Townsend, M. J. Robins, R. K. Robins, W. H. Inskeep and H. Eyring, JACS 93, 1600 (1971).
58. Y. Courtois, P. Fromageot and W. Guschlbauer, EJB 6, 493 (1968).
59. P. A. Hart and J. P. Davis, BBRC 34, 733 (1969).
60. R. E. Schirmer, J. H. Noggle, J. P. Davis and P. A. Hart, JACS 92, 3266 (1970).
61. P. A. Hart and J. P. Davis JACS 93, 753 (1971).
62. J. P. Davis, Tetrahedron 28, 1155 (1972).
63. P. A. Hart and J. P. Davis, JACS 94, 2572 (1972).
64. H. Weiler-Feilchenfeld, G. Zvilichovsky, E. D. Bergmann, H. Berthod, and B. Pullman, in "Conformation of Biological Molecules and Polymer," 5th Jerusalem Symp. Quantum Chem. Biochem. (E. D. Bergmann and B. Pullman, eds.), p. 331. Academic Press, New York, 1973.
65. S. T. Rao and M. Sundaralingam, JACS 92, 4963, (1970).
66. H. Berthod and B. Pullman, BBA 232, 595 (1971).
67. A. D. Broom, M. P. Schweizer and P. O. P. Ts'o, JACS 89, 3612 (1967).
68. D. W. Miles, M. J. Robins, R. K. Robins, M. W. Winkley and H. Eyring, JACS 91, 831 (1969).
69. J. Pitha, Bchem 9, 3678 (1970).
70. S. I. Chan, in "The Purines: Theory and Experiment," 4th Jerusalem Symp. Quantum Chem. Biochem. (E. D. Bergmann and B. Pullman, eds.), p. 46. Academic Press, New York, 1972.
71. P. F. Torrence, J. A. Waters and B. Witkop, JACS 94, 3638 (1972).
72. N. S. Kondo, K. M. Fang, P. S. Miller and P. O. P. Ts'o, Bchem 11, 1991 (1972).
73. H. Berthod and B. Pullman, BBA 246, 359 (1971).
74. H. Berthod and B. Pullman, in "The Purines: Theory and Experiment," 4th Jerusalem Symp. Quantum Chem. Biochem. (E. D. Bergmann and B. Pullman, eds.), p. 30. Academic Press, New York, 1972.
75. B. Pullman and H. Berthod, in "Conformation of Biological Molecules and Polymers," 5th Jerusalem Symp. Quantum Chem. Biochem. (E. D. Bergmann and B. Pullman, eds.), p. 209. Academic Press, New York, 1973.
76. A. E. V. Haschemeyer and H. M. Sobell, Acta Crystallogr. 19, 125 (1965).
77. U. Thewalt, C. E. Bugg and R. E. Marsh, Acta Crystallogr., Sect. B 26, 1089 (1970).
78. E. Shefter, M. Barlow, R. A. Sparks and K. N. Trueblood, Acta Crystallogr. Sect. B 25, 895 (1969).
79. J. Iball, C. H. Morgan and H. R. Wilson Proc. Roy. Soc., Ser. A 295, 320 (1966).
80. M. Sundaralingam and L. H. Jensen, JMB 13, 914 (1965).
81. A. E. V. Haschemeyer and H. M. Sobell, Acta Crystallogr. 18, 525 (1965).
82. I. C. P. Smith, H. H. Mantsch, R. D. Lapper, R. Deslauriers and T. Schleich, in "Conformation of Biological Molecules and Polymers," 5th Jerusalem Symp. Quantum Chem. Biochem. (E. D. Bergmann and B. Pullman, eds.), p. 381. Academic Press, New York.
83. B. J. Blackburn, R. D. Lapper and I. C. P. Smith, JACS 95, 2873 (1973).
84. R. D. Lapper, H. H. Mantsch and I. C. P. Smith, JACS 95, 2878 (1973).
85. W. A. Klee and S. H. Mudd, Bchem 6, 988 (1967).
86. A. Saran, H. Berthod and B. Pullman, BBA 331, 154 (1973).
87. W. Guschlbauer and Y. Courtois, FEBS Lett. 1, 183 (1968).

88. A. E. V. Haschemeyer and A. Rich, *JMB* **27**, 369 (1967).
89. I. Tinoco Jr., R. C. Davis and S. R. Jaskunas, in "Molecular Association in Biology" (B. Pullman, ed.), p. 77. Academic Press, New York, 1968.
90. A. V. Lakshminarayanan and V. Sasisekharan, *Biopolymers* **8**, 475 (1969).
91. H. R. Wilson, A. Rahman and P. Tollin, *JMB* **46**, 585 (1969).
92. H. R. Wilson and A. Rahman, *JMB* **56**, 129 (1971).
93. F. Jordan and B. Pullman, Theoret. Chim. Acta **9**, 242 (1968).
94. S. Kang, *J. Mol. Struct.* **17**, 127 (1973).
95. F. Jordan, *Biopolymers* **12**, 243 (1973).
96. F. Jordan, *J. Theoret. Biol.* **41**, 375 (1973).
97. R. K. Nanda, R. Tewari, G. Govil and I. C. P. Smith, *Can. J. Chem.* **52**, 371 (1974).
98. J. Rubin, T. Brennan and M. Sundaralingam, *Bchem.* **11**, 3112 (1972).
99. W. Saenger and D. Suck, *Nature (London)* **242**, 610 (1973).
100. D. W. Young, P. Tollin and H. R. Wilson, *Nature (London)* **248**, 513 (1974).
101. H. Berthod and B. Pullman, *FEBS Lett.* **30**, 231 (1973).
102. H. Berthod and B. Pullman, *FEBS Lett.* **33**, 147 (1973).
103. S. T. Rao and M. Sundaralingam, *JACS* **91**, 1210 (1969).
104. J. Kraut and L. H. Jensen, *Acta Crystallogr.* **16**, 79 (1963).
105. M. A. Vishwamitra, B. S. Reddy, G.H-Y. Lin and M. Sundaralingam, *JACS* **93**, 4565 (1971).
106. W. Murayama, N. Nagashima and Y. Shimuzu, *Acta Crystallogr., Sect. B* **25**, 2236 (1969).
107. M. P. Schweizer, A. D. Broom, P. O. P. Ts'o and D. P. Hollis, *JACS* **90**, 1042 (1968).
106. C. D. Barry, A. C. T. North, J. A. Glasel, R. J. P. Williams and A. V. Xavier, *Nature (London)* **232**, 236 (1971).
109. C. D. Barry, J. A. Glasel, A. C. T. North, R. J. P. Williams and A. V. Xavier, *BBRC* **47**, 166 (1972).
110. T-D. Son, W. Guschlbauer and M. Guéron, *JACS* **94**, 7903 (1972).
111. W. Guschlbauer, I. Fric and A. Holy, *EJB* **31**, 1 (1972).
112. T-D. Son and C. Chachaty, *BBA* **335**, 1 (1973).
113. N. Yathindra and M. Sundaralingam, *Biopolymers* **12**, 297 (1973).
114. N. Yathindra and M. Sundaralingam, *Biopolymers* **12**, 2075 (1973).
115. N. Yathindra and M. Sundaralingam, *Biopolymers* **12**, 2261 (1973).
116. W. K. Olson, *Biopolymers* **12**, 1787 (1973).
117. D. Vasilescu, J. N. Lespinasse, F. Camous and R. Cornillon, *FEBS Lett.* **27**, 335 (1972).
118. N. Nagashima, K. Wakabayashi, T. Matzuzaki and Y. Iitaka, *Acta Crystallogr., Sect. B* **30**, 320 (1974).
119. A. Saran, B. Pullman and D. Perahia, *BBA* **299**, 497 (1973).
120. A. V. Lakshminarayanan and V. Sasisekharan, *Biopolymers* **8**, 489 (1969).
121. G. Govil and A. Saran, *J. Theoret. Biol.* **30**, 621 (1971).
122. A. Saran and G. Govil, *J. Theoret. Biol.* **33**, 407 (1971).
123. M. Sundaralingam, S. T. Rao and J. Abola, *Science* **172**, 725 (1971).
124. M. Sundaralingam, S. T. Rao and J. Abola, *JACS* **93**, 7055 (1971).
125. F. E. Hruska, A. A. Smith and J. G. Dalton, *JACS* **93**, 4334 (1971); and references quoted therein.
126. E. Subramanian and D. J. Hunt, *Acta Crystallogr., Sect. B* **26**, 303 (1970).
127. S. Furberg, C. S. Petersen and C. Romming, *Acta Crystallogr.* **18**, 313 (1965).

128. D. J. Hunt and E. Subramanian, *Acta Crystallogr., Sect B* **25**, 2144 (1969).
129. W. Saenger and D. Suck, *Acta Crystallogr., Sect. B* **27**, 1178 (1971).
130. G. H-Y, Lin and M. Sundaralingam, *Acta Crystallogr., Sect. E* **27**, 961 (1971).
131. A. Rahman and H. R. Wilson, *Acta Crystallogr., Sect. B* **26**, 1765 (1970).
132. G. H-Y. Lin, M. Sundaralingam and S. K. Arora, *JACS* **93**, 1235 (1971).
133. E. A. Green, R. Shiono, R. D. Rosentein and D. J. Abraham, *JCS, Ser. D*, 53 (1971); also quoted from Sundaralingam (25).
134. E. Shefter and T. I. Kalman, *BBRC* **32**, 878 (1968); also quoted from Sundaralingam (25).
135. C. H. Schwalbe, W. Saenger and J. Gassman, *BBRC* **44**, 56 (1971).
136. C. H. Schwalbe and W. Saenger, *JMB* **75**, 129 (1973).
137. D. Suck, W. Saenger, P. Main, G. Germain and J-P. Declercq, *BBA* **361**, 257 (1974); also private communication.
138. P. Singh and D. J. Hodgson, *JACS* **96**, 1239 (1974).
139. K. N. Trueblood, P. Horn and V. Luzzati, *Acta Crystallogr.* **14**, 965 (1961).
140. J. L. Sussman, N. C. Seeman, S. H. Kim and H. M. Berman, *JMB* **66**, 403 (1972).
141. R. O. Day, N. C. Seeman, J. M. Rosenberg and A. Rich, *PNAS* **70**, 849 (1973).
142. J. M. Rosenberg, N. C. Seeman, J. J. P. Kim, F. L. Suddath, H. B. Nicholas and A. Rich, *Nature (London)* **243**, 150 (1973).
143. M. Sundaralingam and S. K. Arora, *PNAS* **64**, 1021 (1969).
144. M. Sundaralingam and S. K. Arora, *JMB* **71**, 49 (1972).
145. S. Hecht and M. Sundaralingam, *JACS* **94**, 4314 (1972).
146. T. F. Lai and R. E. Marsh, *Acta Crystallogr., Sect. B* **28**, 1982 (1972).
147. A. R. I. Munns and P. Tollin, *Acta Crystallogr., Sect. B* **26**, 1101 (1970).
148. P. G. Lenhert, *Proc. Roy. Soc., Ser. A* **303**, 45 (1968).
149. S. C. Jain and H. M. Sobell, *JMB* **68**, 1 (1972).
150. M. Sundaralingam, *Acta Crystallogr.* **21**, 495 (1966).
151. O. Kennard, N. W. Isaacs, W. D. S. Motherwell, J. C. Coppola, D. L. Wampler, A. C. Larson and D. G. Watson, *Proc. Roy. Soc., Ser. A* **325**, 401 (1971).
152. D. Suck, P. C. Manor, G. Germain, C. H. Schwalbe, G. Weimann and W. Saenger, *Nature NB* **246**, 161 (1973); also private communication.
153. J. Iball, C. H. Morgan and H. R. Wilson, *Proc. Roy. Soc., Ser. A* **302**, 225 (1968).
154. D. R. Harris and W. M. McIntyre, *Biophys. J.* **4**, 203 (1964).
155. S. W. Hawkinson and C. L. Coulter, *Acta Crystallogr., Sect. B* **27**, 34 (1971).
156. D. Suck, W. Saenger and H. Vorbrüggen, *Nature (London)* **235**, 333 (1972).
157. D. Suck and W. Saenger, *JACS* **94**, 6520 (1972).
158. A. Rahman and H. R. Wilson, *Nature (London)* **232**, 333 (1971).
159. A. Rahman and H. R. Wilson, *Acta Crystallogr., Sect. B* **28**, 2260 (1972).
160. N. Camerman and J. Trotter, *Acta Crystallogr.* **18**, 203 (1965).
161. W. Saenger, D. Suck and K. H. Scheit, *FEBS Lett.* **5**, 262 (1969).
162. W. Saenger and D. Suck, *Acta Crystallogr., Sect. B* **27**, 2105 (1971).
163. D. Suck, W. Saenger and K. Zechmeister, *Acta Crystallogr., Sect. B* **28**, 596 (1972).
164. D. Suck, W. Saenger and J. Hobbs, *BBA* **259**, 157 (1972).
165. C. H. Schwalbe, H. G. Gassen and W. Saenger, *Nature NB* **238**, 171 (1972).
166. C. H. Schwalbe and W. Saenger, *Acta Crystallogr., Sect. B* **29**, 61 (1973).
167. D. W. Young and E. M. Morris, *Acta Crystallogr., Sect. B* **29**, 1259 (1973).
168. U. Thewalt and C. E. Bugg, *Acta Crystallogr., Sect. B* **29**, 1393 (1973).

169. D. Suck, W. Saenger and W. Rhode, *BBA* **361**, 1 (1974); also private communication.
170. C. E. Bugg and R. E. Marsh, *JMB* **25**, 67 (1967).
171. G. Kartha, G. Ambady and M. A. Vishwamitra, *Science* **179**, 495 (1973).
172. M. A. Vishwamitra, B. S. Reddy, M. N. G. James and G. J. B. Williams, *Acta Crystallogr., Sect B* **28**, 1108 (1972).
173. F. Eckstein, W. Saenger and D. Suck, *BBRC* **46**, 964 (1972).
174. W. Saenger, D. Suck and F. Eckstein, *EJB* **46**, 559 (1974); also private communication.
175. N. Camerman, J. K. Fawcett and A. Camerman, *Science* **182**, 1142 (1973).
176. E. Subramanian, J. J. Madden and C. E. Bugg, *BBRC* **50**, 691 (1973).
177. U. Thewalt and C. E. Bugg, *JACS* **94**, 8892 (1972).
178. T. Brennan, C. Weeks, E. Shefter, S. T. Rao and M. Sundaralingam, *JACS* **94**, 8548 (1972).
179. K. Shikata, T. Ueki and T. Mitsui, *Acta Crystallogr., Sect. B* **29**, 31 (1973).
180. S. S. Tavale and H. M. Sobell, *JMB* **48**, 109 (1970).
181. C. E. Bugg and U. Thewalt, *BBRC* **37**, 623 (1969).
182. E. Shefter, *J. Pharm. Sci.* **57**, 1157 (1968).
183. G. Koyama, K. Maeda, H. Umezawa and Y. Iitaka, *Tetrahedron Lett.* **6**, 597 (1966).
184. G. Koyama, H. Umezawa and Y. Iitaka, *Acta Crystallogr., Sect. B* **30**, 1511 (1974).
185. P. Prusiner, T. Brennan and M. Sundaralingam, *Bchem* **12**, 1196 (1973).
186. M. Stroud, *Acta Crystallogr., Sect. B* **29**, 690 (1973).
187. J. Abola and M. Sundaralingam, *Acta Crystallogr., Sect. B* **29**, 697 (1973).
188. W. Saenger, *FEBS Lett.* **10**, 81 (1970).
189. W. Saenger, *JACS* **93**, 3035 (1971).
190. D. Voet and A. Rich, *PNAS* **68**, 1151 (1968).
191. N. Nagashima and Y. Iitaka, *Acta Crystallogr., Sect. B* **24**, 1936 (1968).
192. F. Muller (1971), as quoted in Sundaralingam (25).
193. B. S. Reddy and M. A. Vishwamitra, *Cryst. Struct. Commun.* **2**, 9 (1973).
194. D. W. Young, P. Tollin and H. R. Wilson, *Acta Crystallogr., Sect. B* **25**, 1423 (1969).
195. D. G. Watson, D. J. Sutor and P. Tollin, *Acta Crystallogr.* **19**, 111 (1965).
196. M. Remin and D. Shugar, *BBRC* **48**, 636 (1972).
197. B. Pullman, D. Perahia and A. Saran, *BBA* **269**, 1 (1972).
198. D. C. Hodgkin, J. Lindsey, R. A. Sparks, K. N. Trueblood and J. B. White, *Proc. Roy. Soc., Ser. A* **266**, 494 (1962).
199. C. Brink-Shoemaker, D. W. J. Cruickshank, D. C. Hodgkin, M. J. Camper and D. Pilling, *Proc. Roy. Soc., Ser. A* **278**, 1 (1964).
200. D. C. Rohrer and M. Sundaralingam, *JACS* **92**, 4950 (1970).
201. D. C. Rohrer and M. Sundaralingam, *JACS* **92**, 4956 (1970).
202. E. Shefter, M. P. Kotick and T. J. Bardos, *J. Pharm. Sci.* **5**, 1293 (1967).
203. G. W. Frank, quoted from Sundaralingam (205).
204. S. W. Hawkinson, C. L. Coulter and M. L. Greaves, *Proc. Roy. Soc., Ser. A* **318**, 143 (1970).
205. M. Sundaralingam, *JACS* **93**, 6644 (1971).
206. F. E. Hruska, A. A. Grey and I. C. P. Smith, *JACS* **92**, 214 (1970).
207. D. J. Wood, R. J. Mynott, F. E. Hruska and R. H. Sarma, *FEBS Lett.* **34**, 323 (1973); also private communication.
208. V. Sasisekharan and A. V. Lakshminarayanan, *Biopolymers* **8**, 505 (1969).

209. W. K. Olson and P. J. Flory, *Biopolymers* 8, 25 (1972).
210. W. K. Olson and P. J. Flory, *Biopolymers* 8, 57 (1972).
211. S. D. Stellman, B. Hingerty, S. B. Broyde, E. Subramanian, T. Sato and R. Langridge, *Biopolymers* 12, 2731 (1973).
212. D. Perahia, A. Saran and B. Pullman, *in* "Conformation of Biological Molecules and Polymers," 5th Jerusalem Symp. Quantum Chem. Biochem. (E. D. Bergmann and B. Pullman, eds.), p. 225. Academic Press, New York, 1973.
213. D. Perahia, B. Pullman and A. Saran, *BBA* 340, 299 (1974).
214. D. Perahia, B. Pullman and A. Saran, *BBA* 353, 16 (1974).
215. G. Govil, *in* "Conformation of Biological Molecules and Polymers," 5th Jerusalem Symp. Quantum Biochem. (E. D. Bergmann and B. Pullman, eds.), p. 283. Academic Press, New York, 1973.
216. R. Tewari, R. K. Nanda and G. Govil, *J. Theoret. Biol.* 46, 229 (1974).
217. R. K. Nanda, R. Tewari and G. Govil, *Bull. Nat. Inst. Sci., India* 40, 226 (1974).
218. M. D. Newton, *JACS* 95, 256 (1973).
219. E. Clementi and H. Popkie, *Chem. Phys. Lett.* 20, 1 (1973).
220. F. H. C. Crick and J. D. Watson, *Proc. Roy. Soc., Ser. A* 223, 80 (1954).
221. R. Langridge, D. A. Marvin, W. E. Seeds, H. R. Wilson, C. W. Hooper, M. H. F. Wilkins and L. D. Hamilton, *JMB* 2, 38 (1960).
222. S. Arnott, S. D. Dover and A. J. Wonacott, *Acta Crystallogr., Sect. B* 25, 2192 (1969).
223. S. Arnott and D. W. L. Hukins, *BBRC* 47, 1504 (1972).
224. S. Arnott and D. W. L. Hukins, *JMB* 81, 93 (1973).
225. D. A. Marvin, M. Spencer, M. H. F. Wilkins and L. D. Hamilton, *JMB* 3, 547 (1961).
226. W. Fuller, M. H. F. Wilkins, H. R. Wilson and L. D. Hamilton, *JMB* 12, 60 (1965).
227. S. Arnott, M. H. F. Wilkins, W. Fuller and R. Langridge, *JMB* 27, 525 (1967).
228. S. Arnott, M. H. F. Wilkins, W. Fuller and R. Langridge, *JMB* 27, 535 (1967).
229. S. Arnott and D. W. L. Hukins, *JMB* 81, 107 (1973).
230. S. Arnott, D. W. L. Hukins and S. D. Dover, *BBRC* 48, 1392 (1972).
231. E. J. O'Brien and A. W. MacEwan, *JMB* 48, 243 (1970).
232. A. Rich, D. R. Davies, F. H. C. Crick and J. D. Watson, *JMB* 3, 71 (1961).
233. S. K. Mazumdar, W. Saenger and K. H. Scheit, *JMB* 85, 213 (1974).
234. G. Govil and I. C. P. Smith, *Biopolymers* 12, 2589 (1973).
235. G. P. Krieshman and S. I. Chan, *Biopolymers*, 10, 159 (1971).
236. G. C. Lie and E. Clementi, *J. Chem. Phys.* 60, 3005 (1974).
237. Y. Kyogoku and Y. Iitaka, *Acta Crystallogr.* 21, 49 (1966).
238. S. Furberg and J. Solbakk, *Acta Chem. Scand.* 26, 2855 (1972).
239. S. Furberg and J. Solbakk, *Acta Chem. Scand.* 26, 3699 (1972).
240. S. Furberg and J. Solbakk, *Acta Chem. Scand.* 27, 1226 (1973).
241. J. P. Hazel and R. L. Collin, *Acta Crystallogr. B* 28, 2951 (1972).
242. F. S. Ezra and R. L. Collin, *Acta Crystallogr. B* 29, 1398 (1973).
243. L. Giarda, F. Garbassi and M. Calcaterra, *Acta Crystallogr., Sect. B* 29, 1826 (1973).
244. D. Perahia, B. Pullman and A. Saran, *C.R. Acad. Sci., Ser. D* 277, 2257 (1973).
245. W. J. Hehre, R. F. Stewart and J. A. Pople, *J. Chem. Phys.* 51, 2657 (1969).
246. L. Radom and J. A. Pople, *JACS* 92, 4786 (1970).
247. L. Radom and J. A. Pople, *JACS* 93, 289 (1971).
248. W. J. Hehre and J. A. Pople, *JACS* 92, 2191 (1970).

249. W. J. Hehre, W. A. Lathan, R. Ditchfield, M. D. Newton and J. A. Pople, Quantum Chemistry Program Exchange, Program No. 236, Univ. of Indiana, Bloomington, Indiana.
250. M. Tsuboi, S. Takashi, Y. Kyogoku, H. Hayatsu, T. Ukita and M. Kainosho, Science 166, 1504 (1969).
251. P. O. P. Ts'o, N. S. Kondo, M. P. Schweizer and D. P. Hollis, Bchem 8, 997 (1967).
252. N. S. Kondo, H. M. Holmes, L. M. Stempel and P. O. P. Ts'o Bchem. 9, 3479 (1970).
253. S. I. Chan and J. H. Nelson, JACS 91, 168 (1969).
254. B. W. Bangerter and S. I. Chan, JACS 91, 3910 (1969).
255. M. M. Warshaw and I. Tinoco, Jr., JMB 13, 54 (1965).
256. D. Glaubiger, D. A. Lloyd and I. Tinoco, Jr., Biopolymers 6, 409 (1968).
257. M. M. Warshaw and C. R. Cantor, Biopolymers 9, 1079 (1970).
258. W. C. Johnson, M. S. Itzkowitz and I. Tinoco, Jr., Biopolymers 11, 225 (1972).
259. A. J. Alder, L. Grossman and G. D. Fasman, Bchem 7, 3836 (1968).
260. J. Brahms, J. C. Maurizot and A. M. Michelson, JMB 25, 481 (1967).
261. K. N. Fang, N. S. Kondo, P. S. Miller and P. O. P. Ts'o, JACS 93, 6647 (1971).
262. P. S. Miller, K. N. Fang, N. S. Kondo and P. O. P. Ts'o, JACS 93, 6657 (1971).
263. C. Altona, J. H. Van Boom, J. de Jager, H. J. Koeners and G. Van Binst, Nature (London) 247, 558 (1974).
264. S. H. Kim, H. M. Berman, N. C. Seeman and M. D. Newton, Acta Crystallogr., Sect. B 29, 703 (1973).
265. S. Arnott and P. J. Bond, Nature NB 244, 99 (1973).
266. S. Arnott, R. Chandrasekharan and C. M. Marttila, BJ 141, 537 (1974).
267. S. Arnott and E. Selsing, JMB 88, 509 (1974).
268. S. Arnott, R. Chandrasekharan, D. W. L. Hukins, P. J. C. Smith and L. Watts, JMB 88, 523 (1974).
269. J. E. Kilpatrick, K. S. Pitzer and R. Spitzer, JACS 69, 2483 (1947).
270. K. S. Pitzer and W. E. Donath, JACS 81, 3213 (1959).
271. C. Altona, H. R. Buys and E. Havinga, Rech. Trav. Chim., 85, 973 (1966).
272. H. J. Geise, C. Altona and C. Romers, Tetrahedron Lett., 1383 (1967).
273. C. Altona, H. J. Geise and C. Romers, Tetrahedron 24, 13 (1968).
274. C. Altona and A. P. M. Van der Veek, Tetrahedron 24, 4377 (1968).
275. C. Romers, C. Altona, H. R. Buys and E. Havinga, in "Topics in Stereochemistry" (E. L. Eliel and N. L. Allinger, eds.), Vol. 4, p. 39. Wiley (Interscience), New York, 1969.
276. C. Altona, in "Conformational Analysis" (G. Chiurologlu, ed.), p. 1. Academic Press, New York, 1971.
277. F. E. Hruska, D. J. Wood and J. G. Dalton, private communication.
278. L. D. Hall, P. R. Steiner and C. D. Pedersen, Can. J. Chem. 48, 1155 (1970).
279. C. Altona and M. Sundaralingam, JACS 95, 2333 (1973).
280. A. A. Lugovskoi, V. G. Dashevskii and A. I. Kitaigorodskii, Mol. Biol., USSR (Engl. Ed.) 6, 361 (1972).
281. A. A. Lugovskoi, V. G. Dashevskii and A. I. Kitaigorodskii, Mol. Biol., USSR (Engl. Ed.) 6, 494 (1972).
282. G. Govil and A. Saran, J. Theoret. Biol. 33, 399 (1971).
283. D. Perahia, These de Doctorat d'Etat, Université de Paris VI, (1975).
284. J. Konnert, I. Karle and J. Karle, Acta Crystallogr., Sect. B 26, 1089 (1970).
285. W. Saenger and F. Eckstein, Angew. Chem., Int. Engl. Ed. 8, 595 (1969).
286. W. Saenger and F. Eckstein, JACS 92, 4712 (1970).

287. C. L. Coulter and M. L. Greaves, *Science* **169**, 1097 (1970).
288. C. L. Coulter, *JACS* **95**, 570 (1973).
289. K. Watenpaugh, J. Dow, L. H. Jensen and S. Furberg, *Science* **159**, 206 (1968).
290. C. L. Coulter, *Science* **159**, 888 (1968).
291. C. L. Coulter, *Acta Crystallogr.*, Sect. B **25**, 2055 (1969).
292. C. L. Coulter, *Acta Crystallogr.*, Sect. B **26**, 441 (1970).
293. M. Sundaralingam and J. Abola, *Nature NB* **235**, 244 (1972).
294. M. Sundaralingam and J. Abola, *JACS* **94**, 5070 (1972).
295. A. K. Chwang and M. Sundaralingam, *Nature NB* **244**, 136 (1973).
296. A. K. Chwang and M. Sundaralingam, *Acta Crystallogr.*, Sect. B **30**, 1233 (1974).
297. R. D. Lapper, H. H. Mantsch and I. C. P. Smith, *JACS* **94**, 6243 (1972).
298. R. D. Lapper and I. C. P. Smith, *JACS* **95**, 2880 (1973).
299. D. K. Lavallee and C. L. Coulter, *JACS* **95**, 576 (1973).
300. G. Govil and I. C. P. Smith, private communication.
301. F. Jordan, *J. Theoret. Biol.* **41**, 23 (1973).
302. J. N. Lespinasse and D. Vasilescu, *Biopolymers* **13**, 63 (1974).
303. N. Yathindra and M. Sundaralingam, *BBRC* **56**, 119 (1974).
304. A. M. Bryan and P. G. Olafsson, *Biopolymers* **12**, 229 (1973).
305. A. Rabczenko, K. Jankowski and K. Zakrzewska, *BBA* **353**, 1 (1974).
306. W. Bergmann and R. J. Feeney, *JACS* **72**, 2809 (1950).
307. S. S. Cohen, This Series. **5**, 1 (1966).
308. R. J. Suhadolnik, "Nucleoside Antibiotics," Wiley (Interscience). New York, 1970.
309. P. Tougard, *BBRC* **37**, 961 (1969).
310. P. Tougard, *BBA* **319**, 116 (1973).
311. W. Saenger and V. Jacobi, *Angew. Chem., Int. Engl. Ed.* **10**, 187 (1971).
312. W. Saenger, *JACS* **94**, 621 (1972).
313. J. S. Sherfinsky and R. E. Marsh, *Acta Crystallogr.*, Sect. B **29**, 192 (1973).
314. P. Tollin, H. R. Wilson and D. W. Young, *Nature NB* **242**, 49 (1973).
315. P. Tollin, H. R. Wilson and D. W. Young, *Acta Crystallogr.* B **29**, 1641 (1973).
316. A. K. Chwang and M. Sundaralingam, *Nature NB* **243**, 78 (1973).
317. O. Lefebvre-Soubeyran and P. Tougard, *C. R. Acad. Sci., Ser. C* **276**, 403 (1973).
318. P. Tougard and O. Lefebvre-Soubeyran, *Acta Crystallogr.*, Sect. B **30**, 86 (1974).
319. P. Tougard, *Acta Crystallogr.* B **29**, 2227 (1973).
320. D. Suck and W. Saenger, *Acta Crystallogr.* B **29**, 1323 (1973).
321. L. T. Delbaere and M. N. G. James, *Acta Crystallogr.*, B **29**, 2905 (1973).
322. T. Brennan and M. Sundaralingam, *BBRC* **52**, 1348 (1973).
323. W. Guschlbauer and M. Privat de Garilhe, *Bull. Soc. Chim. Biol.* **51**, 1511 (1969) and references quoted therein.
324. R. J. Cushley, K. A. Watanabe and J. J. Fox, *JACS* **89**, 394 (1967).
325. J. G. Dalton, M. Sc. Thesis, University of Manitoba (1971).
326. B. P. Cross and T. Schleich, *Biopolymers* **12**, 2381 (1973).
327. M. Remin and D. Shugar, *JACS* **95**, 8146 (1973).
328. A. Saran, B. Pullman and D. Perahia, *BBA* **349**, 189 (1974).
329a. G. Bunick and D. Voet, *Acta Crystallogr.* B **30**, 1651 (1974).
329b. A. K. Chwang, M. Sundaralingam and S. Hanessian, *Acta Crystallogr.*, Sect. B **30**, 2273 (1974).
330. D. Perahia, B. Pullman and A. Saran, *BBRC* **47**, 1284 (1972).

331. A. Saran, B. Pullman and D. Perahia, *FEBS Lett.* **23**, 332 (1972).
332. H. Berthod and B. Pullman, *C.R. Acad. Sci., Ser. D* **276**, 1767 (1973).
333. D. Perahia, B. Pullman and A. Saran, in "Structure and Conformation of Nucleic Acids and Protein-Nucleic Acid Interactions," 4th Harry Steenbock Symp. (M. Sundaralingam and S. T. Rao, eds.), 1975.
334. N. Yathindra and M. Sundaralingam, *BBA* **308**, 17 (1973).
335. D. W. Miles, D. L. Miles and H. Eyring, *J. Theoret. Biol.* **45**, 577 (1974).
336. C. Giessner-Prettre and B. Pullman, *C.R. Acad. Sci.* **266**, 933 (1968).
337. C. Giessner-Prettre and B. Pullman, *C.R. Acad. Sci.* **268**, 1115 (1969).
338. C. Giessner-Prettre and B. Pullman, *C.R. Acad. Sci.* **270**, 866 (1970).
339. C. Giessner-Prettre and B. Pullman *J. Theoret. Biol.* **27**, 87 (1970).
340. C. Giessner-Prettre and B. Pullman, *J. Theoret. Biol.* **27**, 341 (1970).
341. C. Giessner-Prettre and B. Pullman, *J. Theoret. Biol.* **40**, 441 (1973).
342. C. Giessner-Prettre and B. Pullman, *Intern. J. Quant. Chem. Symp.* **7**, 295 (1973).
343. C. Giessner-Prettre and B. Pullman, *J. Theoret. Biol.* **48**, 425 (1974).
344. A. D. Cross and D. M. Crothers, *Bchem* **10**, 4025 (1971).
345. D. M. Crothers, C. W. Hilbers and R. G. Shulman, *PNAS* **70**, 2899 (1973).
346. D. J. Patel and A. E. Tonelli, *PNAS* **71**, 1945 (1974).
347. R. G. Shulman, C. W., Hilbers, D. R. Kearns, B. R. Reid and Y. P. Wong, *JMB* **78**, 57 (1973).
348. D. R. Lightfoot, K. L. Wong, D. R. Kearns, B. R. Reid and R. G. Shulman, *JMB* **78**, 71 (1973).
349. R. G. Shulman, C. W. Hilbers, Y. P. Wong, K. L. Wong, D. R. Lightfoot, B. R. Reid and D. R. Kearns, *PNAS* **70**, 2024 (1973).
350. K. L. Wong and D. R. Kearns, *Biopolymers* **13**, 371 (1974).
351. M. Kainosho and Y. Kyogoku, *Bchem* **11**, 741 (1972).
352. R. H. Sarma and R. J. Mynott, *JCS, Chem. Commun.*, 977 (1972).
353. R. H. Sarma, C. H. Lee, F. E. Hruska and D. J. Wood, *FEBS Lett.* **36**, 157 (1973).
354. R. H. Sarma and R. J. Mynott, *JACS* **95**, 7470 (1973).
355. R. H. Sarma and R. J. Mynott, in "Conformation of Biological Molecules and Polymers," 5th Jerusalem Symp. Quantum Chem. Biochem. (E. D. Bergmann and B. Pullman, eds.), p. 591. Academic Press, New York, 1973.
356. N. S. Angerman, T. A. Victor, C. L. Bell and S. S. Danyluk, *Bchem* **11**, 2402 (1972).
357. T. R. Krugh and J. W. Neely, *Bchem* **12**, 1775 (1973).
358. T. R. Krugh and J. W. Neely, *Bchem* **12**, 4418 (1973).
359. D. J. Patel, *Bchem* **13**, 1476 (1974).
360. D. J. Patel, *Bchem* **13**, 2388 (1974).
361. G. Melcher, *J. Mol. Evol.* **3**, 121 (1974).
362. C. E. Bugg, in "The Purines: Theory and Experiment," 4th Jerusalem Symp. Quantum Chem. Biochem. (E. D. Bergmann and B. Pullman, eds.), p. 178. Academic Press, New York, 1972.
363. W. D. S. Motherwell and N. W. Isaacs, *JMB* **71**, 231 (1972).
364. W. D. S. Motherwell, L. Riva di Sanseverino and O. Kennard, *JMB* **80**, 405 (1973).
365. J. Caillet and P. Claverie, *Biopolymers* **13**, 601 (1974).
366. N. Yathindra and M. Sundaralingam, *PNAS* **71**, 3325 (1974).

ADDENDUM (for article by R. Brimacombe et al., p. 1)

Since the manuscript for this review was submitted (May 1975), important progress has been made in several areas of "ribosomology." In the following addendum, a short summary is given of some of the more significant recent developments, listed under their appropriate sections.

Section II, A. The number of E. coli ribosomal proteins whose sequences have been determined has now (June 1976) been increased to 27, namely 14 from the small and 13 from the large subunit. More than 4300 amino acids have been sequenced, i.e., 55% of the total present in the E. coli ribosome [see a recent summary (301) for more information].

A search for identical regions among the E. coli ribosomal proteins has been made. The numbers of identical tri-, tetra-, penta- and hexapeptides found in these proteins have been compared with computer-simulated proteins of the same lengths and amino-acid compositions as the ribosomal proteins, but with randomized amino-acid sequences. No significant differences with respect to the number of identical regions were found between the E. coli ribosomal proteins and their computer-simulated isomers (302). In other words: there is no evidence that the proteins present in the E. coli ribosome have originated from a common ancestor. Homologous structures found among them (303) do not occur more frequently than can be expected on a random basis.

Section II, B. Evidence is accumulating that the secondary structure of the 16 S RNA may be much more complicated than was thought earlier (33). RNA complexes have been isolated from both the 5' and 3' regions of the 16 S RNA that have large sections of sequence missing, suggesting that many long-range interactions occur (63, 64, 81, 304). Further, there is evidence that the RNA within the 5'-region of the 16 S RNA is very compactly organized, in the absence of protein, again suggesting the presence of complex tertiary structural interactions (305).

Section III, A. It has been demonstrated (306) that six or seven proteins (namely S3, S5, S9, S12, S13, S18 and possibly S11), not previously classified as primary binding proteins, can associate specifically with 16 S RNA when the latter is prepared by an acetic acid/urea method. RNA prepared in this way has a more open structure, rendering the binding sites of these proteins more accessible.

An approach likely to prove valuable in the investigation of RNA-protein interactions is the further development of techniques for cross-linking protein to RNA. The cross-linking of the 3'-terminus of the 16 S RNA to protein S1 already mentioned (121) has been confirmed, and protein S21 is also involved (307). In addition, it has been shown that S1 will bind reversibly to the 49-nucleotide fragment released from the 3' end of the RNA by colicin E3 (308), and the authors suggest that the role of S1 may be to stabilize this region of the RNA in an open conformation to allow base-pairing to mRNA (see addendum to Section V, A below).

More general methods of cross-linking protein to RNA by direct irradiation with UV light have also been developed. This has led to the identification of a 240-nucleotide stretch of 16 S RNA to which protein S7 can be cross-linked (304), and to the identification of some peptides that are in contact with 16 S RNA in a synthetic S4 · 16 S RNA complex (309). The latter finding is in agreement with studies on tryptic digests of an S4 · 16 S RNA complex (310); a protein fragment of about 155 amino acids was obtained that binds reversibly to 16 S RNA and that contains most of the peptides just mentioned. This portion of protein S4 is therefore presumably the primary binding site.

Section IV. Development of the technique of electron microscopic visualization of antibody markers bound to ribosomal proteins already mentioned (156–158) has advanced to the point where one or more antibody binding sites on the ribosome surface have been identified for each of the 30 S proteins and for many of the 50 S proteins; these results have been incorporated into topographical models of the subparticles (311; see also ref. 301 for review).

Section V, A. In contrast to prokaryotic systems, formylation of the amino group on the aminoacylated eukaryotic initiator tRNA (fMet–tRNA^{fMet}) decreases the ability of the tRNA to serve as a substrate for the initiation factor E IF-3 (312); this factor is equivalent to the prokaryotic factor IF-2.

The hypothesis has already been mentioned (37) that the 3' end of the 16 S RNA can form base-pairs with mRNA during initiation. Evidence that strongly supports this idea has come from the isolation of a stable RNA complex comprising the 49-nucleotide 3' fragment of the 16 S RNA (see above) and a 30-nucleotide fragment from the initiator region of R17 bacteriophage RNA (313). The complex was isolated by mild digestion of a 70 S initiation complex.

Section V, B. Aminoacyl-tRNA synthetases do not attach the amino acids to the same position on tRNA molecules, and three classes can be distinguished, *viz.* those that attach the amino acid to the 2'-position, those that attach it to the 3'-position, and those that attach it to both positions (314–316). This specificity seems to have been conserved during evolution, since, with the exception of tryptophan (316), the same specificity was found in both *E. coli* and yeast.

It has been suggested (283) that EF-Tu carries its own GTPase center. Evidence has recently been reported for a similar center on EF-G (317); affinity-label experiments with GTP analogs show that the majority of the label becomes covalently linked to EF-G within the complex between EF-G, 70 S ribosome, and GTP analog. The low level of reaction with the ribosome seems to be nonspecific (318, 319). Thus, the factor-independent GTPase activity of a 5 S RNA-protein complex (281) and of ribonucleoprotein fragments derived from 50 S (163) seems to be unrelated to the elongation factor-dependent GTPase activities. This conclusion is confirmed by the finding that a totally reconstituted 50 S subparticle lacking 5 S RNA is fully active in EF-G dependent GTPase activity (320).

301. G. Stöffler and H. G. Wittmann, in "Protein Synthesis" (H. Weissbach and S. Pestka, eds.). Academic Press, New York, 1976.
302. B. Wittmann-Liebold and M. Dzionara, *FEBS Lett.* in press.
303. B. Wittmann-Liebold and M. Dzionara, *FEBS Lett.* **61**, 14 (1976).
304. J. Rinke, A. Yuki and R. Brimacombe, *EJB* **64**, 77 (1976).
305. E. Ungewickell and R. A. Garrett, unpublished results.
306. H. K. Hochkeppel, E. Spicer and G. R. Craven, *JMB* **101**, 155 (1976).
307. A. P. Czernilofsky, C. G. Kurland and G. Stöffler, *FEBS Lett.* **58**, 281 (1975).
308. J. Dahlberg and A. E. Dahlberg, *PNAS* **72**, 2940 (1975).
309. B. Ehresmann, J. Reinbolt and J. P. Ebel, *FEBS Lett.* **58**, 106 (1975).
310. C. Schulte, E. Schiltz and R. A. Garrett, *NARes* **2**, 931 (1975).
311. G. W. Tischendorf, H. Zeichhardt and G. Stöffler, *PNAS* **72**, 4820 (1975).
312. R. S. Ranu and I. G. Wool, *Nature (London)* **257**, 616 (1975).
313. J. A. Steitz and K. Jakes, *PNAS* **72**, 4734 (1975).
314. T. H. Fraser and A. Rich, *PNAS* **72**, 3044 (1975).

315. M. Sprinzl and F. Cramer, *PNAS* **72**, 3049 (1975).
316. S. M. Hecht and A. C. Chinault, *PNAS* **73**, 405 (1976).
317. R. C. Marsh, G. Chinali and A. Parmeggiani, *JBC* **250**, 8344 (1975).
318. A. S. Girshovich, E. S. Bochkareva and V. A. Pozdnyakov, *Acta Biol. Med. Germ.* **33**, 639 (1974).
319. A. S. Girshovich, V. A. Pozdnyakov and Y. A. Orchinnikov, *EJB* in press (1976).
320. F. Dohme and K. H. Nierhaus, *PNAS* in press (1976).

Subject Index

D

Deoxyribonucleic acid
  conformation in solution
    anomalies in primary structure, 181–186
    classical forms of double helix, 177–178
    dependence on nucleotide sequence, 178–181
    nature of premelting, 189–202
    superhelical turns, 186–189
  premelting changes in conformation
    early studies, 154–155
    electrochemical analysis, 162–169
    enzymic methods, 171–174
    formaldehyde reaction, 169–170
    optical methods, 155–162
    other techniques, 174–177
    relation between conformation and function, 203–204

E

Escherichia coli
  ribosome
    components, 2–8
    function, 23–36
    protein-RNA interactions, 8–13
    topography of subparticles, 13–23

M

Mononucleotides
  nuclear magnetic resonance studies, 96–97
    hydrogen-bonding effects, 99–100
    proton exchange with solvent, 100–101
    rules for predicting low-field spectra, 101–105
    stacking interactions and ring-current effects, 97–99

N

Nucleic acid(s)
  backbone structure of di- and polynucleotides, 275
    conformational energy maps, 276–291
    geometry of phosphate group, 291–301
    conformation about C4'—C5' bond, 255–256
    α-nucleosides, 270–272
    β-nucleosides, 256–270
    5'-β-nucleotides, 272–275
  conformation of the sugar ring, 301–302
    properties of ribose and deoxyribose, 303–304
    pseudorotational parameters, 302–303
    relation between parameters, 304–308
  glycosyl torsion angle
    rigidity and flexibility of 5'-nucleotides, 251–255
    syn-anti equilibrium in nucleosides, 236–251
  quantum mechanical methods
    basic idea of method of molecular orbitals, 217–218
    classical Hückel approximation, 222
    complete neglect of differential overlap, 224–226
    configuration interaction, 226–227
    extended Hückel theory, 222–223
    localized orbitals, 228
    PCI over localized orbitals method, 228–230
    perturbative configuration interaction, 227–228
    related subjects, 308–310
    self-consistent field method, 218–222
    zero differential overlap approximation, 223–224
  types of torsion angles and definitions, 230–231

mononucleotides, 233–234
nucleosides, 234–236
polynucleotides, 231–233

O

Oligonucleotide fragments
  nuclear magnetic resonance studies
    dimerization and, 110–112
    melting behavior of short helices, 112–113
    simple helices, 105–110

R

Ribosomal RNA
  activities of 50 S subunits with modified 5 S, 80–82
  enzymic activities of 5 S-protein complexes, 83–84
  activity of 50 S subunit lacking 5 S, 79
  complexes of 5 S with protein
    binding, 71–74
    oligonucleotide binding, 78
    protein neighborhood, 74–75
    regions of interaction, 75–78
  primary sequences, 46
    eukaryotic 5 S, 51–54
    eukaryotic 5.8 S, 54–56
    precursors of 5 S and 5.8 S, 56–58
    prokaryotic 5 S, 47–51
  reconstitution of 50 S subunit with modified 5 S, 79–80
  secondary and tertiary structures of 5 S and 5.8 S
    chemical modification, 65–68
    enzymic hydrolysis, 62–65
    oligonucleotide binding, 68–69
    physical characterization, 58–62
    proposed models, 69–71

5 S in binding tRNA to ribosomes, 82–83
function of 5 S, 84
Ribosome
  function
    elongation, 28–35
    initiation, 23–27
    termination, 35–36
  proteins, 2–4
  protein binding sites, 8–9
    16 S RNA, 9–12
    23 S RNA, 12–13
  ribonucleic acid, 4–8
  subparticle topography, 13–14
    30 S, 14–20
    50 S, 20–22
    30–50 S interface, 22–23

T

Transfer RNA
  applications of nuclear magnetic resonance to problems of structure
    denatured conformers, 130–133
    dimerization, 133–134
    effect of aminoacylation, 139–140
    effect of anticodon-loop modifications, 138–139
    interaction with drug's, 141–145
    photocrosslinking, 140–141
    thermal unfolding, 134–138
  high-resolution nuclear magnetic resonance studies
    assignment and interpretation of spectra, 123–130
    evidence for tertiary-structure base-pairs, 117–123
    number of secondary- and tertiary-structure base-pairs, 113–117

Contents of Previous Volumes

Volume 1

"Primer" in DNA Polymerase Reactions
F. J. BOLLUM

The Biosynthesis of Ribonucleic Acid in Animal Systems
R. M. S. SMELLIE

The Role of DNA in RNA Synthesis
JERARD HURWITZ AND J. T. AUGUST

Polynucleotide Phosphorylase
M. GRUNBERG-MANAGO

Messenger Ribonucleic Acid
FRITZ LIPMANN

The Recent Excitement in the Coding Problem
F. H. C. CRICK

Some Thoughts on the Double-Stranded Model of Deoxyribonucleic Acid
AARON BENDICH AND HERBERT S. ROSENKRANZ

Denaturation and Renaturation of Deoxyribonucleic Acid
J. MARMUR, R. ROWND, AND C. L. SCHILDKRAUT

Some Problems Concerning the Macromolecular Structure of Ribonucleic Acids
A. S. SPIRIN

The Structure of DNA as Determined by X-Ray Scattering Techniques
VITTORIO LUZZATI

Molecular Mechanisms of Radiation Effects
A. WACKER

AUTHOR INDEX—SUBJECT INDEX

Volume 2

Nucleic Acids and Information Transfer
LIEBE F. CAVALIERI AND BARBARA H. ROSENBERG

Nuclear Ribonucleic Acid
HENRY HARRIS

Plant Virus Nucleic Acids
  Roy Markham

The Nucleases of *Escherichia coli*
  I. R. Lehman

Specificity of Chemical Mutagenesis
  David R. Krieg

Column Chromatography of Oligonucleotides and Polynucleotides
  Matthys Staehelin

Mechanism of Action and Application of Azapyrimidines
  J. Skoda

The Function of the Pyrimidine Base in the Ribonuclease Reaction
  Herbert Witzel

Preparation, Fractionation, and Properties of sRNA
  G. L. Brown

Author Index—Subject Index

Volume 3

Isolation and Fractionation of Nucleic Acids
  K. S. Kirby

Cellular Sites of RNA Synthesis
  David M. Prescott

Ribonucleases in Taka-Diastase: Properties, Chemical Nature, and Applications
  Fujio Egami, Kenji Takahashi, and Tsuneko Uchida

Chemical Effects of Ionizing Radiations on Nucleic Acids and Related Compounds
  Joseph J. Weiss

The Regulation of RNA Synthesis in Bacteria
  Frederick C. Neidhardt

Actinomycin and Nucleic Acid Function
  E. Reich and I. H. Goldberg

De Novo Protein Synthesis *in Vitro*
  B. Nisman and J. Pelmont

Free Nucleotides in Animal Tissues
  P. Mandel

Author Index—Subject Index

Volume 4

Fluorinated Pyrimidines
  Charles Heidelberger

Genetic Recombination in Bacteriophage
  E. Volkin

DNA Polymerases from Mammalian Cells
  H. M. Keir

The Evolution of Base Sequences in Polynucleotides
  B. J. McCarthy

Biosynthesis of Ribosomes in Bacterial Cells
  Syozo Osawa

5-Hydroxymethylpyrimidines and Their Derivatives
  T. L. V. Ulbricht

Amino Acid Esters of RNA, Nucleotides, and Related Compounds
  H. G. Zachau and H. Feldmann

Uptake of DNA by Living Cells
  L. Ledoux

Author Index—Subject Index

Volume 5

Introduction to the Biochemistry of D-Arabinosyl Nucleosides
  Seymour S. Cohen

Effects of Some Chemical Mutagens and Carcinogens on Nucleic Acids
  P. D. Lawley

Nucleic Acids in Chloroplasts and Metabolic DNA
  Tatsuichi Iwamura

Enzymatic Alteration of Macromolecular Structure
  P. R. Srinivasan and Ernest Borek

Hormones and the Synthesis and Utilization of Ribonucleic Acids
  J. R. Tata

Nucleoside Antibiotics
  JACK J. FOX, KYOICHI A. WATANABE, AND ALEXANDER BLOCH

Recombination of DNA Molecules
  CHARLES A. THOMAS, JR.

  Appendix I. Recombination of a Pool of DNA Fragments with Complementary Single-Chain Ends
    G. S. WATSON, W. K. SMITH, AND CHARLES A. THOMAS, JR.

  Appendix II. Proof That Sequences of A, C, G, and T Can Be Assembled to Produce Chains of Ultimate Length, Avoiding Repetitions Everywhere
    A. S. FRAENKEL AND J. GILLIS

The Chemistry of Pseudouridine
  ROBERT WARNER CHAMBERS

The Biochemistry of Pseudouridine
  EUGENE GOLDWASSER AND ROBERT L. HEINRIKSON

AUTHOR INDEX—SUBJECT INDEX

Volume 6

Nucleic Acids and Mutability
  STEPHEN ZAMENHOF

Specificity in the Structure of Transfer RNA
  KIN-ICHIRO MIURA

Synthetic Polynucleotides
  A. M. MICHELSON, J. MASSOULIÉ, AND W. GUSCHLBAUER

The DNA of Chloroplasts, Mitochondria, and Centrioles
  S. GRANICK AND AHARON GIBOR

Behavior, Neural Function, and RNA
  H. HYDÉN

The Nucleolus and the Synthesis of Ribosomes
  ROBERT P. PERRY

The Nature and Biosynthesis of Nuclear Ribonucleic Acids
  G. P. GEORGIEV

Replication of Phage RNA
  CHARLES WEISSMANN AND SEVERO OCHOA

AUTHOR INDEX—SUBJECT INDEX

Volume 7

Autoradiographic Studies on DNA Replication in Normal and Leukemic Human Chromosomes
  FELICE GAVOSTO

Proteins of the Cell Nucleus
  LUBOMIR S. HNILICA

The Present Status of the Genetic Code
  CARL R. WOESE

The Search for the Messenger RNA of Hemoglobin
  H. CHANTRENNE, A. BURNY, AND G. MARBAIX

Ribonucleic Acids and Information Transfer in Animal Cells
  A. A. HADJIOLOV

Transfer of Genetic Information during Embryogenesis
  MARTIN NEMER

Enzymatic Reduction of Ribonucleotides
  AGNE LARSSON AND PETER REICHARD

The Mutagenic Action of Hydroxylamine
  J. H. PHILLIPS AND D. M. BROWN

Mammalian Nucleolytic Enzymes and Their Localization
  DAVID SHUGAR AND HALINA SIERAKOWSKA

AUTHOR INDEX—SUBJECT INDEX

Volume 8

Nucleic Acids—The First Hundred Years
  J. N. DAVIDSON

Nucleic Acids and Protamine in Salmon Testes
  GORDON H. DIXON AND MICHAEL SMITH

Experimental Approaches to the Determination of the Nucleotide Sequences of Large Oligonucleotides and Small Nucleic Acids
  ROBERT W. HOLLEY

Alterations of DNA Base Composition in Bacteria
  G. F. GAUSE

Chemistry of Guanine and Its Biologically Significant Derivatives
  ROBERT SHAPIRO

Bacteriophage ϕX174 and Related Viruses
  ROBERT L. SINSHEIMER

The Preparation and Characterization of Large Oligonucleotides
  GEORGE W. RUSHIZKY AND HERBERT A. SOBER

Purine N-Oxides and Cancer
  GEORGE BOSWORTH BROWN

The Photochemistry, Photobiology, and Repair of Polynucleotides
  R. B. SETLOW

What Really Is DNA? Remarks on the Changing Aspects of a
    Scientific Concept
  ERWIN CHARGAFF

Recent Nucleic Acid Research in China
  TIEN-HSI CHENG AND ROY H. DOI

AUTHOR INDEX—SUBJECT INDEX

Volume 9

The Role of Conformation in Chemical Mutagenesis
  B. SINGER AND H. FRAENKEL-CONRAT

Polarographic Techniques in Nucleic Acid Research
  E. PALEČEK

RNA Polymerase and the Control of RNA Synthesis
  JOHN P. RICHARDSON

Radiation-Induced Alterations in the Structure of Deoxyribonucleic
    Acid and Their Biological Consequences
  D. T. KANAZIR

Optical Rotatory Dispersion and Circular Dichroism of Nucleic Acids
  JEN TSI YANG AND TATSUYA SAMEJIMA

The Specificity of Molecular Hybridization in Relation to Studies
    on Higher Organisms
  P. M. B. WALKER

Quantum-Mechanical Investigations of the Electronic Structure of
    Nucleic Acids and Their Constituents
  BERNARD PULLMAN AND ALBERTE PULLMAN

The Chemical Modification of Nucleic Acids
  N. K. KOCHETKOV AND E. I. BUDOWSKY

AUTHOR INDEX—SUBJECT INDEX

Volume 10

Induced Activation of Amino Acid Activating Enzymes by Amino Acids and tRNA
  ALAN H. MEHLER

Transfer RNA and Cell Differentiation
  NOBORU SUEOKA AND TAMIKO KANO-SUEOKA

N^6-(Δ^2-Isopentenyl)adenosine: Chemical Reactions, Biosynthesis, Metabolism, and Significance to the Structure and Function of tRNA
  ROSS H. HALL

Nucleotide Biosynthesis from Preformed Purines in Mammalian Cells: Regulatory Mechanisms and Biological Significance
  A. W. MURRAY, DAPHNE C. ELLIOTT, AND M. R. ATKINSON

Ribosome Specificity of Protein Synthesis in Vitro
  ORIO CIFERRI AND BRUNO PARISI

Synthetic Nucleotide-peptides
  ZOE A. SHABAROVA

The Crystal Structures of Purines, Pyrimidines and Their Intermolecular Complexes
  DONALD VOET AND ALEXANDER RICH

AUTHOR INDEX—SUBJECT INDEX

Volume 11

The Induction of Interferon by Natural and Synthetic Polynucleotides
  CLARENCE COLBY, JR.

Ribonucleic Acid Maturation in Animal Cells
  R. H. BURDON

Liporibonucleoprotein as an Integral Part of Animal Cell Membranes
  V. S. SHAPOT AND S. YA. DAVIDOVA

Uptake of Nonviral Nucleic Acids by Mammalian Cells
  PUSHPA M. BHARGAVA AND G. SHANMUGAM

The Relaxed Control Phenomenon
  ANN M. RYAN AND ERNEST BOREK

Molecular Aspects of Genetic Recombination
  CEDRIC I. DAVERN

Principles and Practices of Nucleic Acid Hybridization
  DAVID E. KENNELL

Recent Studies Concerning the Coding Mechanism
  THOMAS H. JUKES AND LILA GATLIN

The Ribosomal RNA Cistrons
  M. L. BIRNSTIEL, M. CHIPCHASE, AND J. SPEIRS

Three-Dimensional Structure of tRNA
  FRIEDRICH CRAMER

Current Thoughts on the Replication of DNA
  ANDREW BECKER AND JERARD HURWITZ

Reaction of Aminoacyl-tRNA Synthetases with Heterologous tRNA's
  K. BRUCE JACOBSON

On the Recognition of tRNA by Its Aminoacyl-tRNA Ligase
  ROBERT W. CHAMBERS

AUTHOR INDEX—SUBJECT INDEX

Volume 12

Ultraviolet Photochemistry as a Probe of Polyribonucleotide Conformation
  A. J. LOMANT AND JACQUES R. FRESCO

Some Recent Developments in DNA Enzymology
  MEHRAN GOULIAN

Minor Components in Transfer RNA: Their Characterization, Location, and Function
  SUSUMU NISHIMURA

The Mechanism of Aminoacylation of Transfer RNA
  ROBERT B. LOFTFIELD

Regulation of RNA Synthesis
  EKKEHARD K. F. BAUTZ

The Poly(dA-dT) of Crab
  M. LASKOWSKI, SR.

The Chemical Synthesis and the Biochemical Properties of Peptidyl-tRNA
  YEHUDA LAPIDOT AND NATHAN DE GROOT

SUBJECT INDEX

Volume 13

Reactions of Nucleic Acids and Nucleoproteins with Formaldehyde
  M. Ya. Feldman

Synthesis and Functions of the -C-C-A Terminus of Transfer RNA
  Murray P. Deutscher

Mammalian RNA Polymerases
  Samson T. Jacob

Poly(adenosine diphosphate ribose)
  Takashi Sugimura

The Stereochemistry of Actinomycin Binding to DNA and Its Implications in Molecular Biology
  Henry M. Sobell

Resistance Factors and Their Ecological Importance to Bacteria and to Man
  M. H. Richmond

Lysogenic Induction
  Ernest Borek and Ann Ryan

Recognition in Nucleic Acids and the Anticodon Families
  Jacques Ninio

Translation and Transcription of the Tryptophan Operon
  Fumio Imamoto

Lymphoid Cell RNA's and Immunity
  A. Arthur Gottlieb

Subject Index

Volume 14

DNA Modification and Restriction
  Werner Arber

Mechanism of Bacterial Transformation and Transfection
  Nihal K. Notani and Jane K. Setlow

DNA Polymerases II and III of *Escherichia coli*
  Malcolm L. Gefter

The Primary Structure of DNA
  Kenneth Murray and Robert W. Old

RNA-Directed DNA Polymerase—Properties and Functions in Oncogenic RNA Viruses and Cells
  Maurice Green and Gray F. Gerard

Subject Index

Volume 15

Information Transfer in Cells Infected by RNA Tumor Viruses and Extension to Human Neoplasia
D. GILLESPIE, W. C. SAXINGER, AND R. C. GALLO

Mammalian DNA Polymerases
F. J. BOLLUM

Eukaryotic RNA Polymerases and the Factors That Control Them
B. B. BISWAS, A. GANGULY, AND D. DAS

Structural and Energetic Consequences of Noncomplementary Base Oppositions in Nucleic Acid Helices
A. J. LOMANT AND JACQUES R. FRESCO

The Chemical Effects of Nucleic Acid Alkylation and Their Relation to Mutagenesis and Carcinogenesis
B. SINGER

Effects of the Antibiotics Netropsin and Distamycin A on the Structure and Function of Nucleic Acids
CHRISTOPH ZIMMER

SUBJECT INDEX

Volume 16

Initiation of Enzymic Synthesis of Deoxyribonucleic Acid by Ribonucleic Acid Primers
ERWIN CHARGAFF

Transcription and Processing of Transfer RNA Precursors
JOHN D. SMITH

Bisulfite Modification of Nucleic Acids and Their Constituents
HIKOYA HAYATSU

The Mechanism of the Mutagenic Action of Hydroxylamines
E. I. BUDOWSKY

Diethyl Pyrocarbonate in Nucleic Acid Research
L. EHRENBERG, I. FEDORCSÁK, AND F. SOLYMOSY

SUBJECT INDEX

Volume 17

The Enzymic Mechanism of Guanosine 5′,3′-Polyphosphate Synthesis
 Fritz Lipmann and Jose Sy

Effects of Polyamines on the Structure and Reactivity of tRNA
 Ted T. Sakai and Seymour S. Cohen

Information Transfer and Sperm Uptake by Mammalian Somatic Cells
 Aaron Bendich, Ellen Borenfreund, Steven S. Witkins, Delia Beju, and Paul J. Higgins

Studies on the Ribosome and Its Components
 Pnina Spitnik-Elson and David Elson

Classical and Postclassical Modes of Regulation of the Synthesis of Degradative Bacterial Enzymes
 Boris Magasanik

Characteristics and Significance of the Polyadenylate Sequence in Mammalian Messenger RNA
 George Brawerman

Polyadenylate Polymerases
 Mary Edmonds and Mary Ann Winters

Three-Dimensional Structure of Transfer RNA
 Sung-Hou Kim

Insights into Protein Biosynthesis and Ribosome Function through Inhibitors
 Sidney Pestka

Interaction with Nucleic Acids of Carcinogenic and Mutagenic N-Nitroso Compounds
 W. Lijinsky

Biochemistry and Physiology of Bacterial Ribonuclease
 Alok K. Datta and Salil K. Niyogi

Subject Index